THING

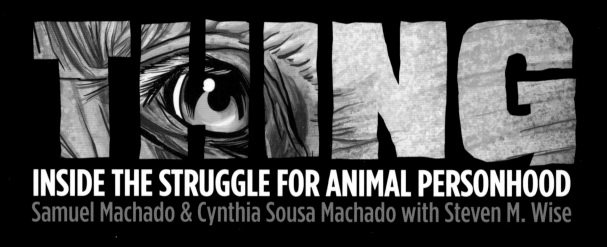

THING

INSIDE THE STRUGGLE FOR ANIMAL PERSONHOOD

Samuel Machado & Cynthia Sousa Machado with Steven M. Wise

◯ **ISLAND**PRESS | Washington | Covelo

Library of Congress Control Number: 2022948611

All Island Press books are printed on environmentally responsible materials.

Manufactured in the United States of America
10 9 8 7 6 5 4 3 2 1

Keywords: animal rights, autonomy, Bronx Zoo, chimpanzees, civil rights, cognition, common law, communication, dolphin, elephant, empathy, habeas corpus, Happy the elephant, nonhuman animals, Nonhuman Rights Project, orca, personhood, rights of nature, sanctuary, self-awareness, zoo

Dedicated to our parents, Eloy, Gilbert, Martha and Norma, and our grandmothers, Carmen, Paula and Vinita for teaching us to be humans, and to all of our alloparents like uncle Eduardo (Lalo) who was always there, aunt Mary who gives more than she has.

CONTENTS

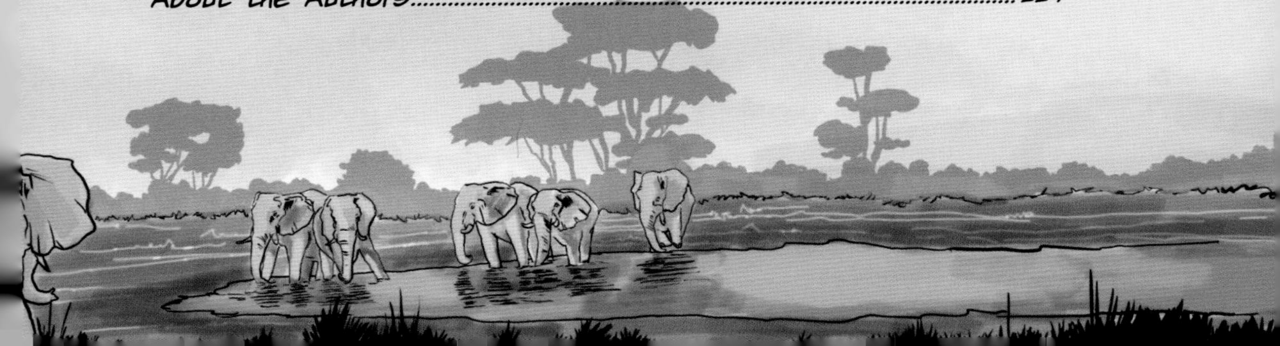

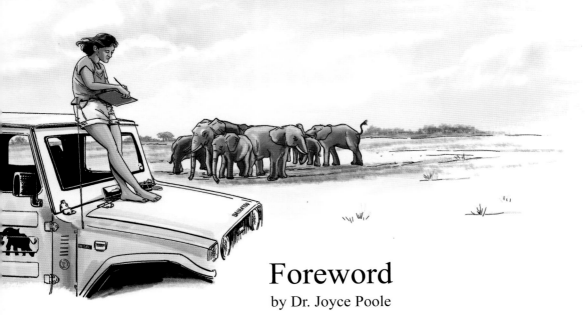

Foreword

by Dr. Joyce Poole

Cofounder and Scientific Director of ElephantVoices

When I was 11 years old, my mother took me to a talk by Dr. Jane Goodall. I knew then what I wanted to be. Today, I have had the pleasure of spending a large part of my life among elephants. Throughout a career studying elephant behavior and communication, I've come to see that they have so many of our best qualities as a species. They are extremely intelligent, communicative, and sensitive in ways we may not fully fathom. But most impressively, they are autonomous, meaning they are capable of self-determined behavior. And if for no other reason, we ought to acknowledge that a life of captivity and isolation is no life for one of these magnificent animals.

Thing: Inside the Struggle for Animal Personhood follows the battle to move Happy, a fifty-year-old Asian elephant, from the Bronx Zoo to sanctuary. For almost a decade, I have been an advocate for Happy's release, submitting affidavits on her behalf to New York courts with the help of the Nonhuman Rights Project and its President and Founder, Steven M. Wise. Sam Machado and Cynthia Sousa Machado beautifully illustrate the shared anatomy between elephants and humans that makes it possible for Happy to possess self-awareness, empathy, and many other capacities erroneously considered to be uniquely human.

We are not the only beings with personalities and minds capable of reasoning. Nor are we the only beings with emotions like happiness, sadness, fear, and despair. And we are certainly not the only beings capable of giving and receiving love. In her natural environment with her birth family, Happy would have been in the middle of a collection of aunts, sisters, brothers, and of course, her own mother. All of these social connections would have contributed to the sophisticated social fabric of Happy's life. The loss of any of them would send ripples of grief and sadness through her entire family, as it would to any of us.

Happy would also have learned to exert her voice while grazing across Thailand, engaging in decision-making, debate, and disagreement with her family. Elephants like Happy petition, argue, and persuade each other into making decisions just like we do. In this book, Sam and Cynthia articulate the complex web of signals, rumbles, and pointing behavior that elephants use to communicate preferences and convince each other of what to do and where to go. Deprived of this rich social life, Happy is instead locked in her one-and-a-half-acre enclosure, separated from her family, and will never know an elephant's life. Sanctuary is the closest Happy will ever come to living a meaningful life of her own.

Happy deserves better. Any autonomous being of any species deserves better than a life on display in captivity. Even under the best of circumstances, the guidelines that govern Happy's confinement, the Association of Zoos and Aquariums (AZA) Standards for Elephant Management and Care, and the Animal Welfare Act, are woefully inadequate. I hope that this groundbreaking graphic novel will bring others to see, as I have, that we cannot continue to treat autonomous beings as if they exist for our education or entertainment. Happy deserves to be free.

THING

CHAPTER 1: THE CASE FOR PERSONHOOD

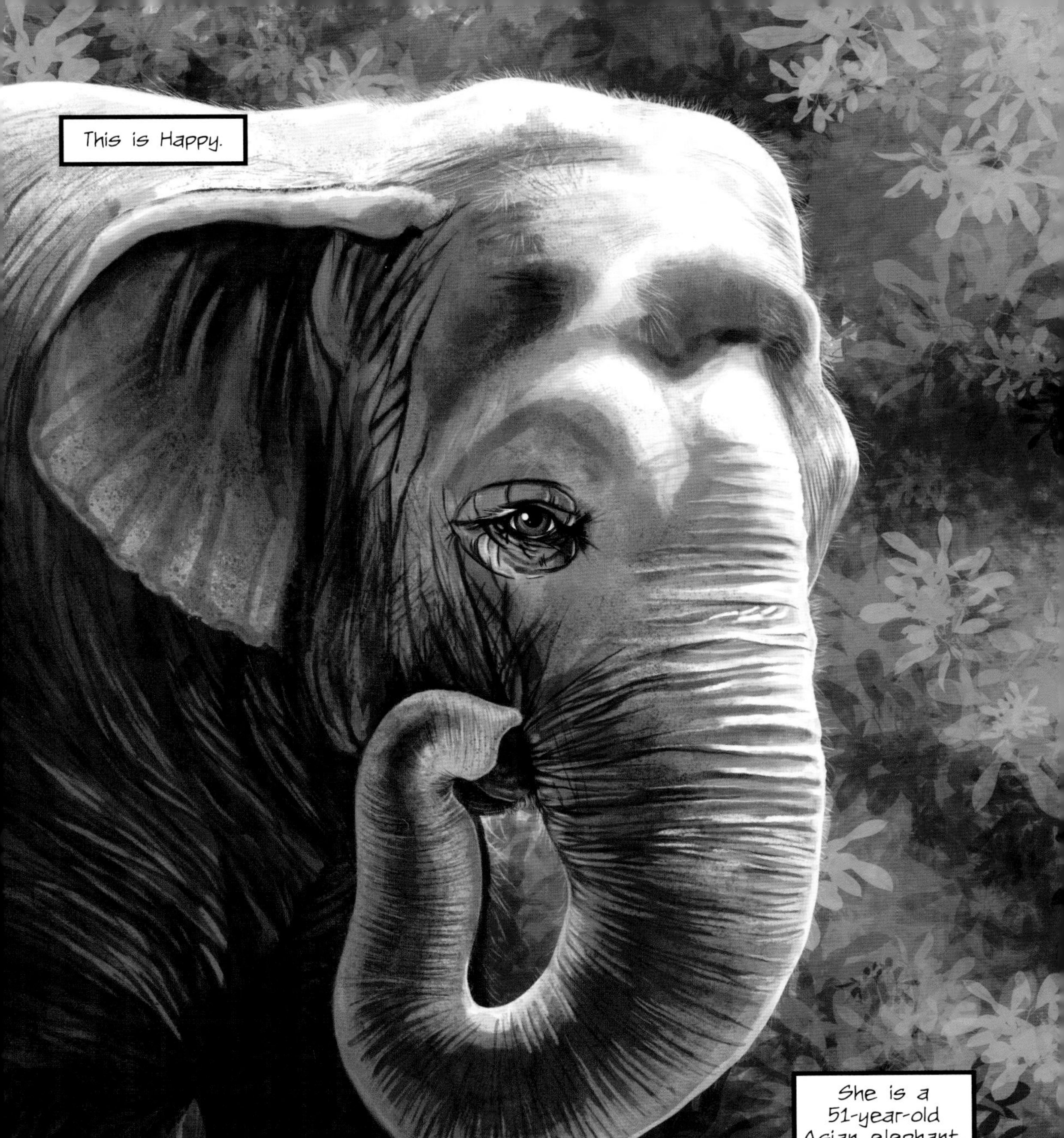
This is Happy.
She is a 51-year-old Asian elephant.

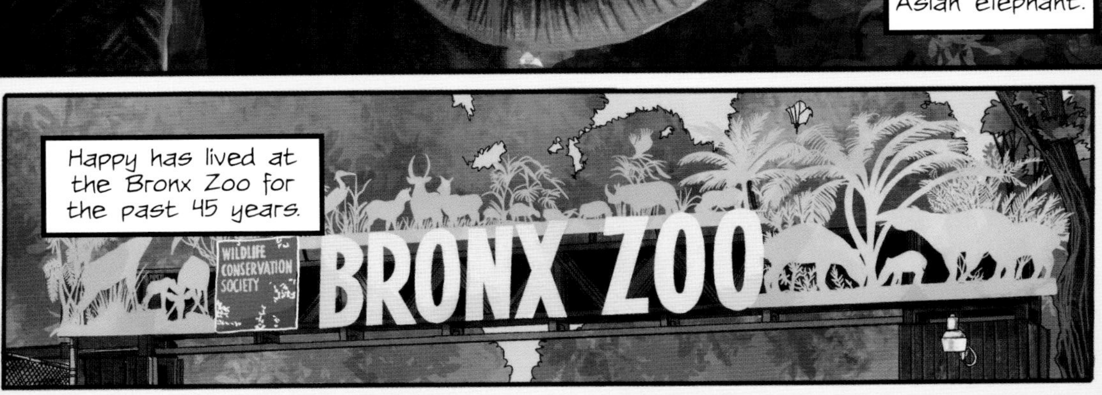
Happy has lived at the Bronx Zoo for the past 45 years.
WILDLIFE CONSERVATION SOCIETY
BRONX ZOO

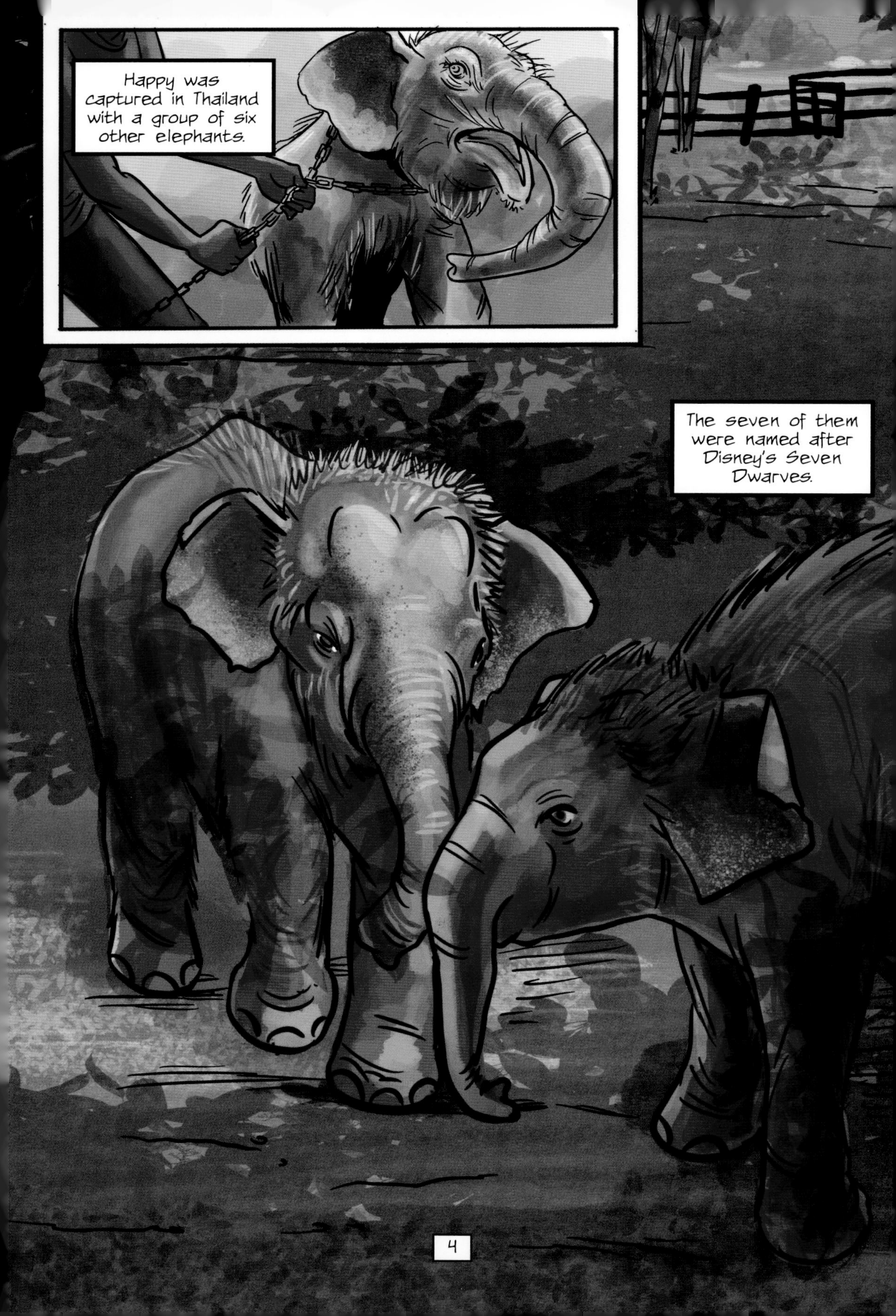
Happy was captured in Thailand with a group of six other elephants.
The seven of them were named after Disney's Seven Dwarves.

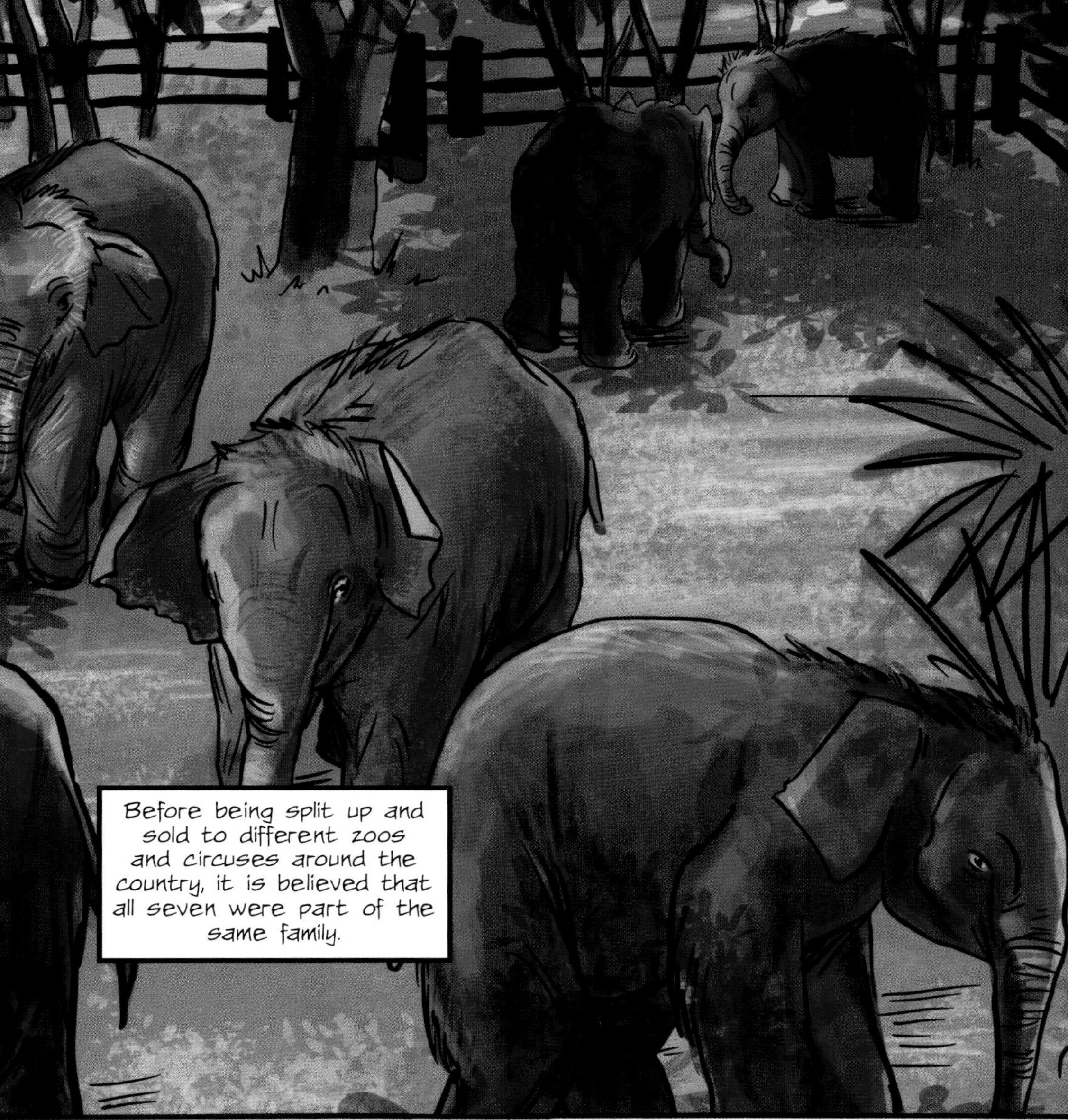
Before being split up and sold to different zoos and circuses around the country, it is believed that all seven were part of the same family.

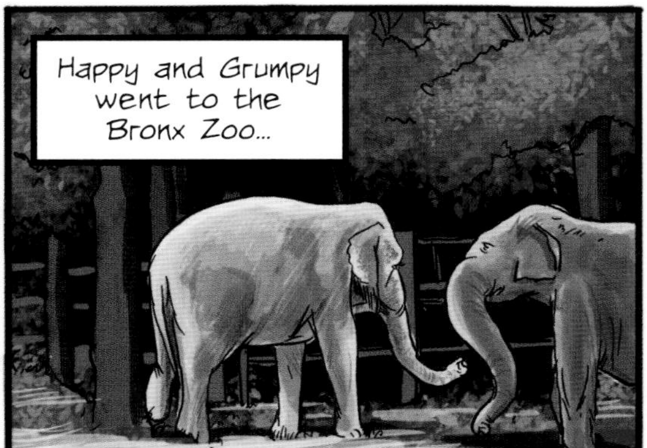
Happy and Grumpy went to the Bronx Zoo...

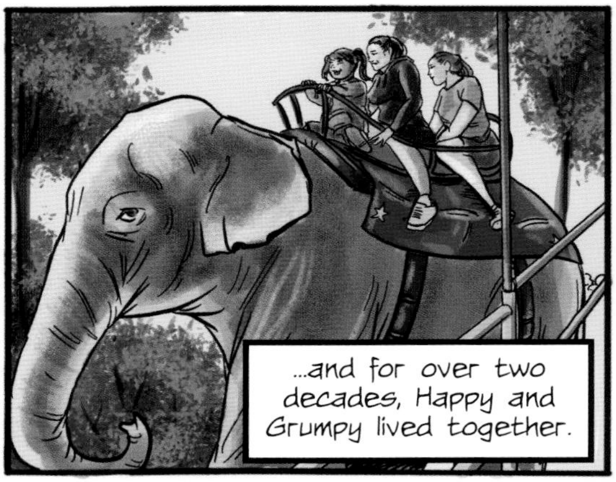
...and for over two decades, Happy and Grumpy lived together.

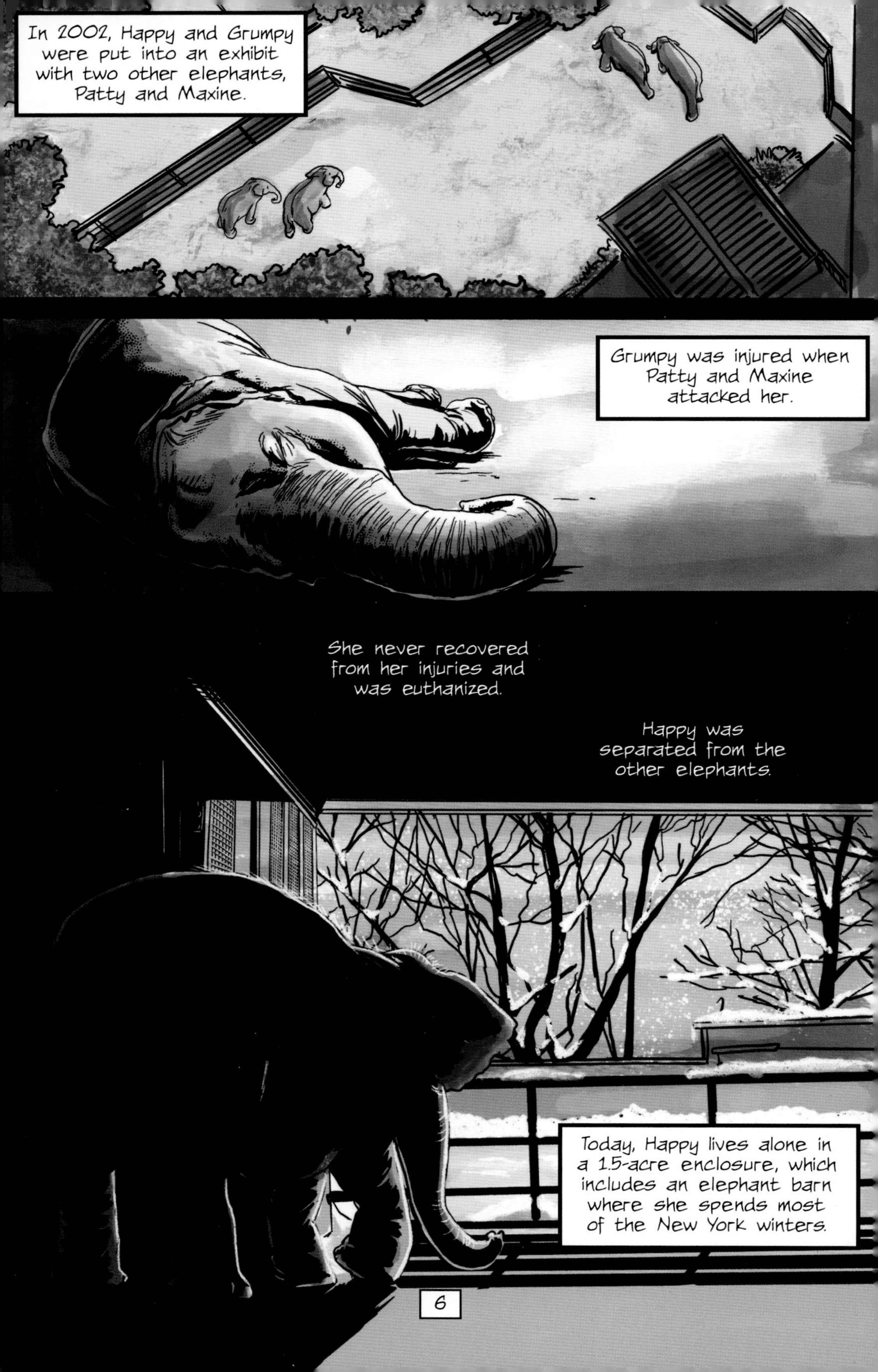
In 2002, Happy and Grumpy were put into an exhibit with two other elephants, Patty and Maxine.
Grumpy was injured when Patty and Maxine attacked her.
She never recovered from her injuries and was euthanized.
Happy was separated from the other elephants.
Today, Happy lives alone in a 1.5-acre enclosure, which includes an elephant barn where she spends most of the New York winters.

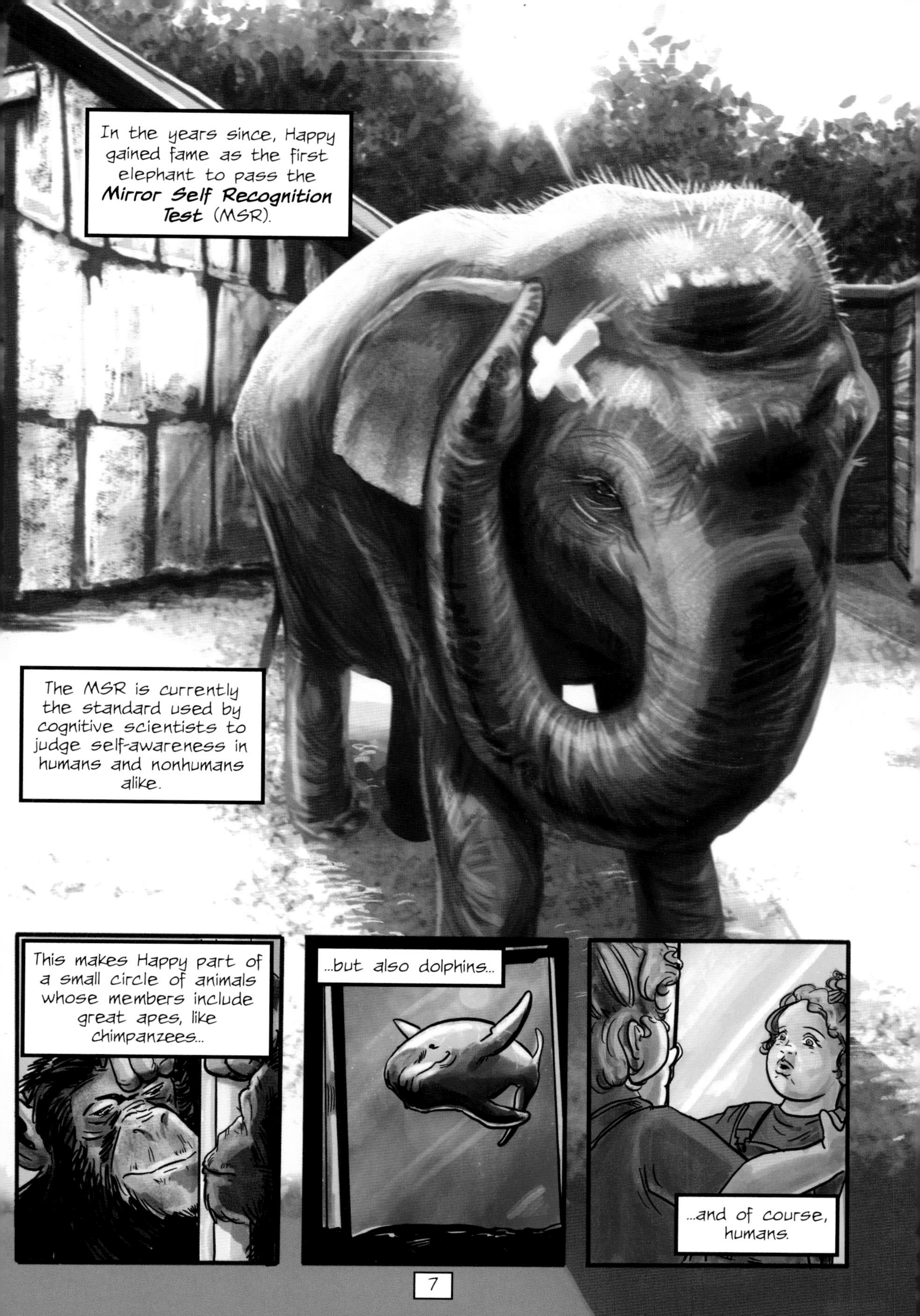
In the years since, Happy gained fame as the first elephant to pass the ***Mirror Self Recognition Test*** (MSR).
The MSR is currently the standard used by cognitive scientists to judge self-awareness in humans and nonhumans alike.
This makes Happy part of a small circle of animals whose members include great apes, like chimpanzees...
...but also dolphins...
...and of course, humans.

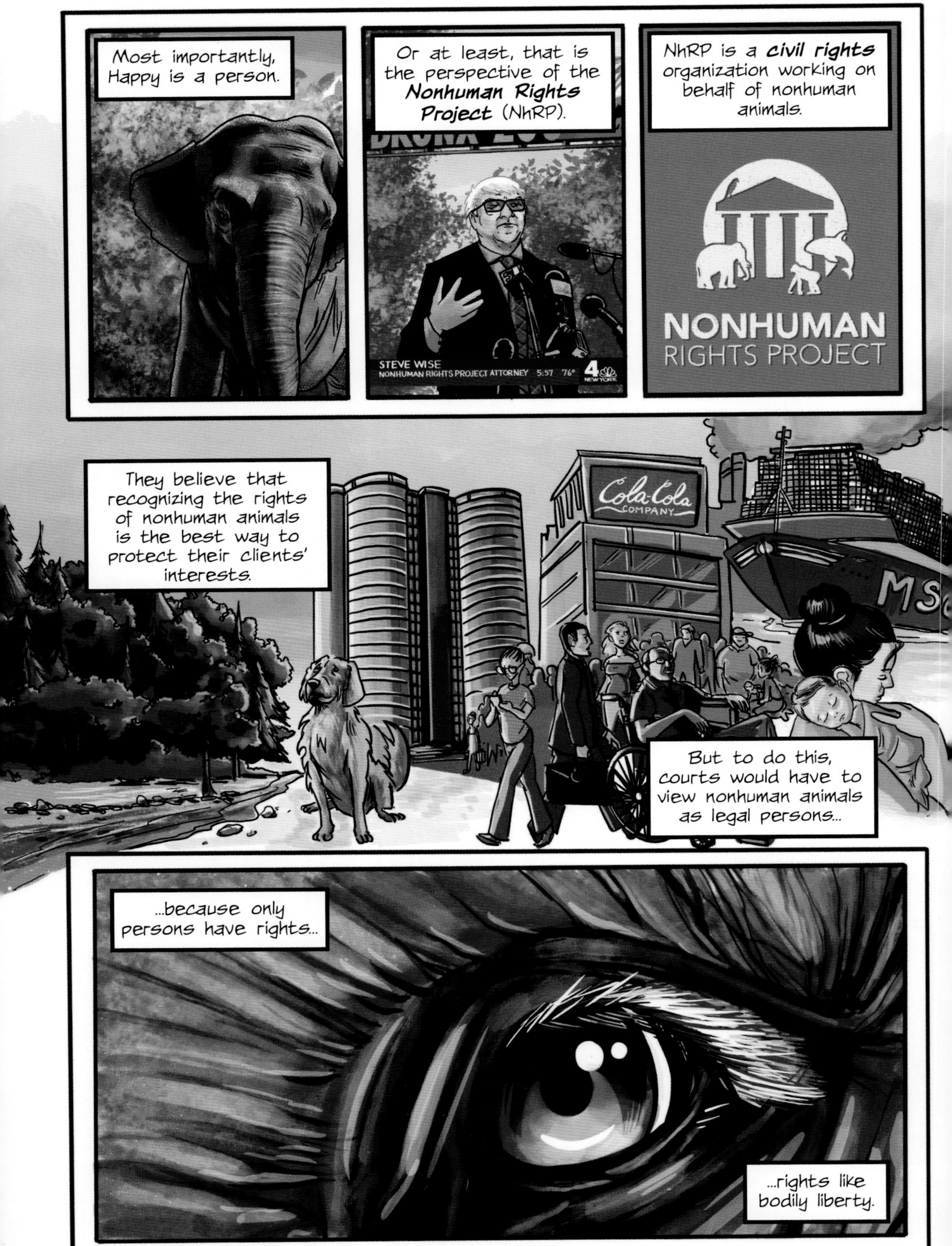
Most importantly, Happy is a person.
Or at least, that is the perspective of the **Nonhuman Rights Project** (NhRP).
STEVE WISE
NONHUMAN RIGHTS PROJECT ATTORNEY 5:57 76°
NhRP is a **civil rights** organization working on behalf of nonhuman animals.
NONHUMAN RIGHTS PROJECT
They believe that recognizing the rights of nonhuman animals is the best way to protect their clients' interests.
Cola-Cola COMPANY
But to do this, courts would have to view nonhuman animals as legal persons...
...because only persons have rights...
...rights like bodily liberty.

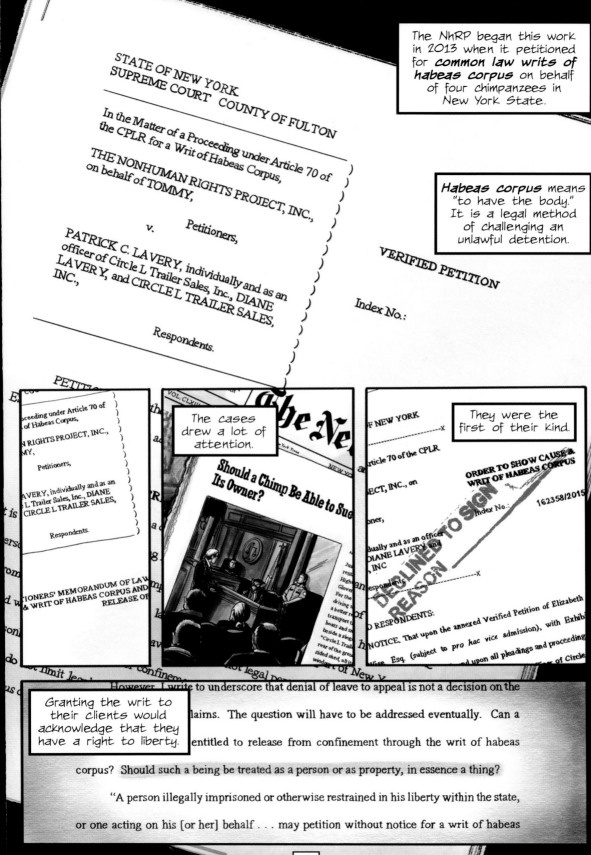

The NhRP began this work in 2013 when it petitioned for **common law writs of habeas corpus** on behalf of four chimpanzees in New York State.

Habeas corpus means "to have the body." It is a legal method of challenging an unlawful detention.

STATE OF NEW YORK
SUPREME COURT COUNTY OF FULTON
_____
In the Matter of a Proceeding under Article 70 of
the CPLR for a Writ of Habeas Corpus,

THE NONHUMAN RIGHTS PROJECT, INC.,
on behalf of TOMMY,

              Petitioners,

     v.

PATRICK C. LAVERY, individually and as an
officer of Circle L Trailer Sales, Inc., DIANE
LAVERY, and CIRCLE L TRAILER SALES,
INC.,

              Respondents.

VERIFIED PETITION

Index No.:

The cases drew a lot of attention.

Should a Chimp Be Able to Sue Its Owner?

They were the first of their kind.

ORDER TO SHOW CAUSE & WRIT OF HABEAS CORPUS

Index No.: 162358/2015

DECLINED TO SIGN — REASON

However, I write to underscore that denial of leave to appeal is not a decision on the

Granting the writ to their clients would acknowledge that they have a right to liberty.

claims. The question will have to be addressed eventually. Can a

entitled to release from confinement through the writ of habeas

corpus? Should such a being be treated as a person or as property, in essence a thing?

"A person illegally imprisoned or otherwise restrained in his liberty within the state,

or one acting on his [or her] behalf . . . may petition without notice for a writ of habeas

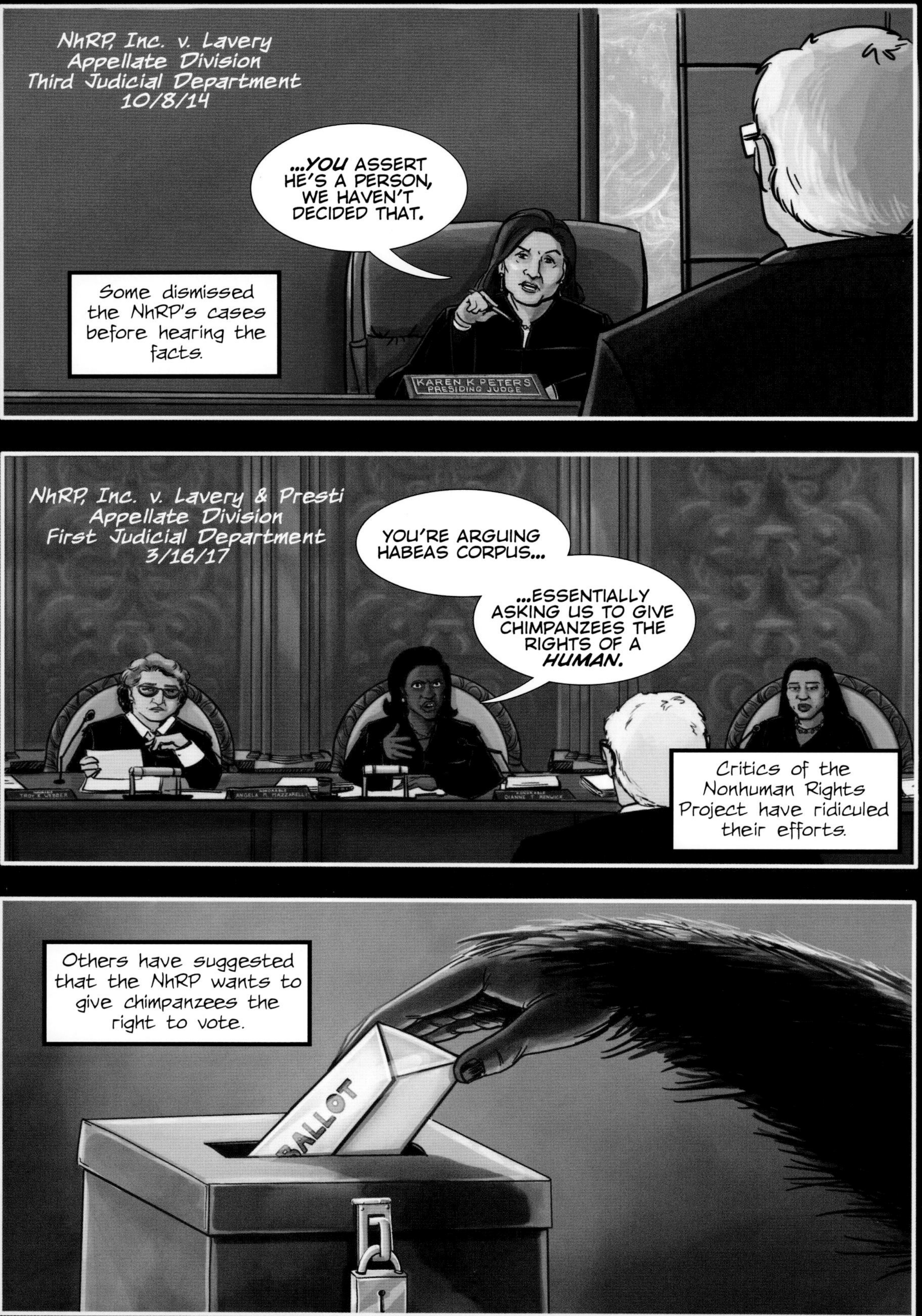
NhRP, Inc. v. Lavery
Appellate Division
Third Judicial Department
10/8/14
...YOU ASSERT HE'S A PERSON, WE HAVEN'T DECIDED THAT.
Some dismissed the NhRP's cases before hearing the facts.
KAREN K PETERS
PRESIDING JUDGE
NhRP, Inc. v. Lavery & Presti
Appellate Division
First Judicial Department
3/16/17
YOU'RE ARGUING HABEAS CORPUS...
...ESSENTIALLY ASKING US TO GIVE CHIMPANZEES THE RIGHTS OF A HUMAN.
Critics of the Nonhuman Rights Project have ridiculed their efforts.
Others have suggested that the NhRP wants to give chimpanzees the right to vote.
BALLOT

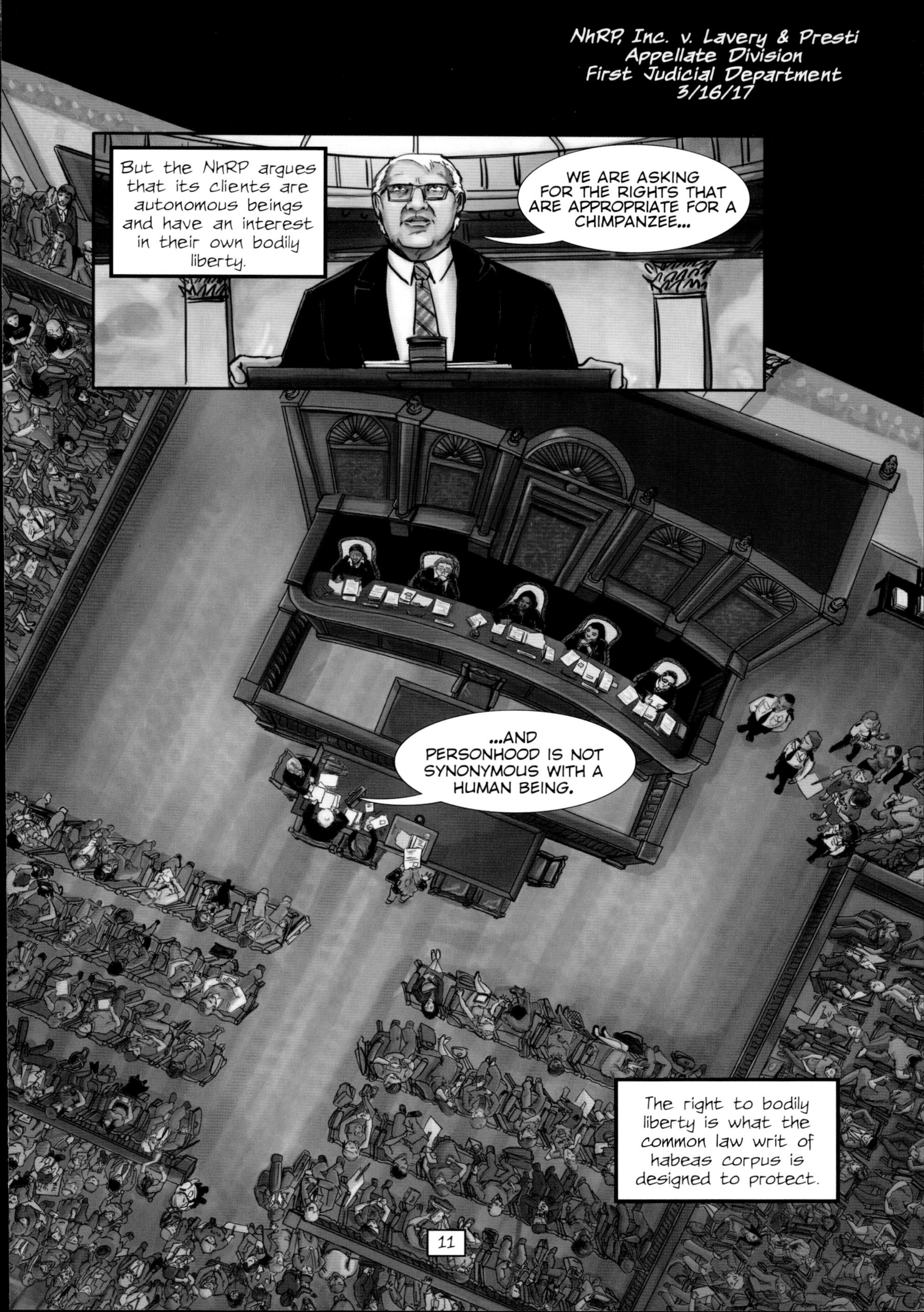
NhRP, Inc. v. Lavery & Presti
Appellate Division
First Judicial Department
3/16/17
But the NhRP argues that its clients are autonomous beings and have an interest in their own bodily liberty.
WE ARE ASKING FOR THE RIGHTS THAT ARE APPROPRIATE FOR A CHIMPANZEE...
...AND PERSONHOOD IS NOT SYNONYMOUS WITH A HUMAN BEING.
The right to bodily liberty is what the common law writ of habeas corpus is designed to protect.

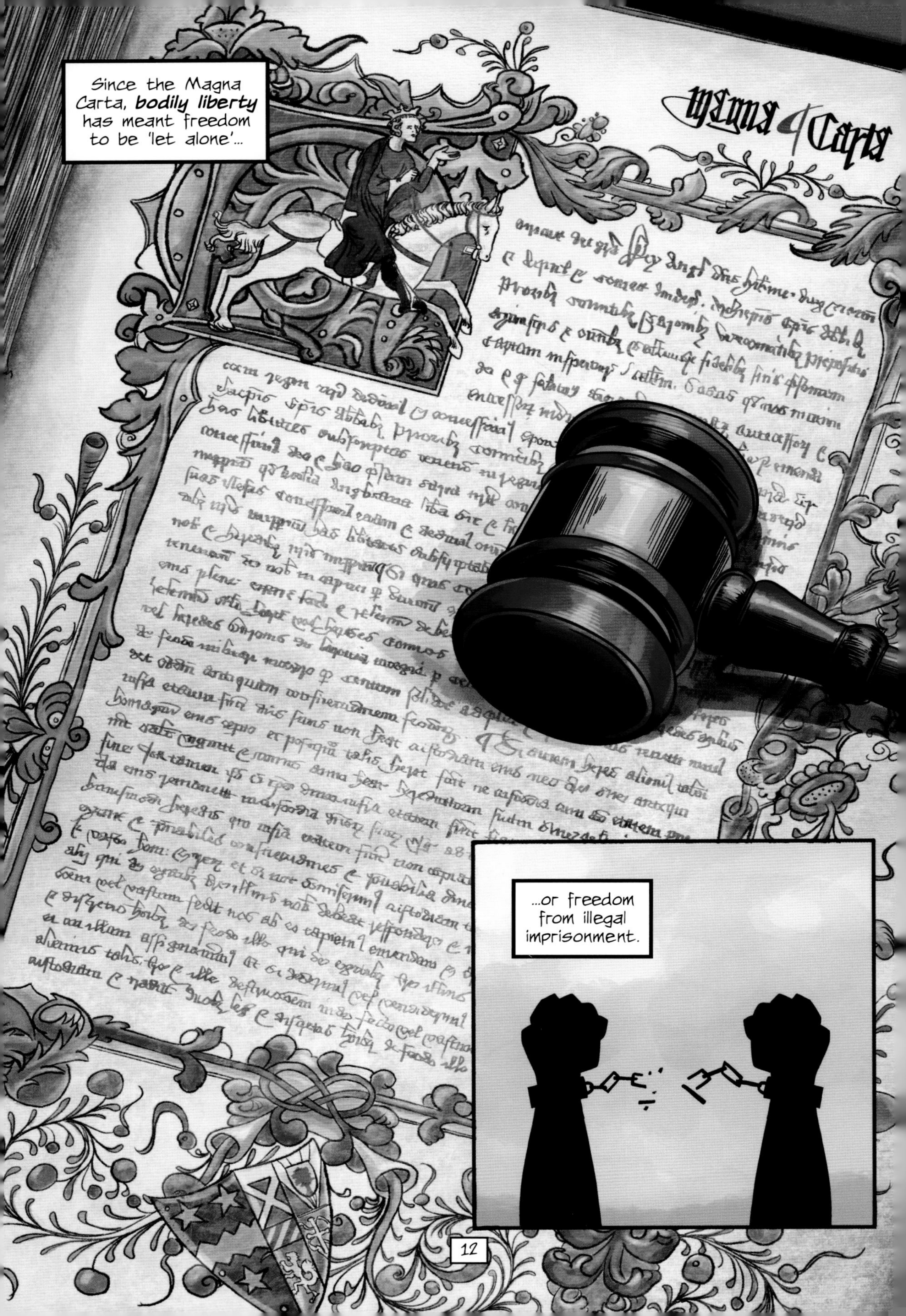
Since the Magna Carta, **bodily liberty** has meant freedom to be 'let alone'...
Magna Carta
...or freedom from illegal imprisonment.

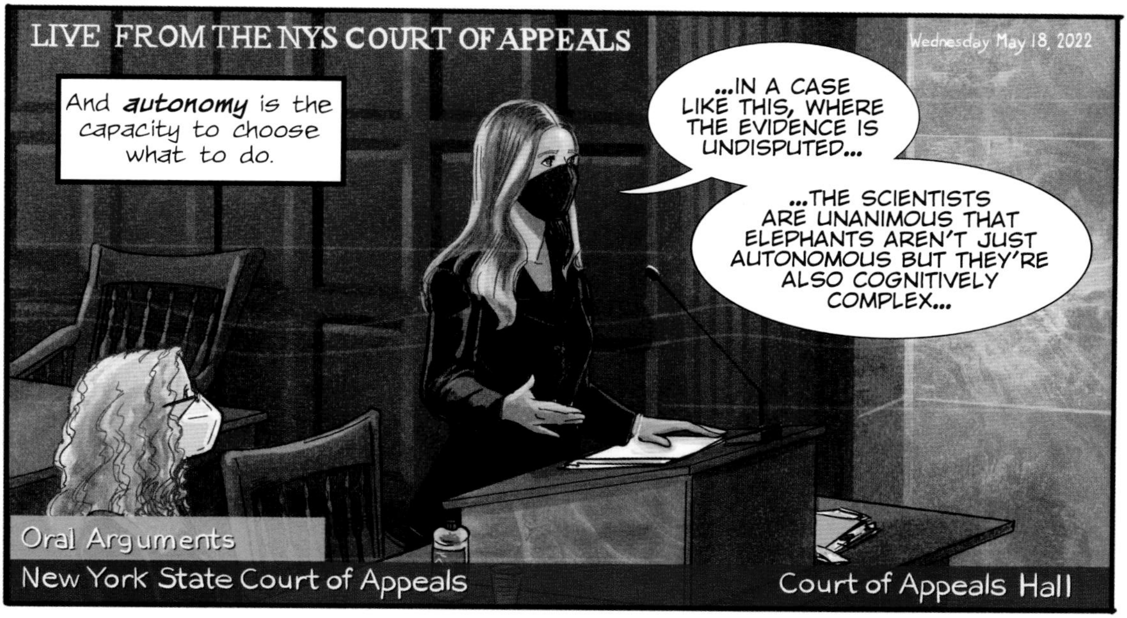
LIVE FROM THE NYS COURT OF APPEALS
Wednesday May 18, 2022
And ***autonomy*** is the capacity to choose what to do.
...IN A CASE LIKE THIS, WHERE THE EVIDENCE IS UNDISPUTED...
...THE SCIENTISTS ARE UNANIMOUS THAT ELEPHANTS AREN'T JUST AUTONOMOUS BUT THEY'RE ALSO COGNITIVELY COMPLEX...
Oral Arguments
New York State Court of Appeals
Court of Appeals Hall

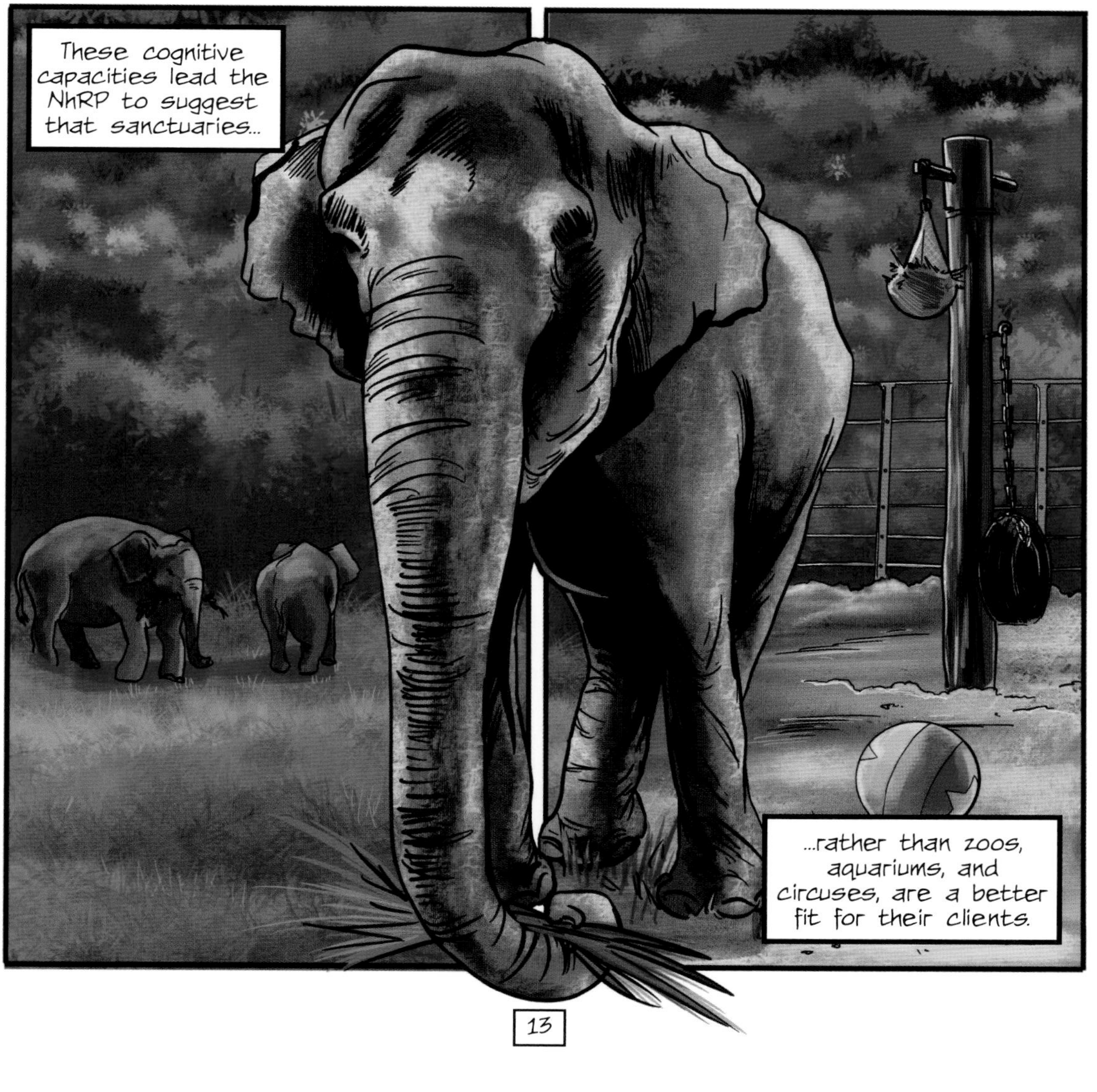
These cognitive capacities lead the NhRP to suggest that sanctuaries...
...rather than zoos, aquariums, and circuses, are a better fit for their clients.

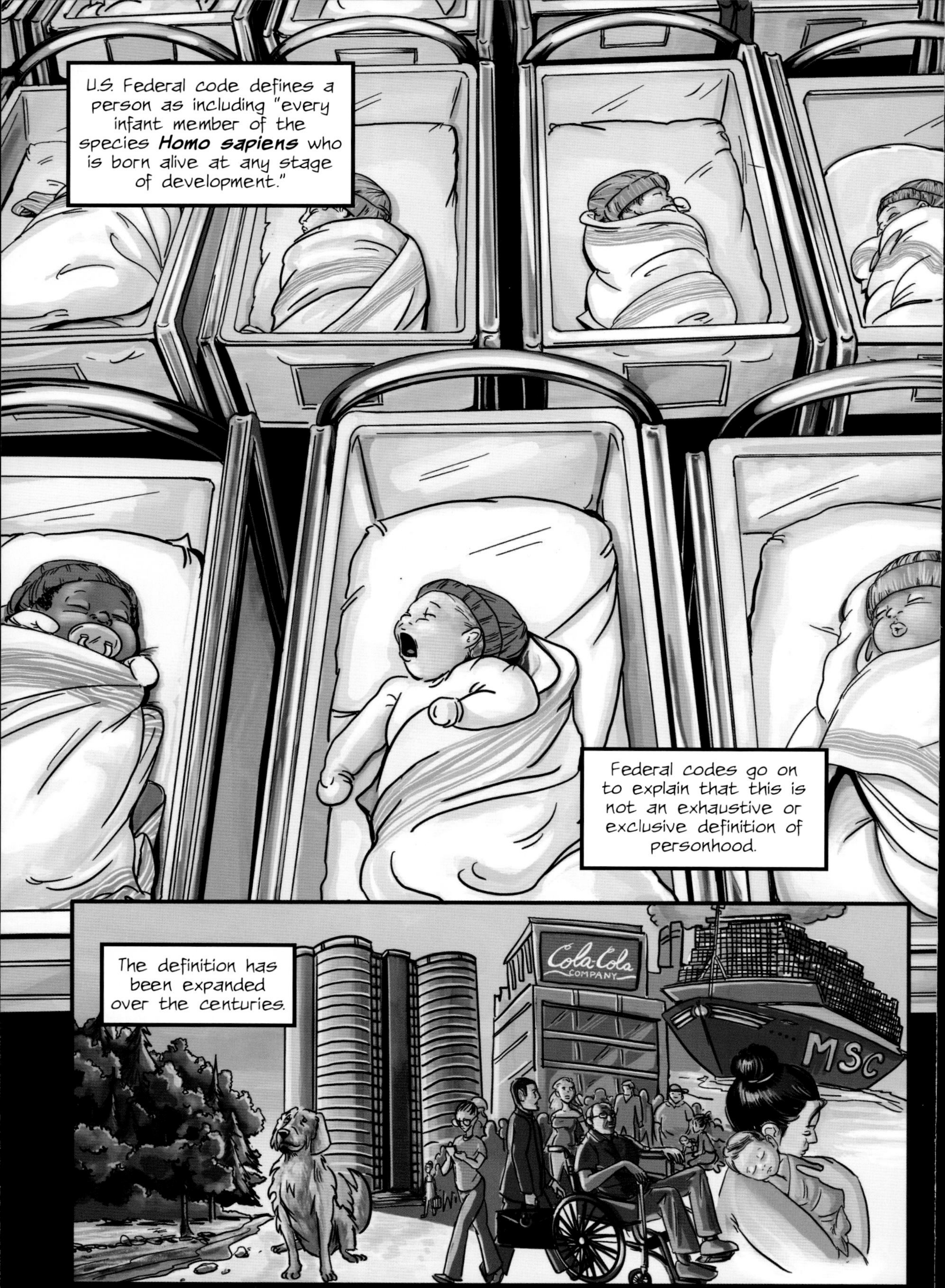
U.S. Federal code defines a person as including "every infant member of the species **Homo sapiens** who is born alive at any stage of development."
Federal codes go on to explain that this is not an exhaustive or exclusive definition of personhood.
The definition has been expanded over the centuries.
Cola-Cola
COMPANY
MSC

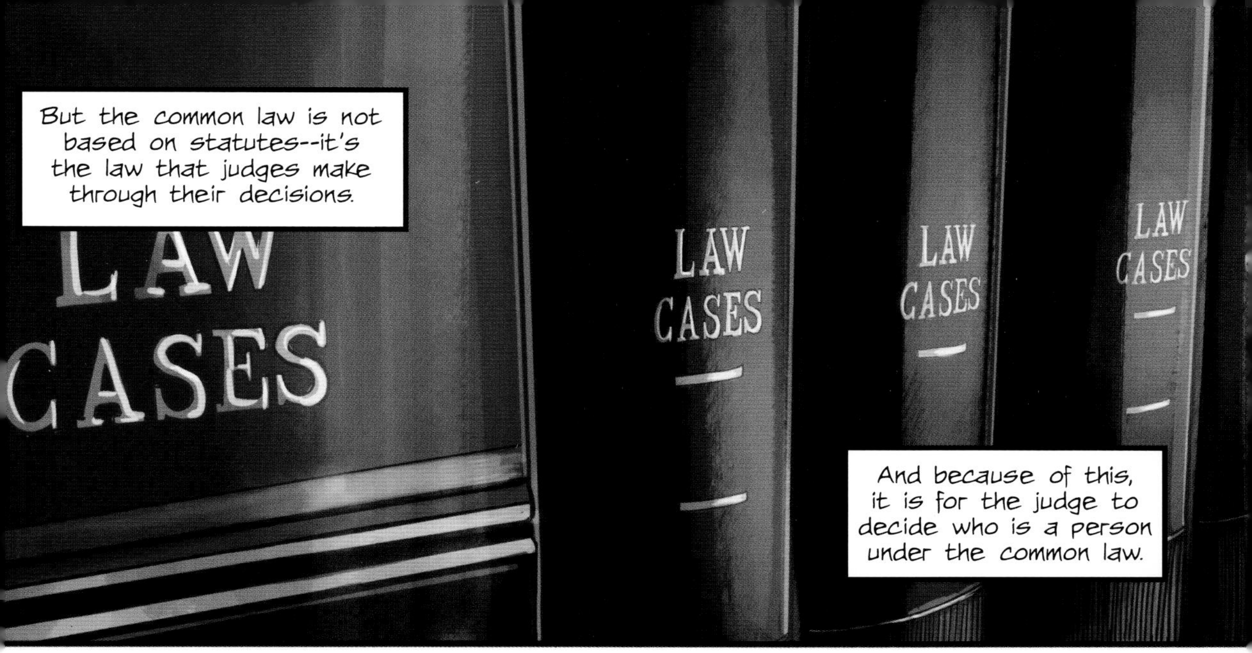
But the common law is not based on statutes--it's the law that judges make through their decisions.
LAW CASES
LAW CASES
LAW CASES
LAW CASES
And because of this, it is for the judge to decide who is a person under the common law.

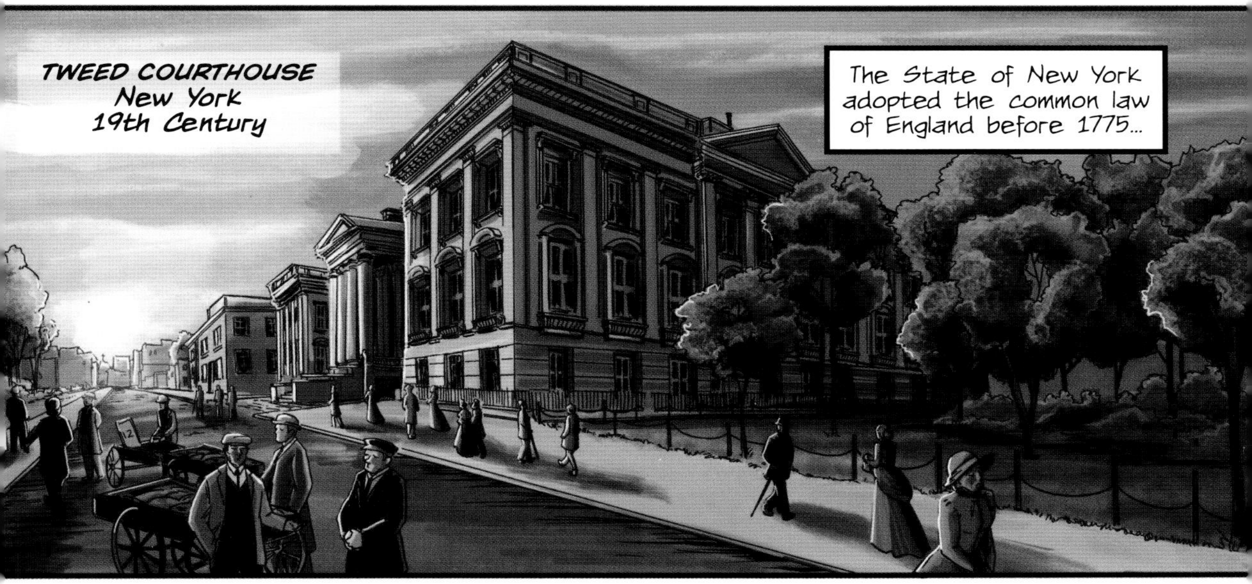
TWEED COURTHOUSE New York 19th Century
The State of New York adopted the common law of England before 1775...

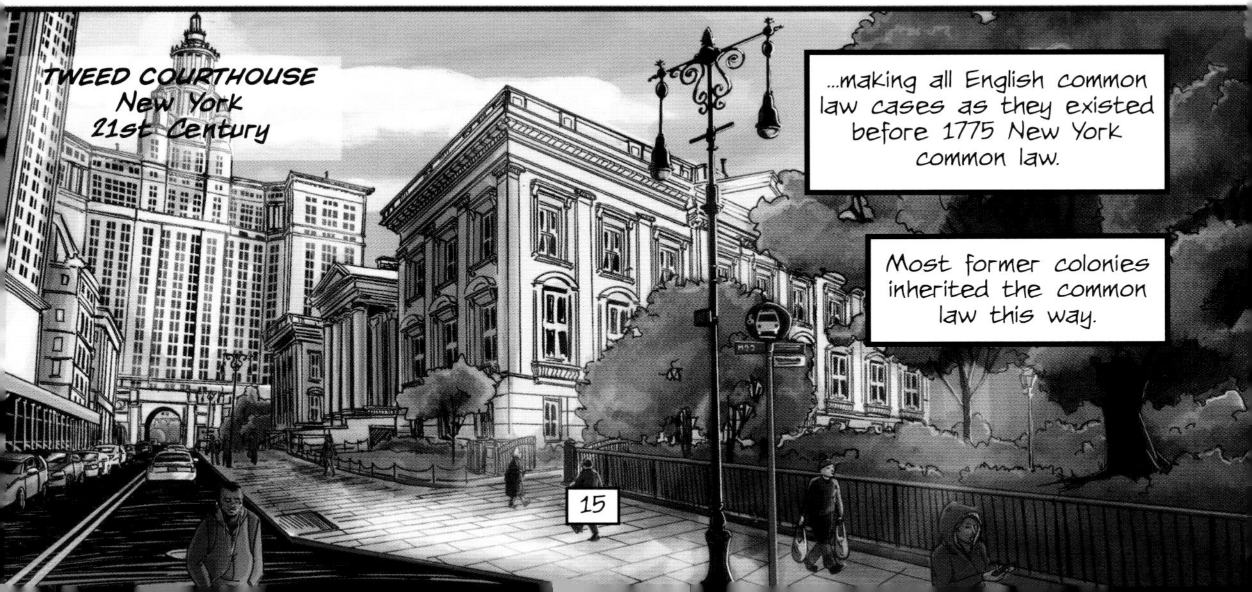
TWEED COURTHOUSE New York 21st Century
...making all English common law cases as they existed before 1775 New York common law.
Most former colonies inherited the common law this way.

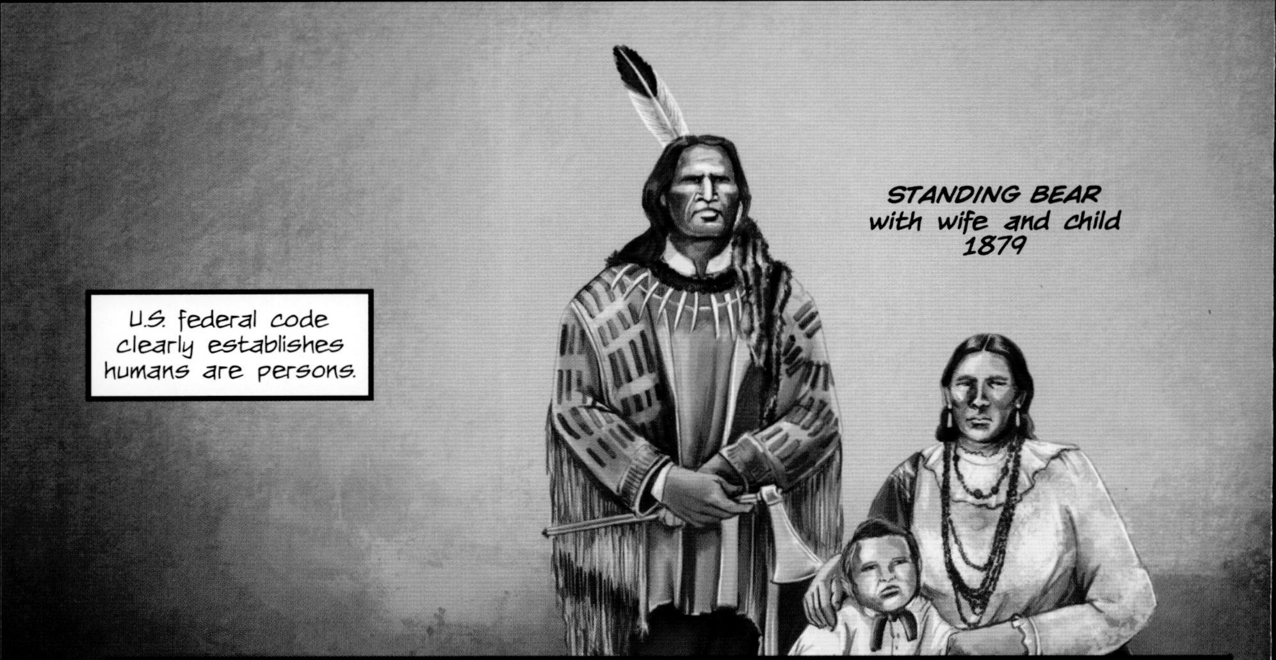
U.S. federal code clearly establishes humans are persons.
STANDING BEAR
with wife and child
1879

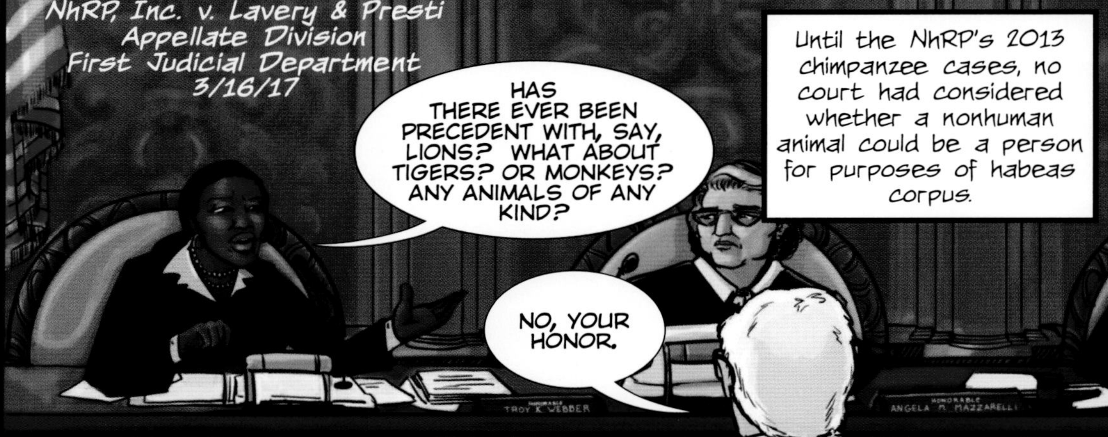
NhRP, Inc. v. Lavery & Presti
Appellate Division
First Judicial Department
3/16/17
HAS THERE EVER BEEN PRECEDENT WITH, SAY, LIONS? WHAT ABOUT TIGERS? OR MONKEYS? ANY ANIMALS OF ANY KIND?
Until the NhRP's 2013 chimpanzee cases, no court had considered whether a nonhuman animal could be a person for purposes of habeas corpus.
NO, YOUR HONOR.
TROY K. WEBBER
ANGELA M. MAZZARELLI

When the First Judicial Department upheld the lower court's ruling against NhRP's chimpanzee clients...

Needless to say, unlike human beings, chimpanzees cannot bear any legal duties, submit to societal responsibilities or be held legally accountable for their actions. In our view, it is this incapability to bear any legal responsibilities and societal duties that renders it inappropriate to confer upon chimpanzees the legal rights – such as the fundamental right to liberty protected by the writ of habeas corpus – that have been afforded to human beings.

...critics of the decision observed that many children and disabled humans would not fit this definition of person.

Judge Eugene Fahey's opinion in the final chimpanzee case was critical of the decisions of New York Appellate Courts.

The reliance on a paradigm that determines entitlement to a court decision based on whether the party is considered a "person" or relegated to the category of a "thing" amounts to a refusal to confront a manifest injustice. Whether a being has the right to seek freedom from confinement through the writ of habeas corpus should not be treated as a simple either/or proposition. The evolving nature of life makes clear that chimpanzees and humans exist on a continuum of living beings. Chimpanzees share at least 96% of their DNA with humans. They are autonomous, intelligent creatures. To solve this dilemma, we have to recognize its complexity and confront it.

He described mistakes made by the New York courts as "...a refusal to confront a manifest injustice."

JUDGE FAHEY
Court of Appeals
State of New York
5/8/18

JUSTICE TUITT
NhRP v. Breheny
State of New York
Supreme Court
Bronx County
1/6/20

IN GOD WE TRUST

In 2019, Justice Alison Tuitt gave time for counsel on both sides to make their arguments regarding the merits of Happy's case.

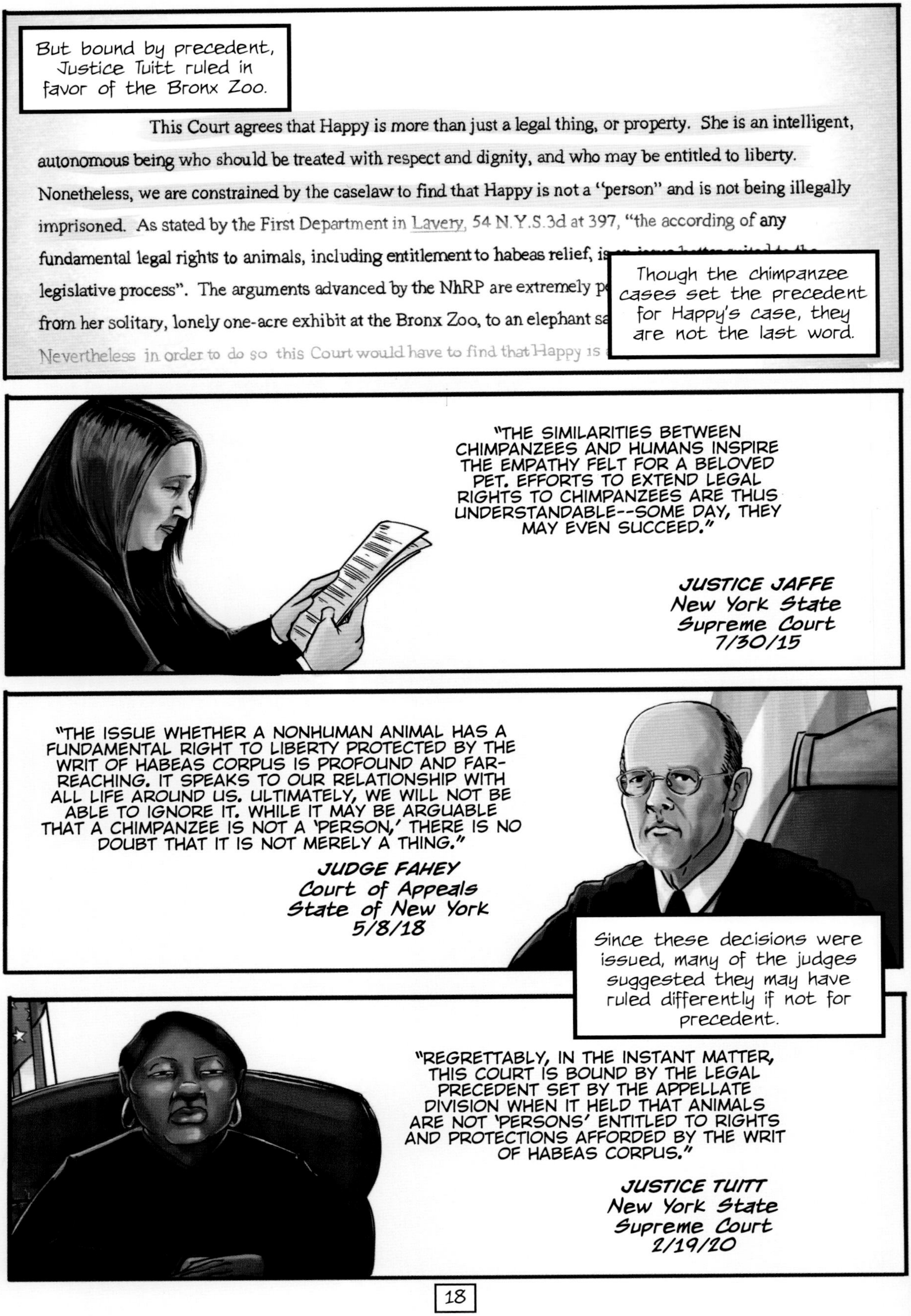
But bound by precedent, Justice Tuitt ruled in favor of the Bronx Zoo.
This Court agrees that Happy is more than just a legal thing, or property. She is an intelligent, autonomous being who should be treated with respect and dignity, and who may be entitled to liberty. Nonetheless, we are constrained by the caselaw to find that Happy is not a "person" and is not being illegally imprisoned. As stated by the First Department in Lavery, 54 N.Y.S.3d at 397, "the according of any fundamental legal rights to animals, including entitlement to habeas relief, is
legislative process". The arguments advanced by the NhRP are extremely
from her solitary, lonely one-acre exhibit at the Bronx Zoo, to an elephant
Nevertheless in order to do so this Court would have to find that Happy is
Though the chimpanzee cases set the precedent for Happy's case, they are not the last word.
"THE SIMILARITIES BETWEEN CHIMPANZEES AND HUMANS INSPIRE THE EMPATHY FELT FOR A BELOVED PET. EFFORTS TO EXTEND LEGAL RIGHTS TO CHIMPANZEES ARE THUS UNDERSTANDABLE--SOME DAY, THEY MAY EVEN SUCCEED."
JUSTICE JAFFE
New York State
Supreme Court
7/30/15
"THE ISSUE WHETHER A NONHUMAN ANIMAL HAS A FUNDAMENTAL RIGHT TO LIBERTY PROTECTED BY THE WRIT OF HABEAS CORPUS IS PROFOUND AND FAR-REACHING. IT SPEAKS TO OUR RELATIONSHIP WITH ALL LIFE AROUND US. ULTIMATELY, WE WILL NOT BE ABLE TO IGNORE IT. WHILE IT MAY BE ARGUABLE THAT A CHIMPANZEE IS NOT A 'PERSON,' THERE IS NO DOUBT THAT IT IS NOT MERELY A THING."
JUDGE FAHEY
Court of Appeals
State of New York
5/8/18
Since these decisions were issued, many of the judges suggested they may have ruled differently if not for precedent.
"REGRETTABLY, IN THE INSTANT MATTER, THIS COURT IS BOUND BY THE LEGAL PRECEDENT SET BY THE APPELLATE DIVISION WHEN IT HELD THAT ANIMALS ARE NOT 'PERSONS' ENTITLED TO RIGHTS AND PROTECTIONS AFFORDED BY THE WRIT OF HABEAS CORPUS."
JUSTICE TUITT
New York State
Supreme Court
2/19/20

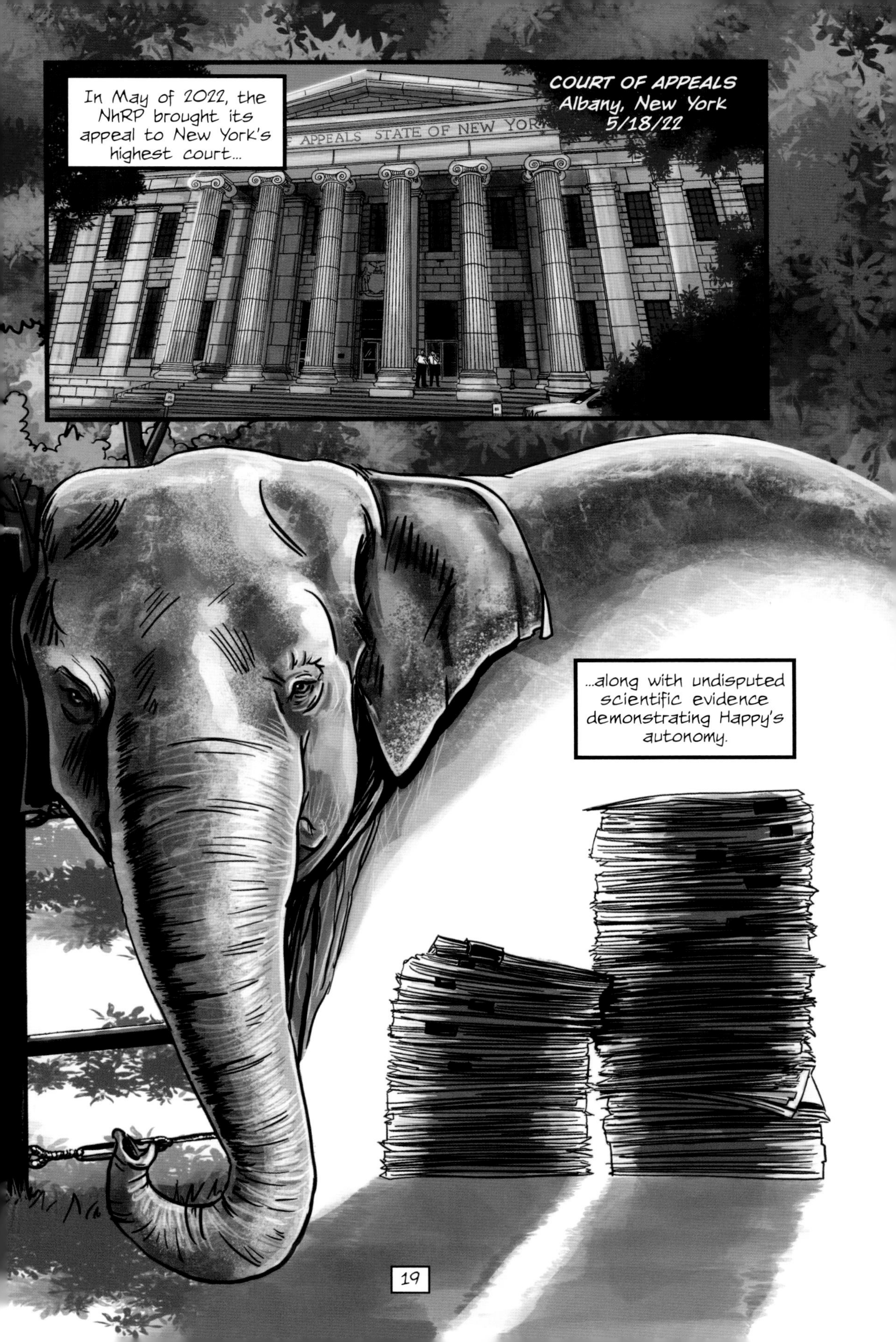

In May of 2022, the NhRP brought its appeal to New York's highest court...
COURT OF APPEALS
Albany, New York
5/18/22
STATE OF NEW YORK
...along with undisputed scientific evidence demonstrating Happy's autonomy.

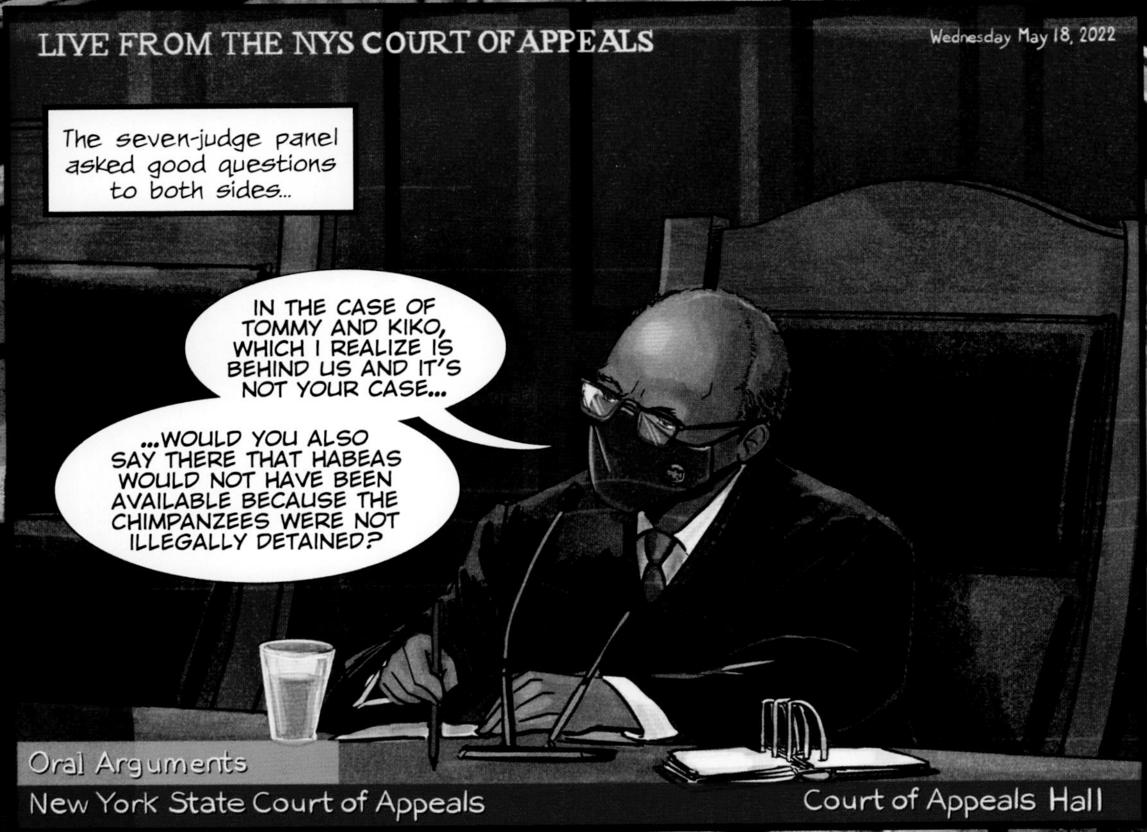
LIVE FROM THE NYS COURT OF APPEALS
Wednesday May 18, 2022
The seven-judge panel asked good questions to both sides...
IN THE CASE OF TOMMY AND KIKO, WHICH I REALIZE IS BEHIND US AND IT'S NOT YOUR CASE...
...WOULD YOU ALSO SAY THERE THAT HABEAS WOULD NOT HAVE BEEN AVAILABLE BECAUSE THE CHIMPANZEES WERE NOT ILLEGALLY DETAINED?
Oral Arguments
New York State Court of Appeals
Court of Appeals Hall

...but ultimately ruled that only humans could use the writ of habeas corpus and that animals could not be legal persons.

- 2 - No. 52

habeas corpus relief on behalf of Happy, an elephant residing at the Bronx Zoo, in order to secure her transfer to an elephant sanctuary. Because the writ of habeas corpus is intended to protect the liberty right of *human beings* to be free of unlawful confinement, it has no applicability to Happy, a nonhuman animal who is not a "person" subjected to illegal detention. Thus, while no one disputes that elephants are intelligent beings deserving of proper care and compassion, the courts below properly granted the motion to dismiss the petition for a writ of habeas corpus, and we therefore affirm.

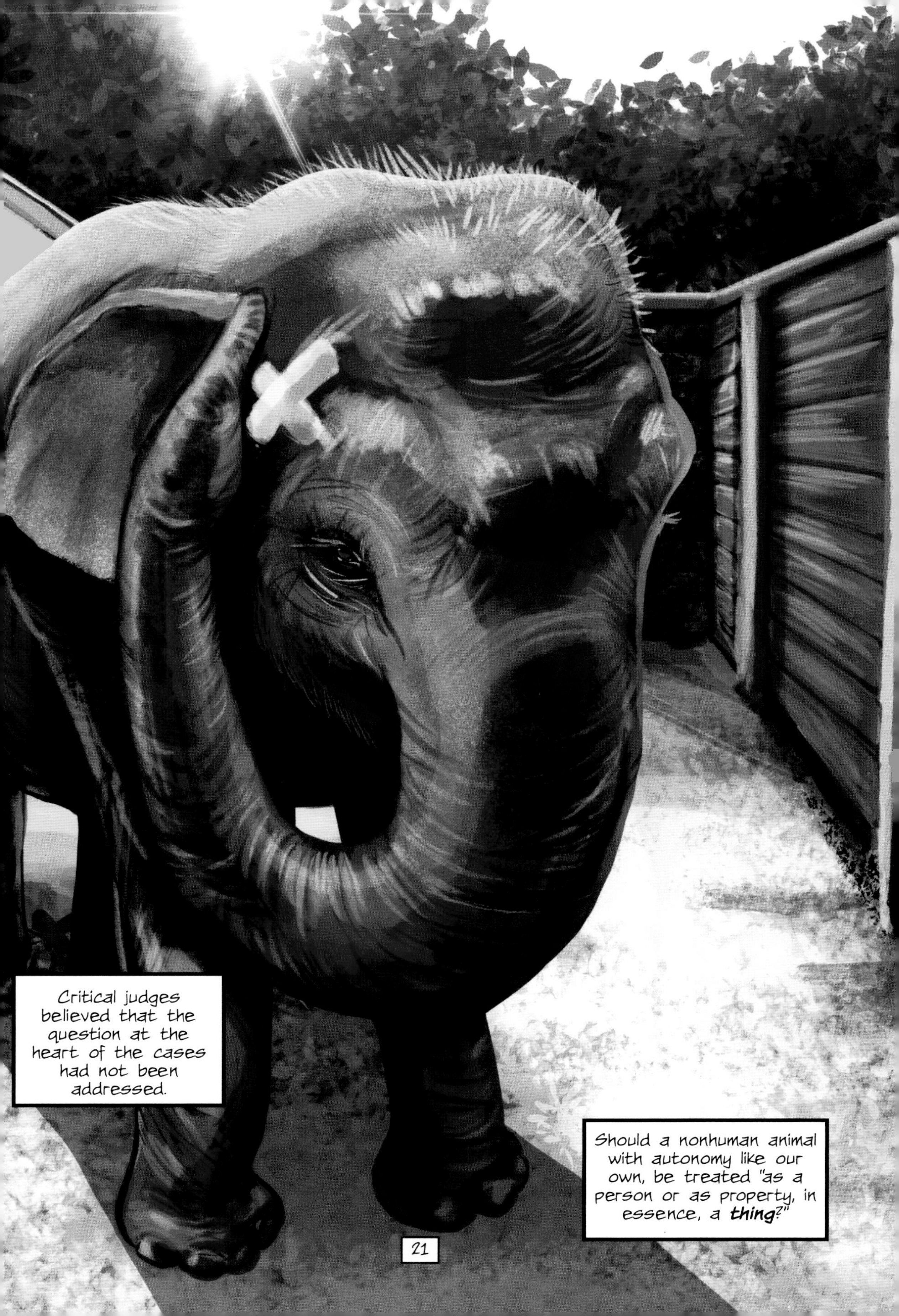
Critical judges believed that the question at the heart of the cases had not been addressed.
Should a nonhuman animal with autonomy like our own, be treated "as a person or as property, in essence, a ***thing***?"

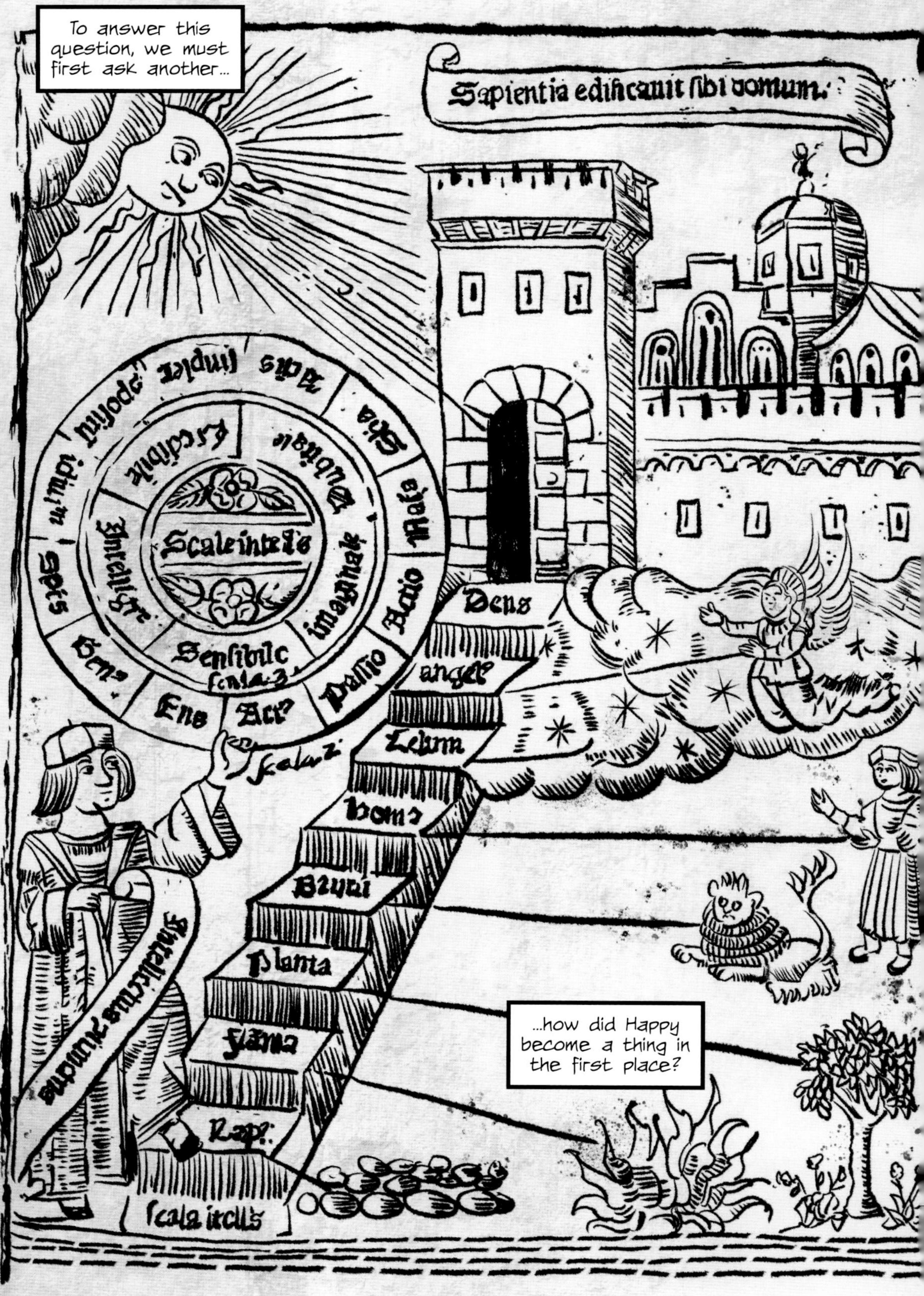
To answer this question, we must first ask another...
...how did Happy become a thing in the first place?

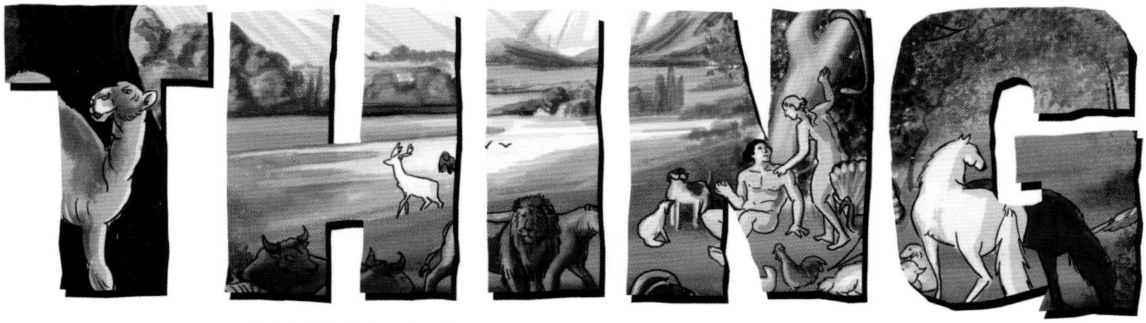
THING

CHAPTER 2: THE GREAT CHAIN OF BEING

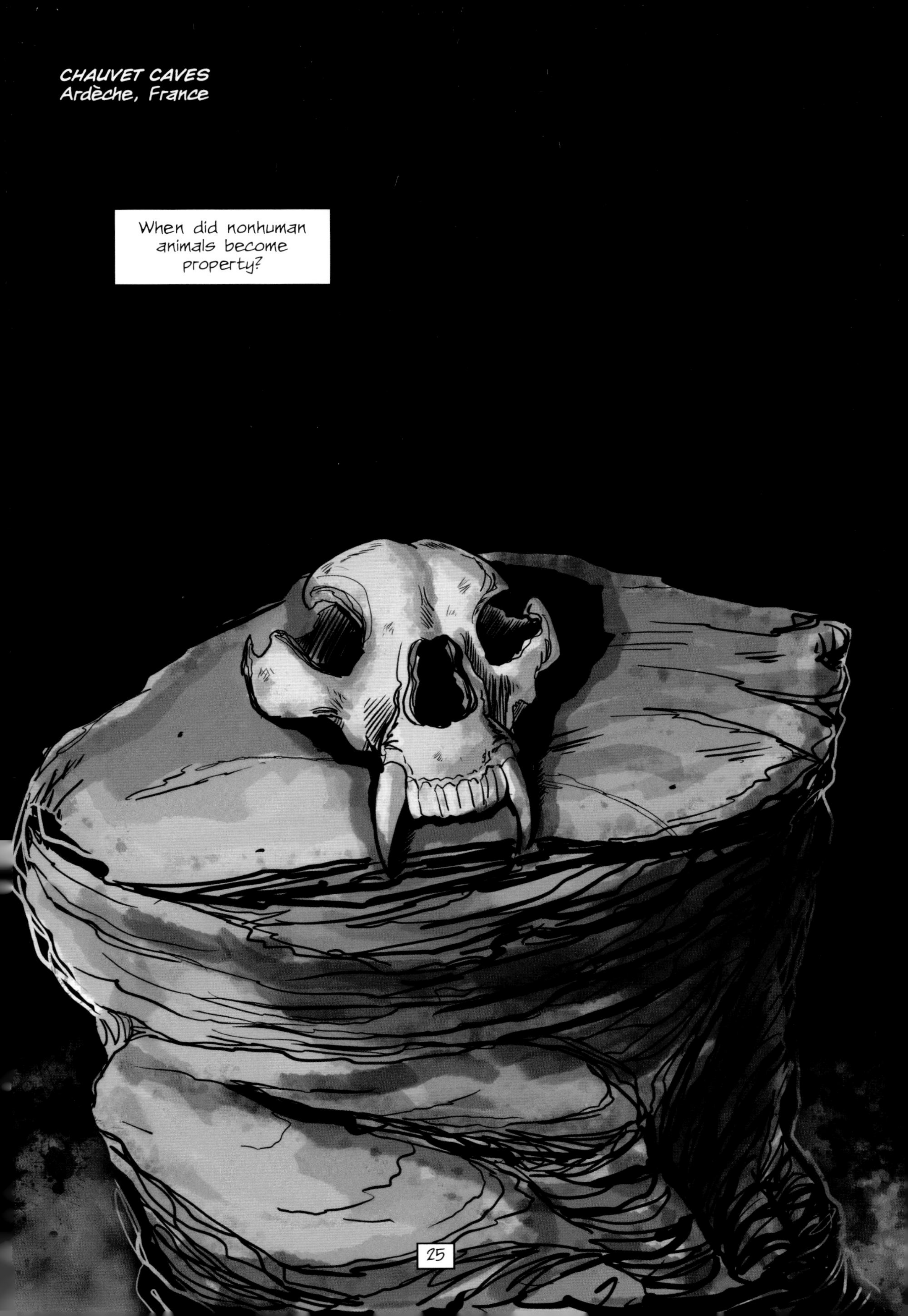
CHAUVET CAVES
Ardèche, France
When did nonhuman animals become property?

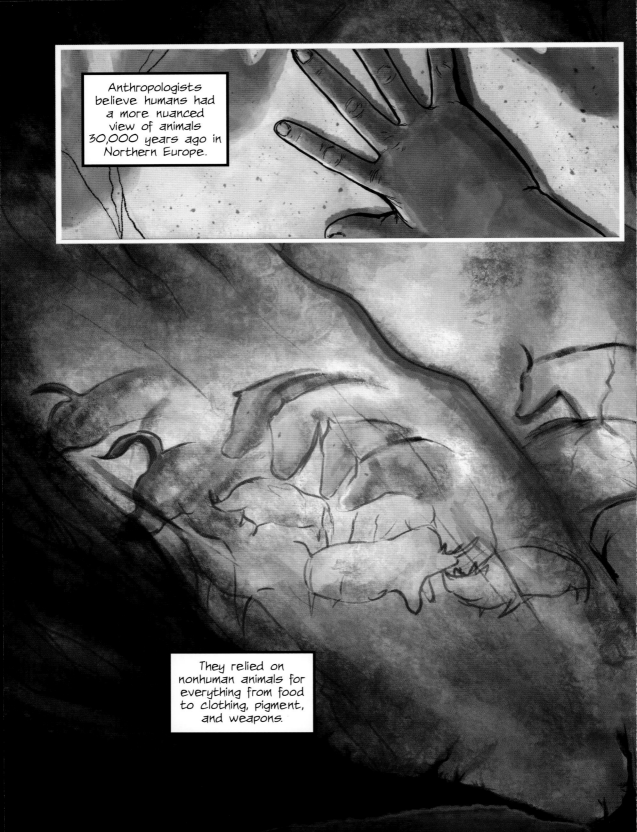

Anthropologists believe humans had a more nuanced view of animals 30,000 years ago in Northern Europe.

They relied on nonhuman animals for everything from food to clothing, pigment, and weapons.

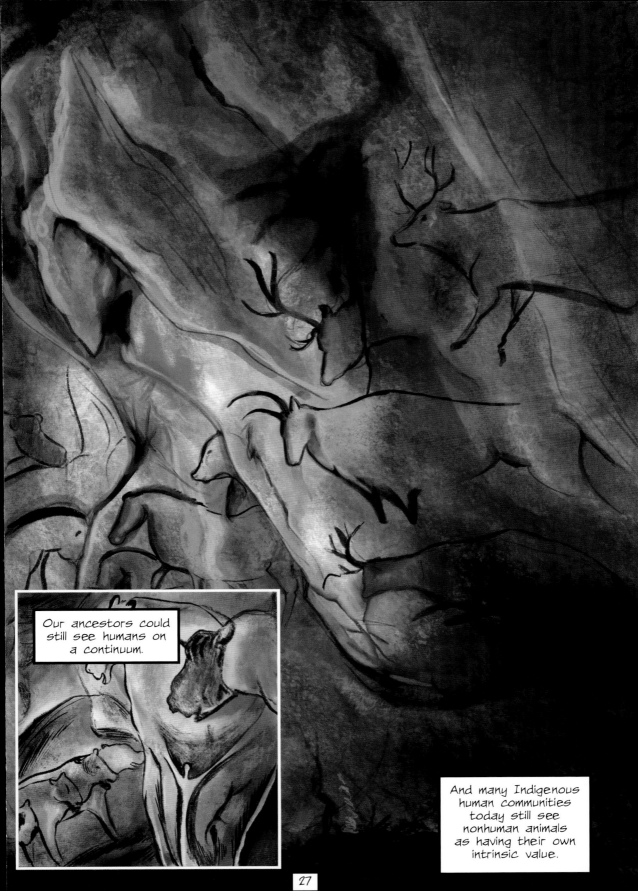

Our ancestors could still see humans on a continuum.

And many Indigenous human communities today still see nonhuman animals as having their own intrinsic value.

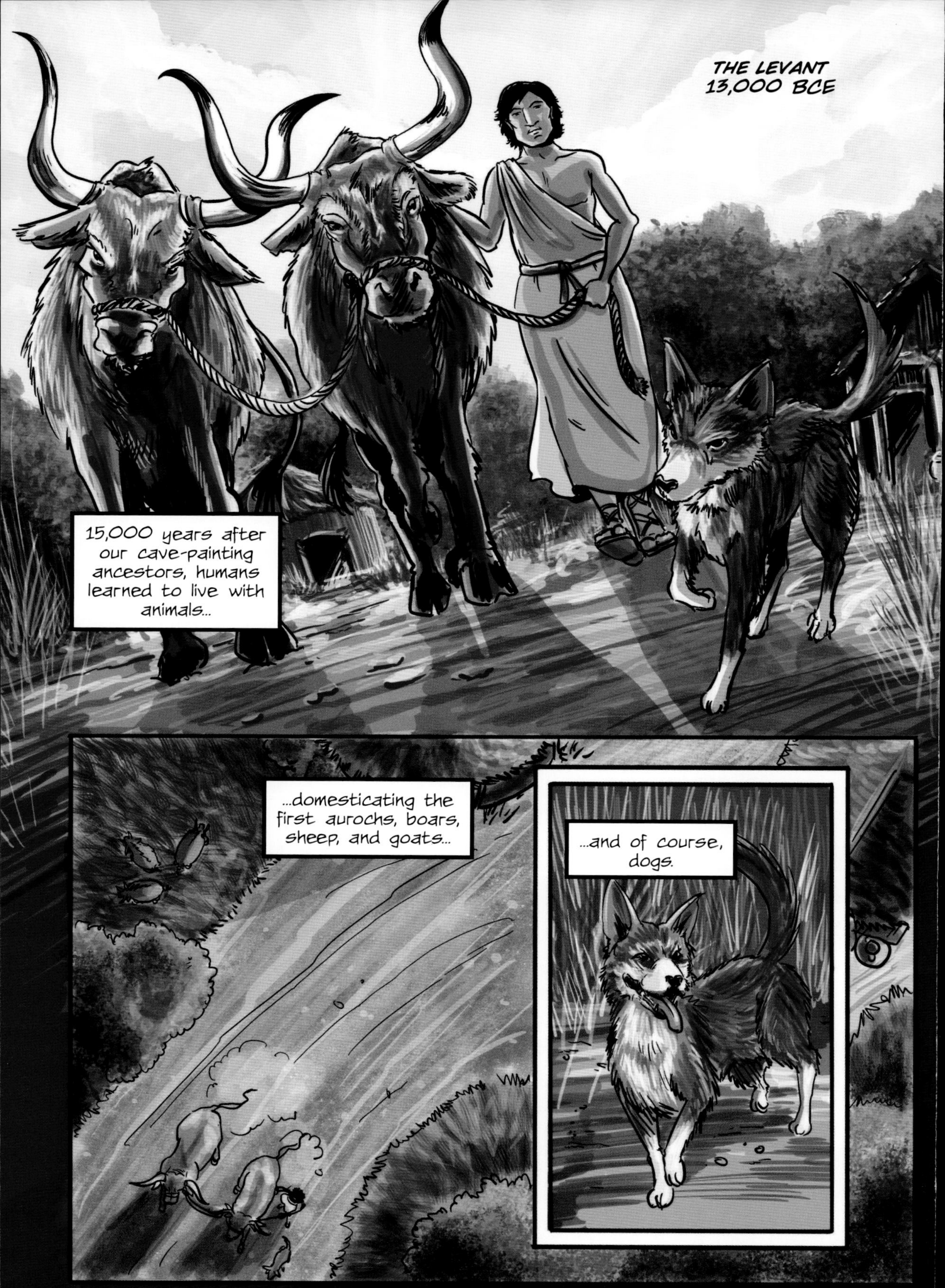
THE LEVANT
13,000 BCE
15,000 years after our cave-painting ancestors, humans learned to live with animals...
...domesticating the first aurochs, boars, sheep, and goats...
...and of course, dogs.

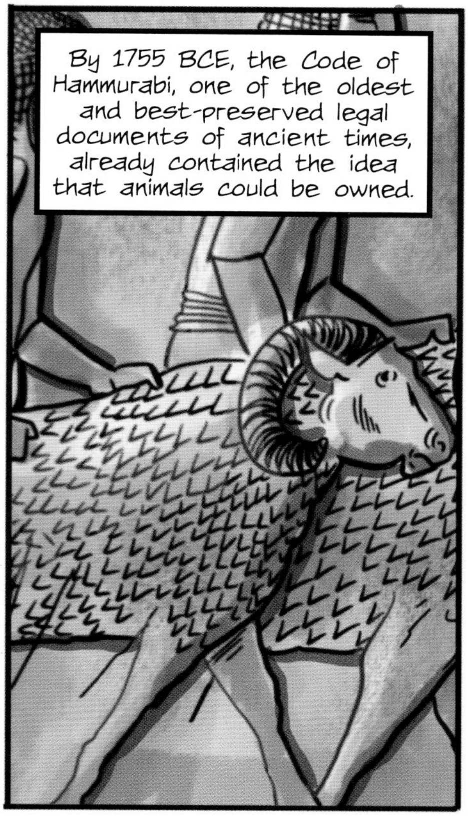
By 1755 BCE, the Code of Hammurabi, one of the oldest and best-preserved legal documents of ancient times, already contained the idea that animals could be owned.

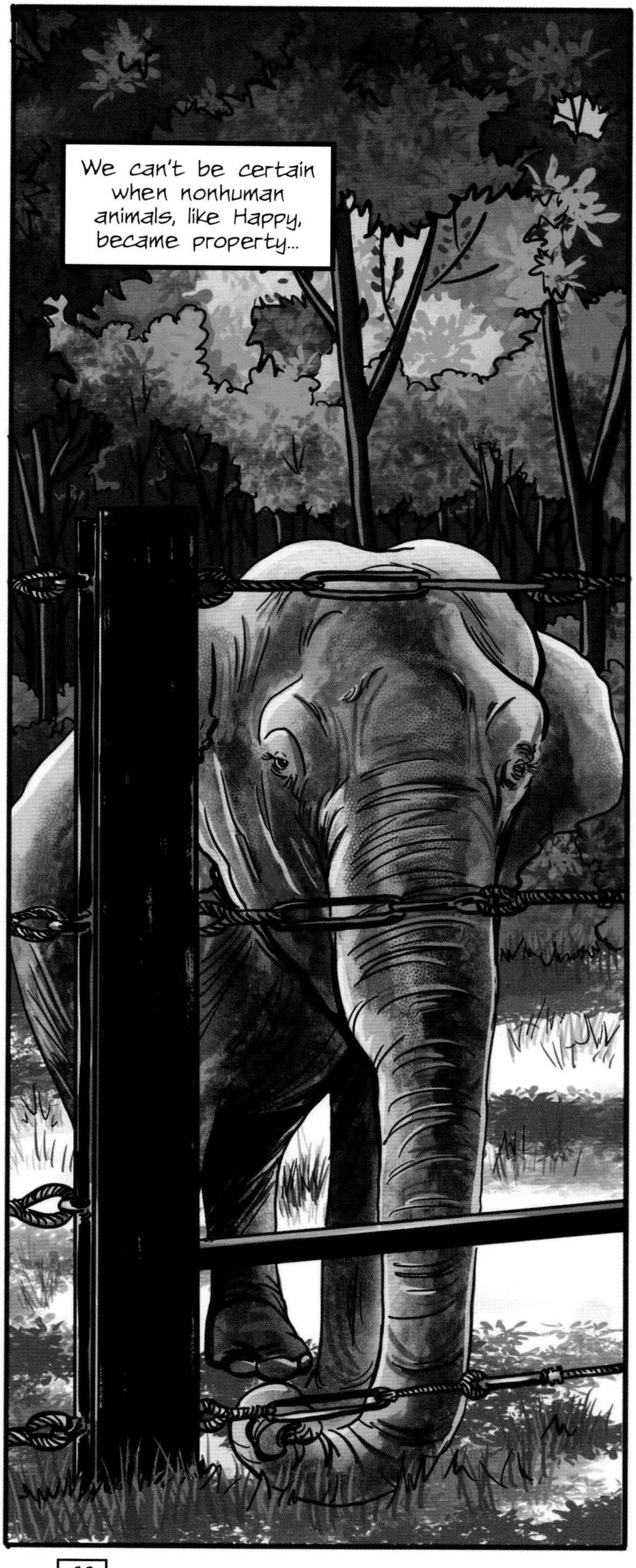
We can't be certain when nonhuman animals, like Happy, became property...

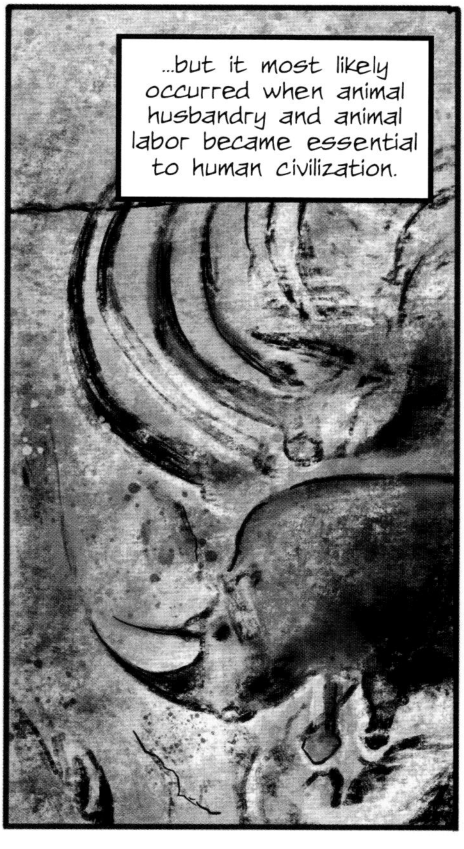
...but it most likely occurred when animal husbandry and animal labor became essential to human civilization.

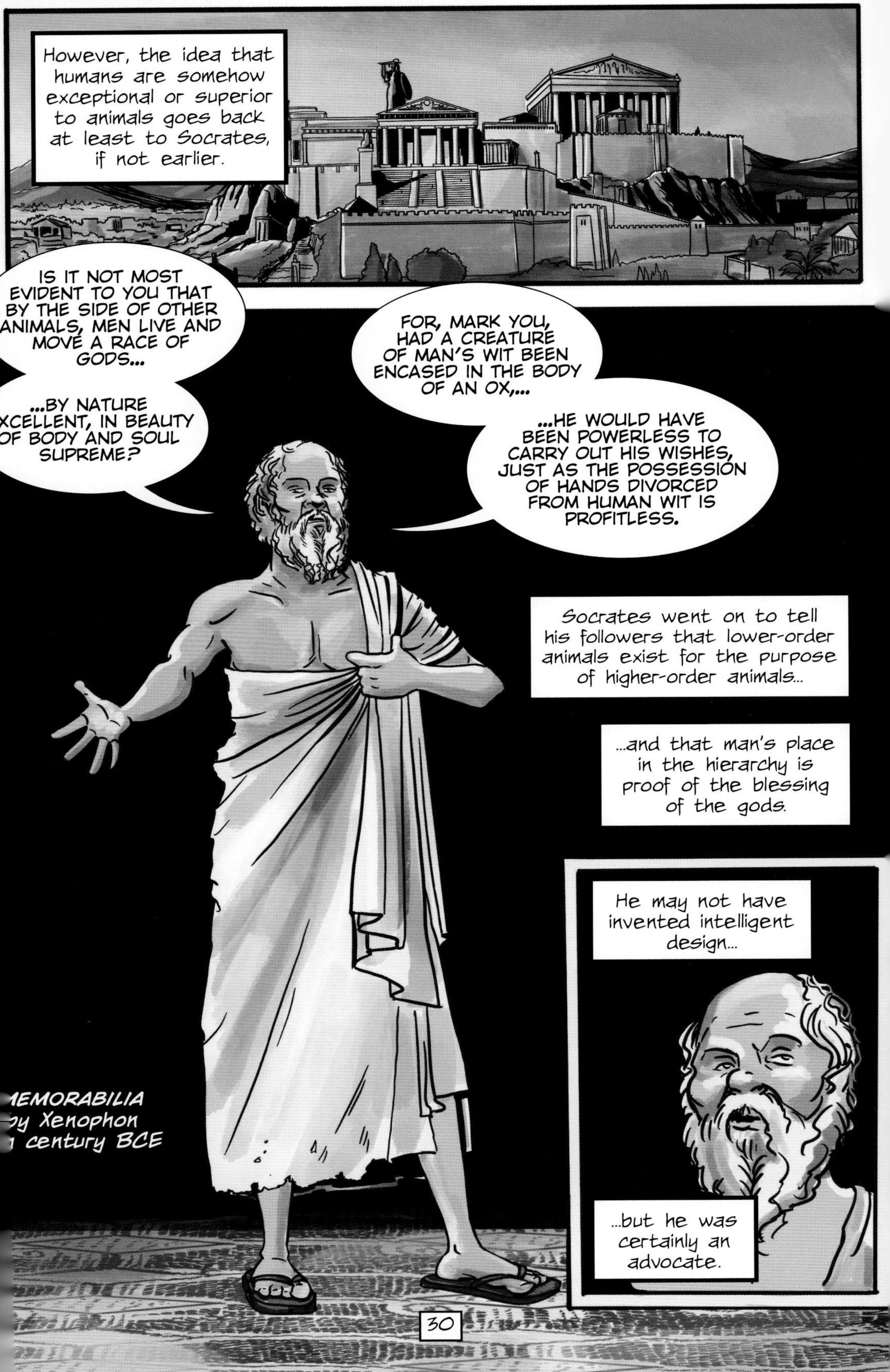
However, the idea that humans are somehow exceptional or superior to animals goes back at least to Socrates, if not earlier.
IS IT NOT MOST EVIDENT TO YOU THAT BY THE SIDE OF OTHER ANIMALS, MEN LIVE AND MOVE A RACE OF GODS...
...BY NATURE XCELLENT, IN BEAUTY OF BODY AND SOUL SUPREME?
FOR, MARK YOU, HAD A CREATURE OF MAN'S WIT BEEN ENCASED IN THE BODY OF AN OX,...
...HE WOULD HAVE BEEN POWERLESS TO CARRY OUT HIS WISHES, JUST AS THE POSSESSION OF HANDS DIVORCED FROM HUMAN WIT IS PROFITLESS.
Socrates went on to tell his followers that lower-order animals exist for the purpose of higher-order animals...
...and that man's place in the hierarchy is proof of the blessing of the gods.
MEMORABILIA
Xenophon
century BCE
He may not have invented intelligent design...
...but he was certainly an advocate.

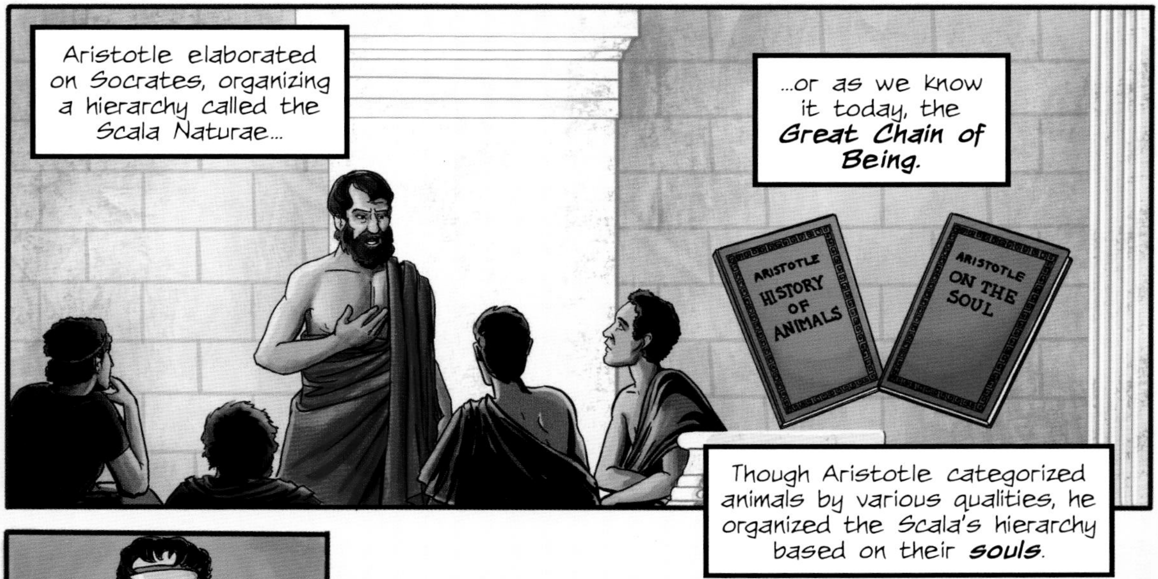
Aristotle elaborated on Socrates, organizing a hierarchy called the Scala Naturae...
...or as we know it today, the ***Great Chain of Being.***
ARISTOTLE HISTORY OF ANIMALS
ARISTOTLE ON THE SOUL
Though Aristotle categorized animals by various qualities, he organized the Scala's hierarchy based on their ***souls***.

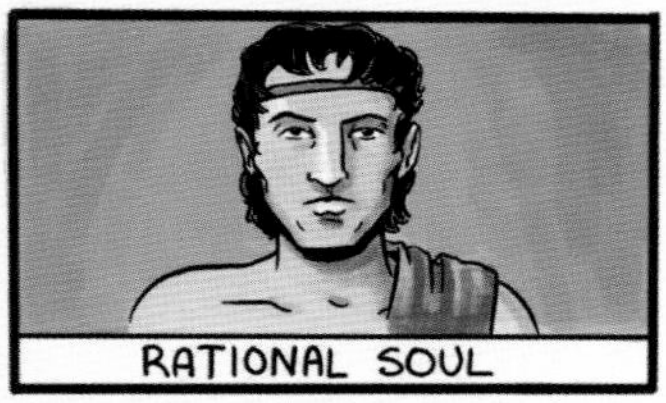
RATIONAL SOUL

Rational souls are possessed by humans and make us capable of thoughts and reflection, making reason, intellect, and justice unique to us, although we can still tap into our instinctual or sensitive soul.

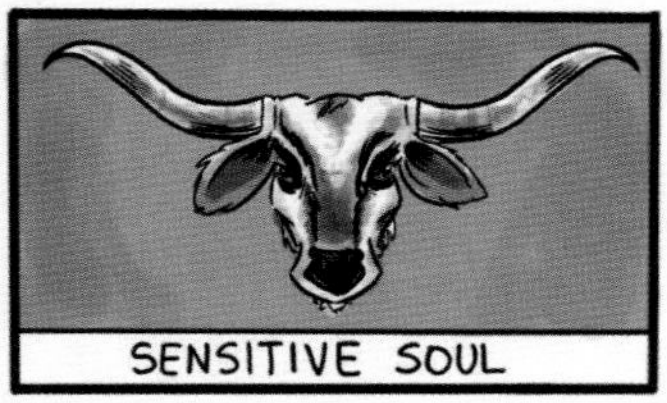
SENSITIVE SOUL

Sensitive souls are possessed by animals, making them capable of mobility and sensation, hunger, a reproductive drive, and instinct. Animals with sensitive souls still have access to vegetative souls.

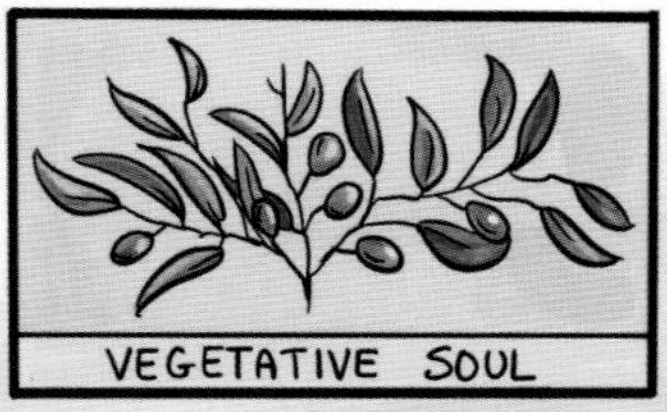
VEGETATIVE SOUL

Vegetative souls allow plants to reproduce and grow. This type of soul is the simplest and contains no deeper drives.

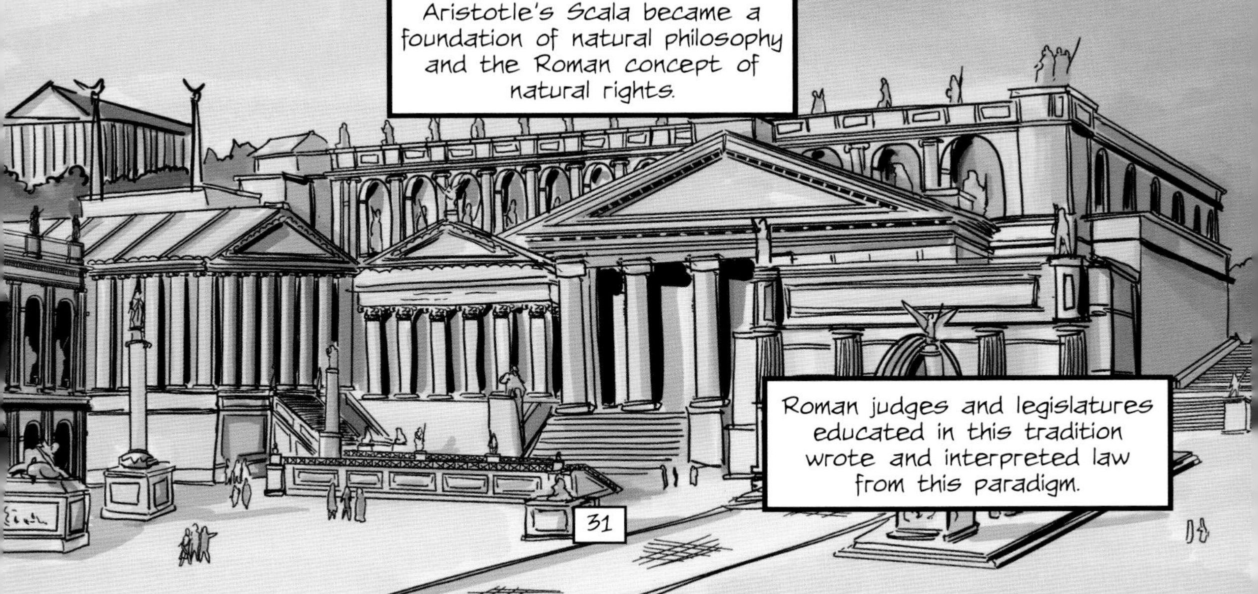
Aristotle's Scala became a foundation of natural philosophy and the Roman concept of natural rights.
Roman judges and legislatures educated in this tradition wrote and interpreted law from this paradigm.

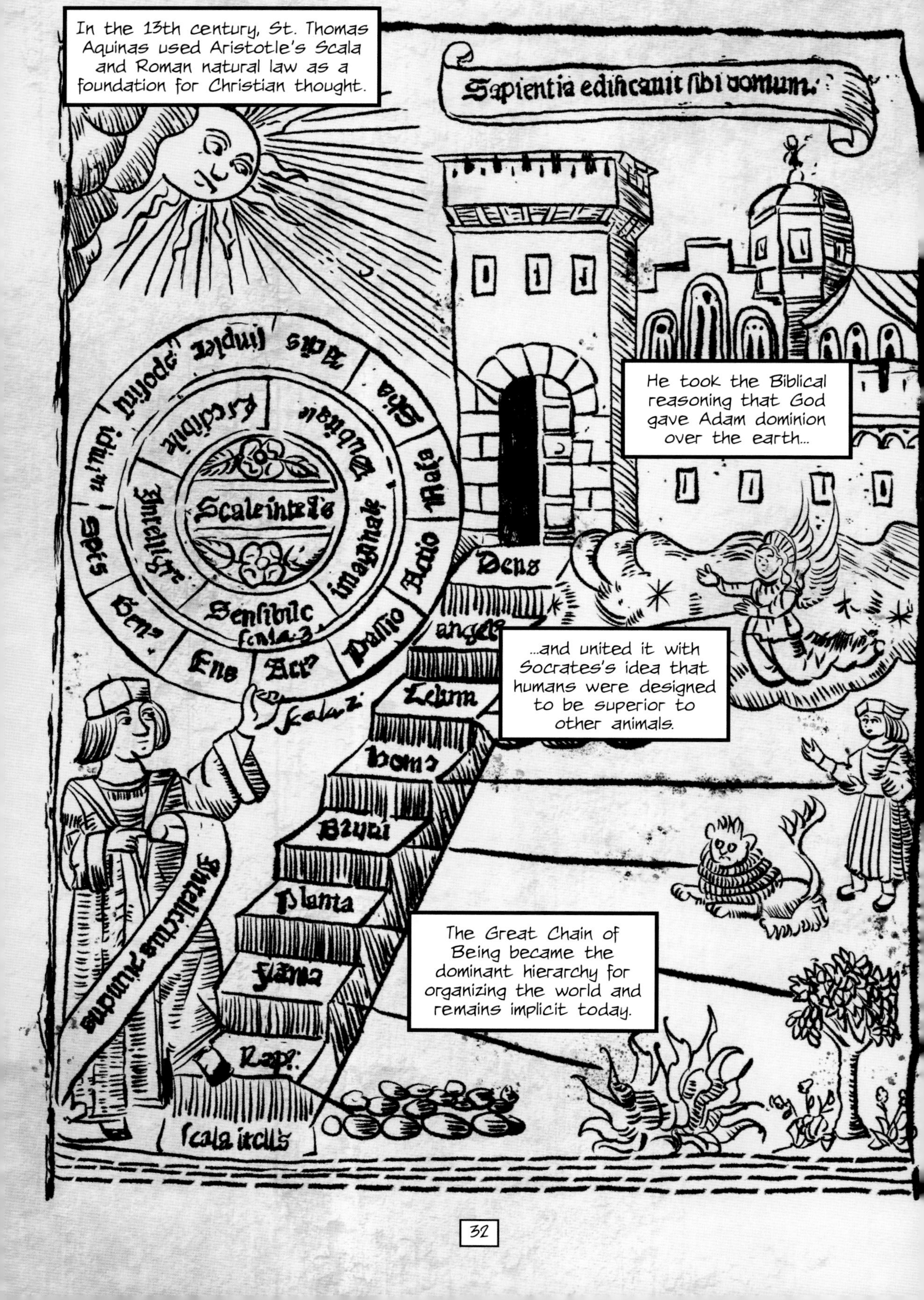
In the 13th century, St. Thomas Aquinas used Aristotle's Scala and Roman natural law as a foundation for Christian thought.
Sapientia edificauit sibi domum.
He took the Biblical reasoning that God gave Adam dominion over the earth...
...and united it with Socrates's idea that humans were designed to be superior to other animals.
The Great Chain of Being became the dominant hierarchy for organizing the world and remains implicit today.

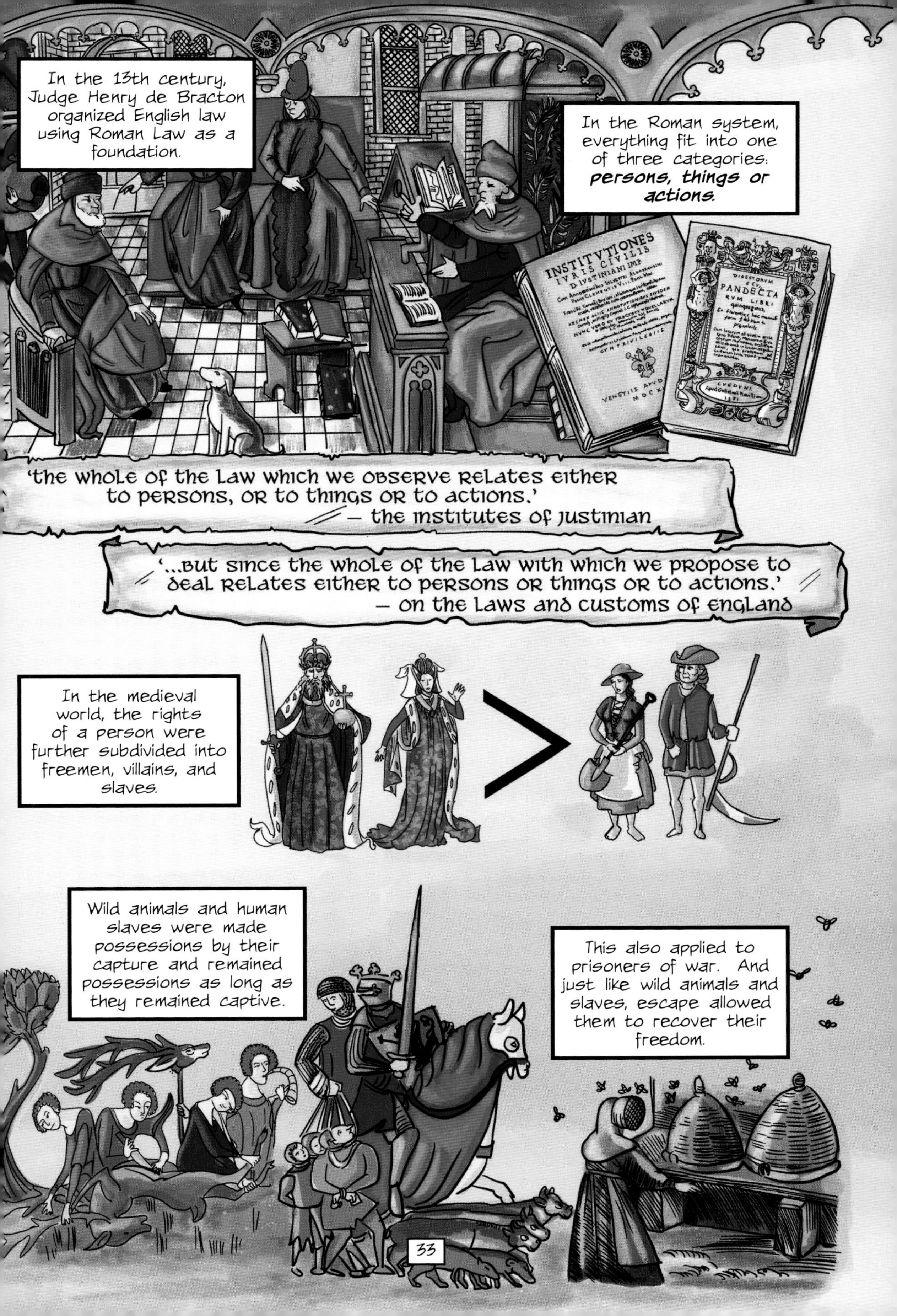
In the 13th century, Judge Henry de Bracton organized English law using Roman Law as a foundation.
In the Roman system, everything fit into one of three categories: ***persons, things or actions.***
INSTITVTIONES IVRIS CIVILIS
PANDECTA
'the whole of the law which we observe relates either to persons, or to things or to actions.' – the institutes of justinian
'...but since the whole of the law with which we propose to deal relates either to persons or things or to actions.' – on the laws and customs of england
In the medieval world, the rights of a person were further subdivided into freemen, villains, and slaves.
Wild animals and human slaves were made possessions by their capture and remained possessions as long as they remained captive.
This also applied to prisoners of war. And just like wild animals and slaves, escape allowed them to recover their freedom.

The first animal welfare laws didn't arrive until the 19th century...

...first proposed by Lord Erskine, the Chancellor of England.

IF FORTUNATELY ADOPTED BY YOUR LORDSHIPS, WILL ENACT THIS LAW AS A SPONTANEOUS RULE, IN THE MIND OF EVERY MAN WHO READS IT...

...WHICH WILL MAKE EVERY HUMAN BOSOM A SANCTUARY AGAINST CRUELTY...

...AND CONSECRATE PERHAPS IN ALL NATIONS AND IN ALL AGES...

...THAT JUST AND ETERNAL PRINCIPLE WHICH BINDS THE WHOLE LIVING WORLD IN ONE HARMONIOUS CHAIN, UNDER THE DOMINION OF ENLIGHTENED MAN,...

...THE LORD AND GOVERNOR OF ALL.

"WHEREAS IT PLEASED ALMIGHTY GOD TO SUBDUE TO THE DOMINION, USE AND COMFORT OF MAN, THE STRENGTH AND FACULTIES OF MANY USEFUL ANIMALS, AND TO PROVIDE OTHERS FOR HIS FOOD--AND WHEREAS THE ABUSE OF THAT DOMINION BY CRUEL AND OPPRESSIVE TREATMENT OF SUCH ANIMALS IS NOT ONLY HIGHLY UNJUST AND IMMORAL, BUT MOST PERNICIOUS IN ITS EXAMPLE, HAVING AN EVIDENT TENDENCY TO HARDEN THE HEART AGAINST THE NATURAL FEELING OF HUMANITY."

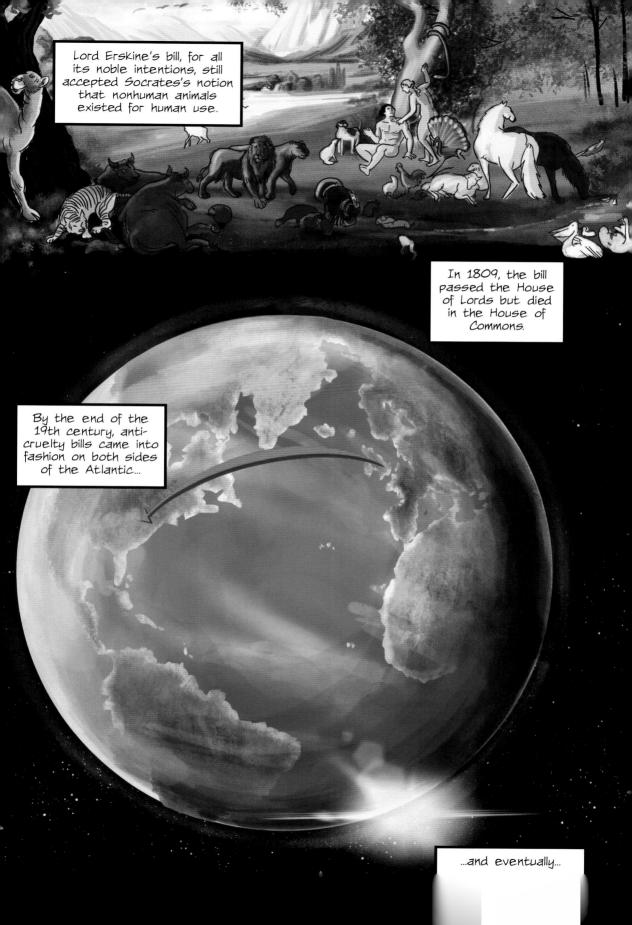

Lord Erskine's bill, for all its noble intentions, still accepted Socrates's notion that nonhuman animals existed for human use.

In 1809, the bill passed the House of Lords but died in the House of Commons.

By the end of the 19th century, anti-cruelty bills came into fashion on both sides of the Atlantic...

...and eventually...

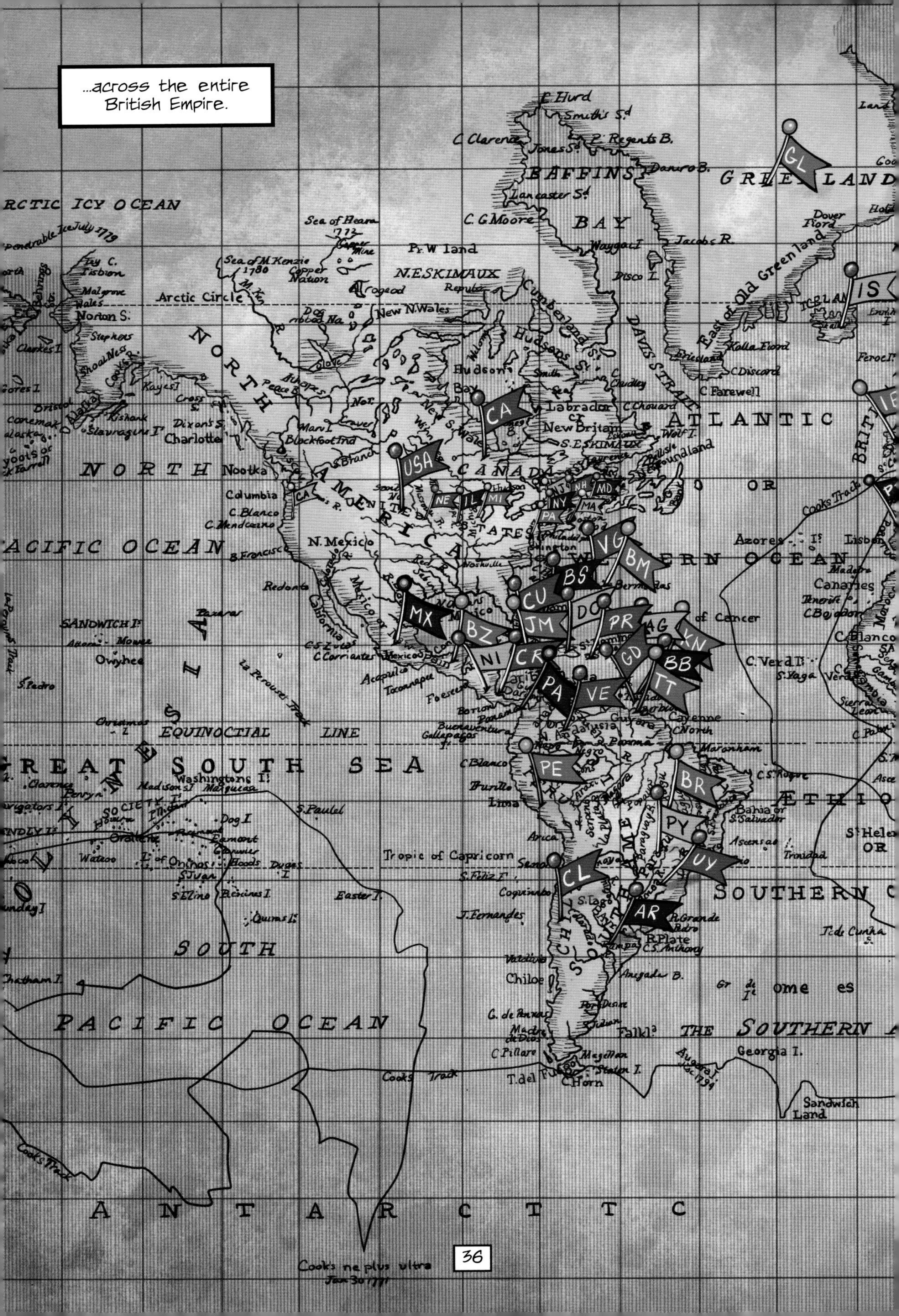
...across the entire British Empire.
GL
GREENLAND
BAFFINS BAY
Lancaster St
C. G Moore
Jacobs R.
Disco I.
Cumberland St
DAVIS STRAIT
Hudsons St
Hudson Bay
Labrador
New Britain
Wolf I.
Newfoundland
C Farewell
East of Old Greenland
ICELAND
IS
ATLANTIC
WESTERN OCEAN
Azores Is
Canaries
C. Verd Is
Madeira
Tropic of Cancer
ARCTIC ICY OCEAN
Arctic Circle
Sea of M. Kenzie 1780
Copper Nation
N. ESKIMAUX
New N. Wales
Norton S.
NORTH AMERICA
UNITED STATES
CANADA
Nootka
Columbia
C. Blanco
C. Mendocino
N. Mexico
S. Francisco
California
Acapulco
NORTH PACIFIC OCEAN
SANDWICH Is
Owyhee
EQUINOCTIAL LINE
GREAT SOUTH SEA
Washingtons Is
SOCIETY Is
Tropic of Capricorn
Easter I.
J. Fernandes
SOUTH PACIFIC OCEAN
Chiloe
Lima
Falkld
Georgia I.
C Horn
T. del Fuego
THE SOUTHERN
SOUTHERN
Sandwich Land
Cooks Track
ANTARCTIC
Cooks ne plus ultra Jan 30 1774
CA
USA
CA
NE
IL
MI
NJ
NH
MD
NY
MA
PA
VG
BM
BS
CU
DO
MX
JM
PR
AG
KN
BZ
NI
CR
GD
BB
PA
VE
TT
PE
BR
PY
CL
UY
AR

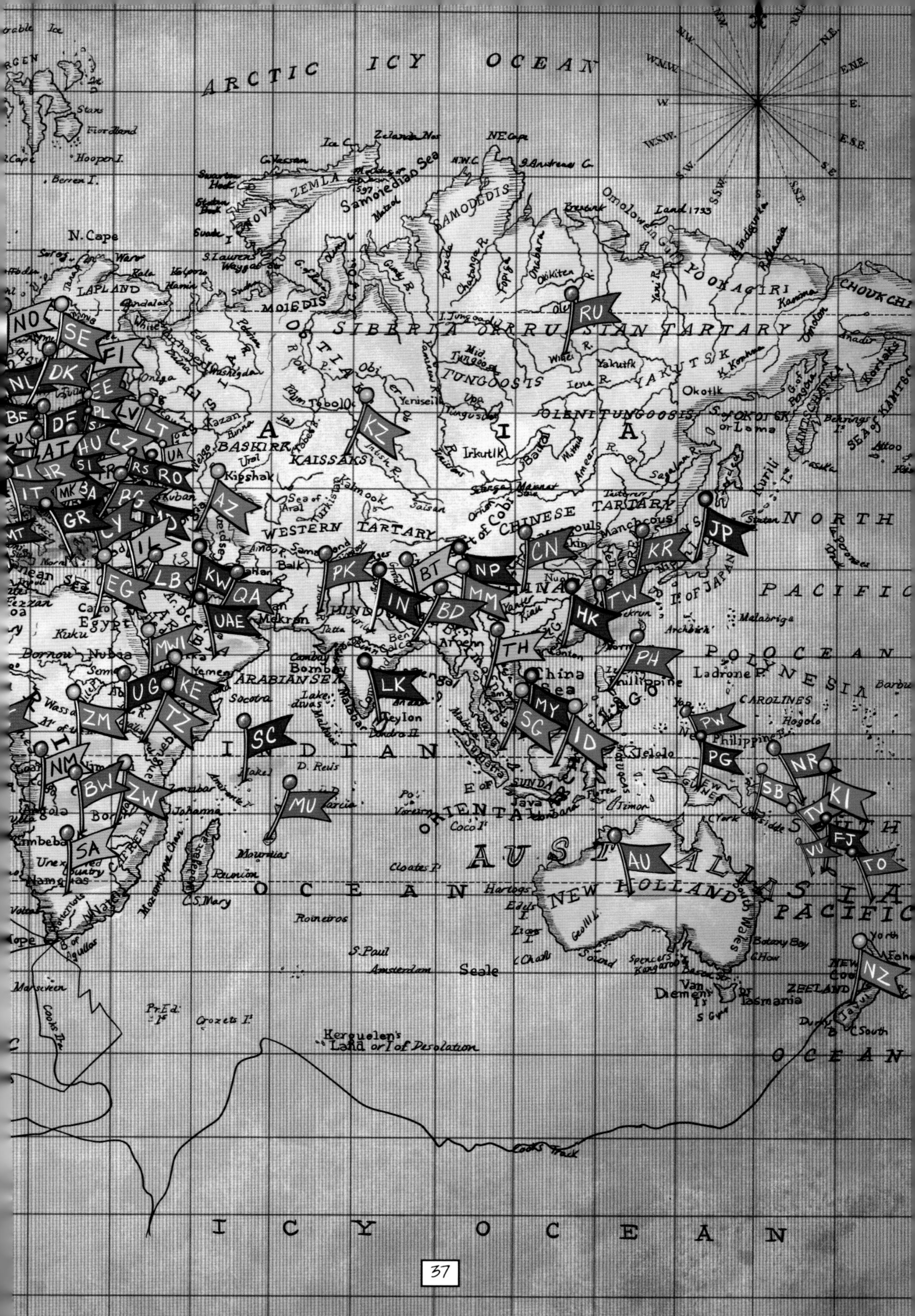
ARCTIC ICY OCEAN
N.W.
N.E.
W.N.W.
E.N.E.
W.
E.
W.S.W.
E.S.E.
S.W.
S.E.
S.S.W.
S.S.E.
NOVA ZEMLA
Samoiediao Sea
SAMOIEDIS
N.Cape
LAPLAND
MOIEDIS
SIBERIA OR RUSSIAN TARTARY
YOOKAGIRI
CHOUKCHI
TUNGOOSIS
YAKUTSK
OLENI TUNGOOSIS
A
I
A
BASKIRK
KAISSAKS
Kipshak
Sea of Aral
WESTERN TARTARY
CHINESE TARTARY
NORTH
PACIFIC
Egypt
Nubia
ARABIAN SEA
Socotra
Bombay
Ceylon
China Sea
INDIAN
POLYNESIA
CAROLINES
Ladrone I.
ORIENTAL
AUSTRALASIA
NEW HOLLAND
PACIFIC
OCEAN
Tasmania
NEW ZEELAND
Kerguelen's Land or I. of Desolation
Cooks Track
ICY OCEAN
RU
NO
SE
FI
DK
NL
EE
BE
DE
PL
LV
LT
LU
AT
HU
CZ
UA
SI
HR
RS
RO
IT
MK
BA
BG
GR
CY
MT
IL
EG
LB
KW
QA
AZ
KZ
UAE
PK
IN
BT
NP
MM
BD
CN
KR
JP
TW
HK
TH
PH
LK
MY
SG
ID
MWI
UG
KE
TZ
ZM
SC
MU
NM
BW
ZW
SA
PW
PG
NR
KI
SB
TV
FJ
VU
TO
AU
NZ

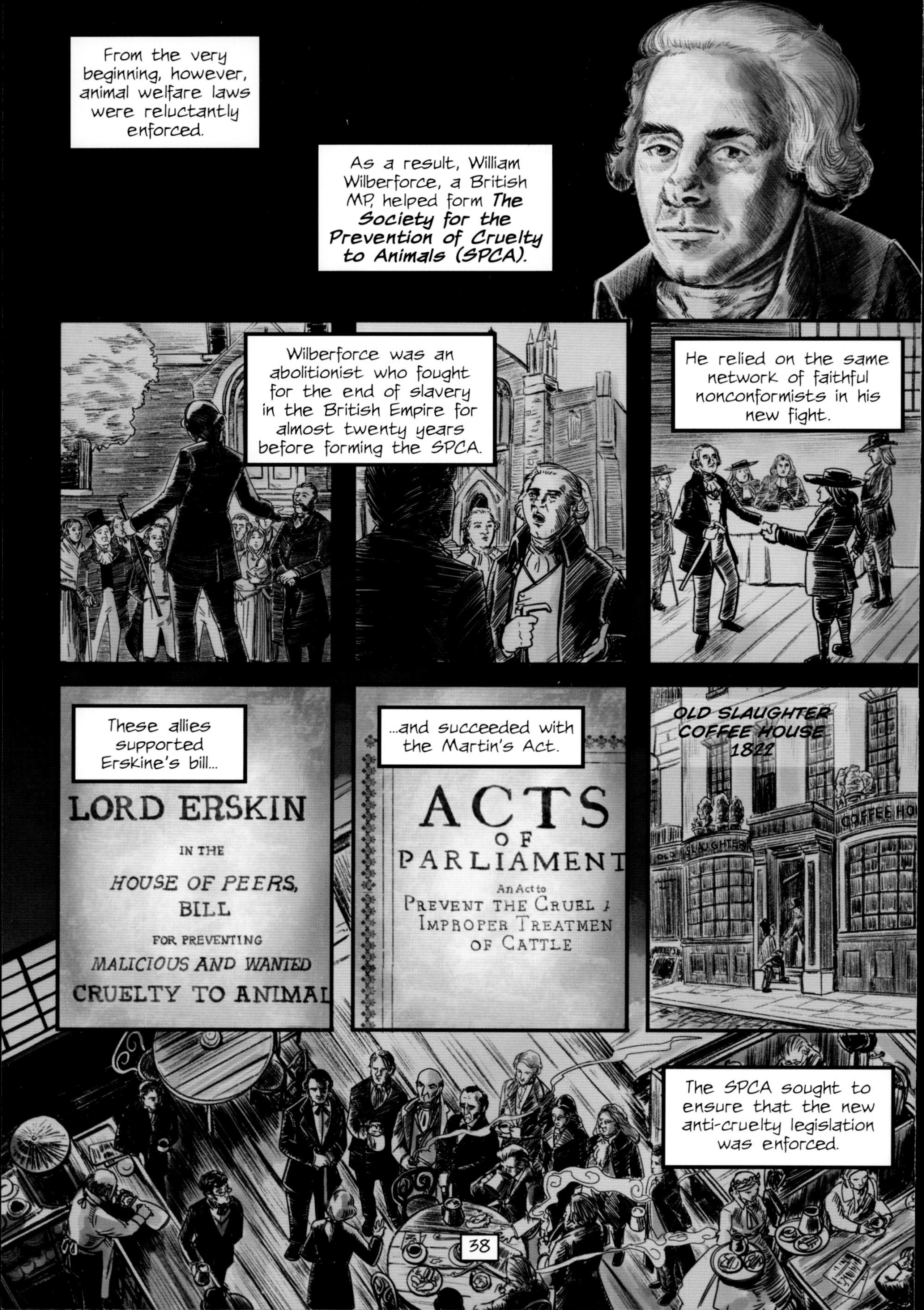

From the very beginning, however, animal welfare laws were reluctantly enforced.

As a result, William Wilberforce, a British MP, helped form *The Society for the Prevention of Cruelty to Animals (SPCA).*

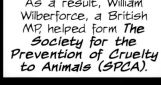

Wilberforce was an abolitionist who fought for the end of slavery in the British Empire for almost twenty years before forming the SPCA.

He relied on the same network of faithful nonconformists in his new fight.

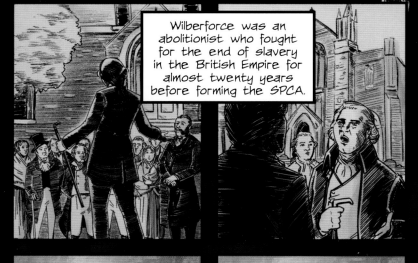

These allies supported Erskine's bill...

LORD ERSKIN

IN THE

HOUSE OF PEERS,

BILL

FOR PREVENTING

MALICIOUS AND WANTED

CRUELTY TO ANIMAL

...and succeeded with the Martin's Act.

ACTS

OF

PARLIAMENT

An Act to

PREVENT THE CRUEL /

IMPROPER TREATMEN

OF CATTLE

OLD SLAUGHTER COFFEE HOUSE 1822

The SPCA sought to ensure that the new anti-cruelty legislation was enforced.

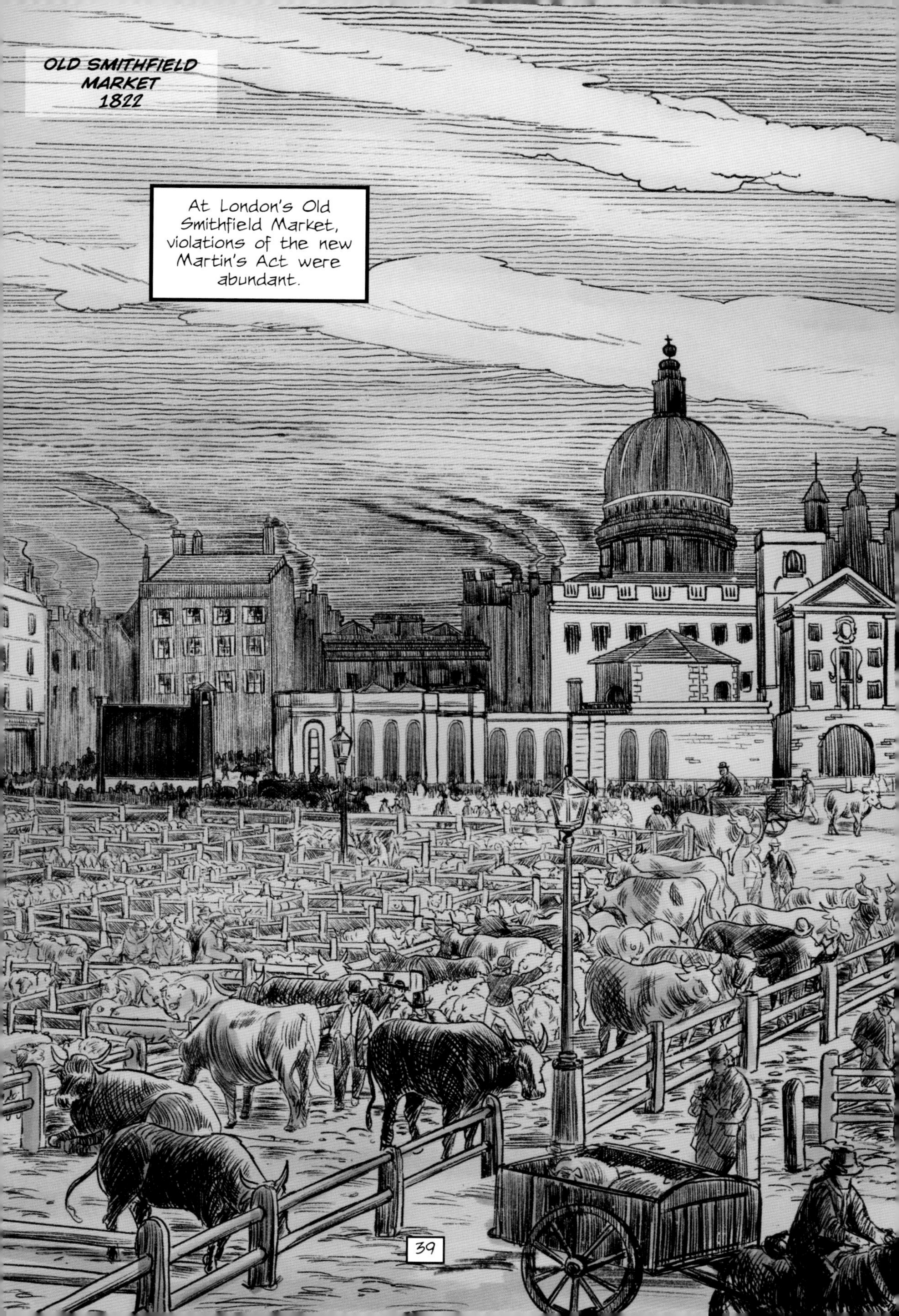
OLD SMITHFIELD MARKET 1822
At London's Old Smithfield Market, violations of the new Martin's Act were abundant.

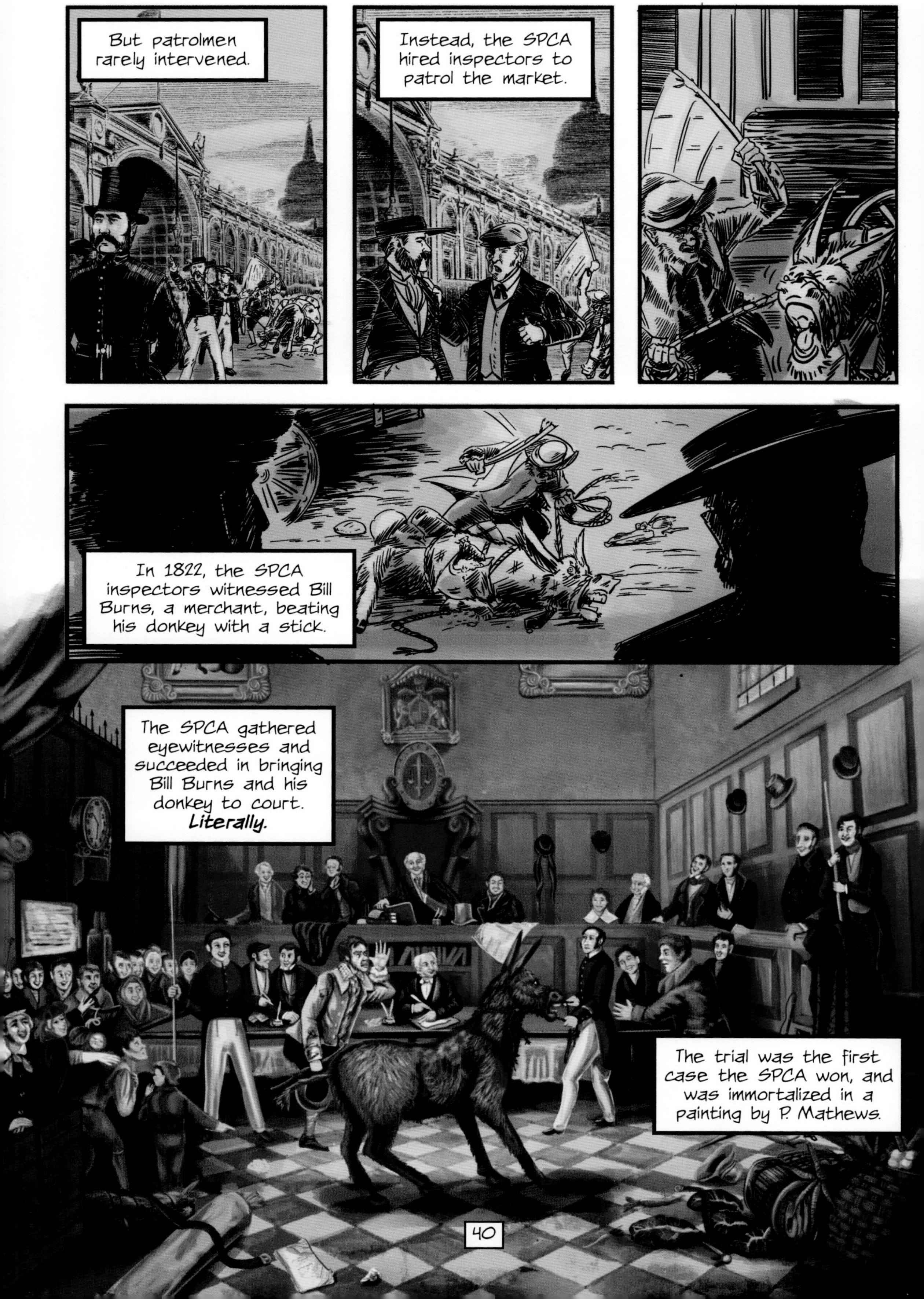
But patrolmen rarely intervened.
Instead, the SPCA hired inspectors to patrol the market.
In 1822, the SPCA inspectors witnessed Bill Burns, a merchant, beating his donkey with a stick.
The SPCA gathered eyewitnesses and succeeded in bringing Bill Burns and his donkey to court. ***Literally.***
The trial was the first case the SPCA won, and was immortalized in a painting by P. Mathews.

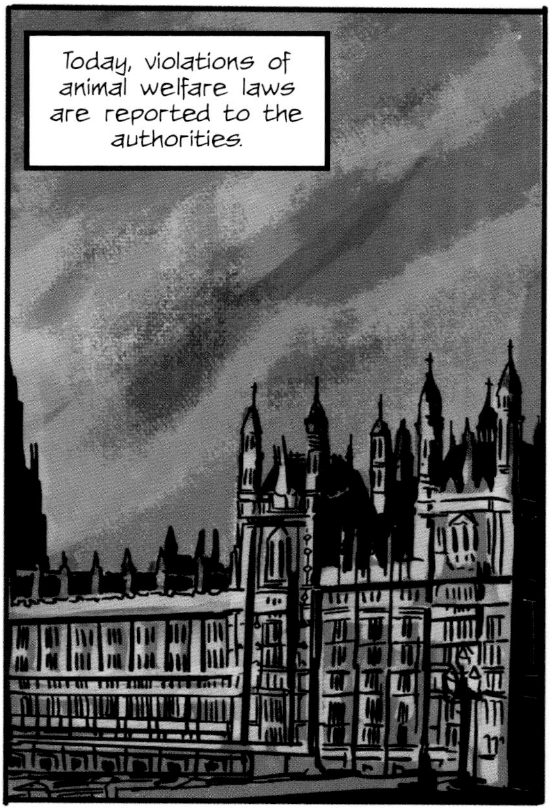
Today, violations of animal welfare laws are reported to the authorities.

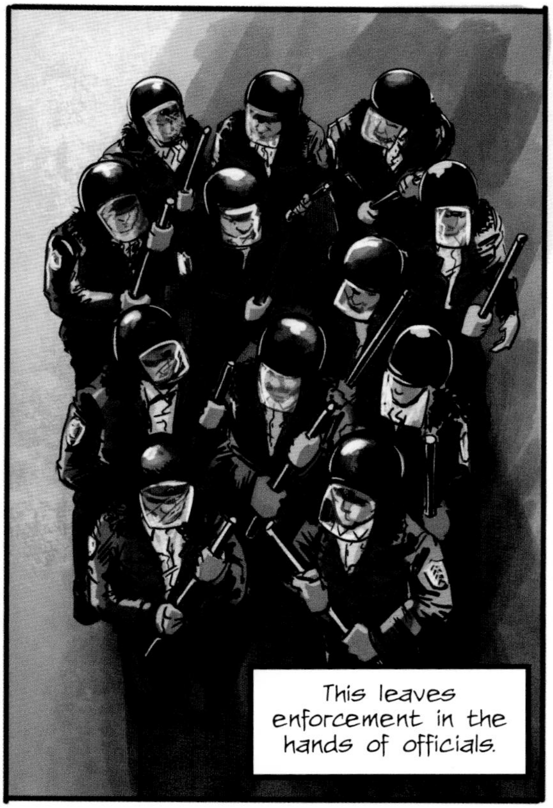
This leaves enforcement in the hands of officials.

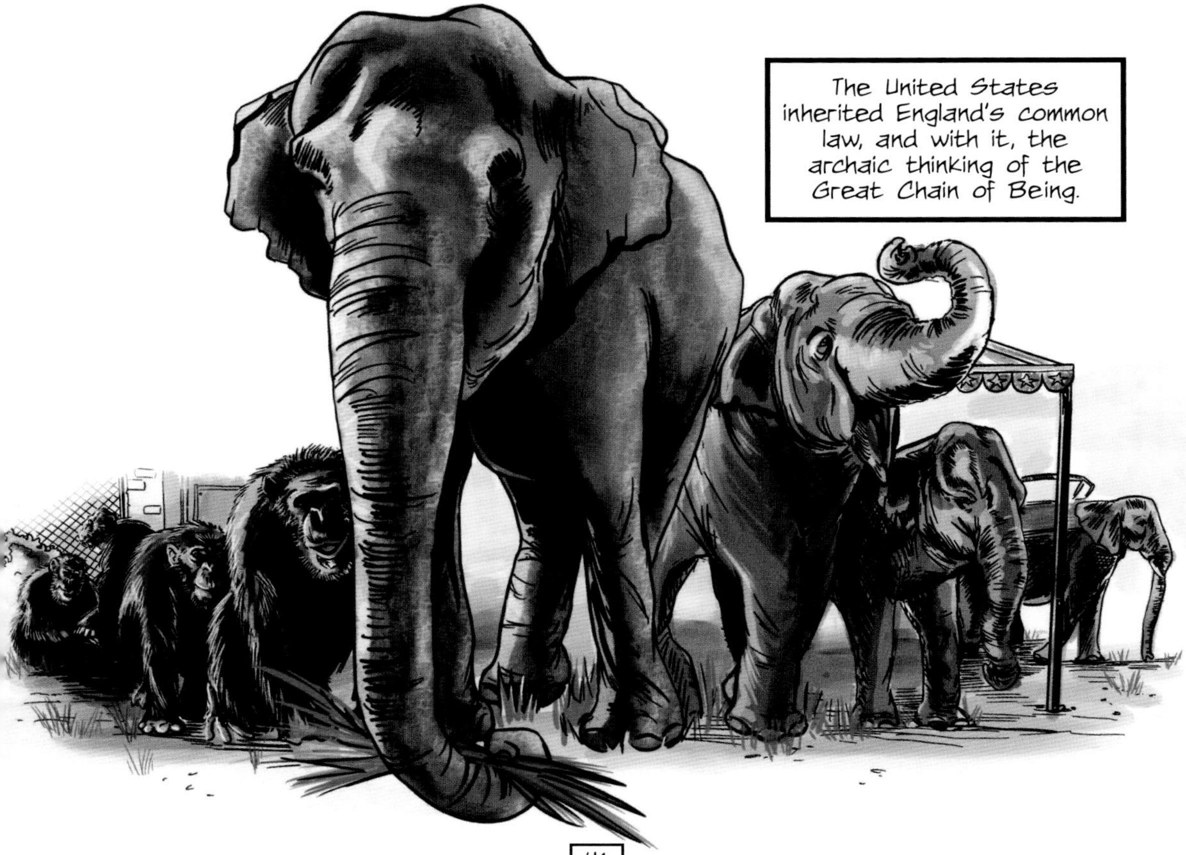
The United States inherited England's common law, and with it, the archaic thinking of the Great Chain of Being.

But the origins of this thinking emerged with human civilization.
P.T. BARNUM'S GREATEST SHOW ON EARTH
THE GIANT AFRICAN ELEPHANT
JUMBO

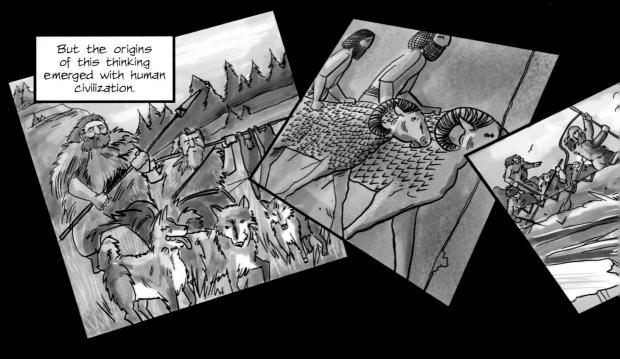

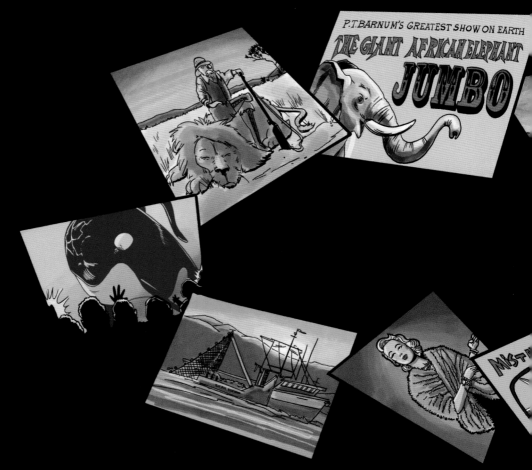

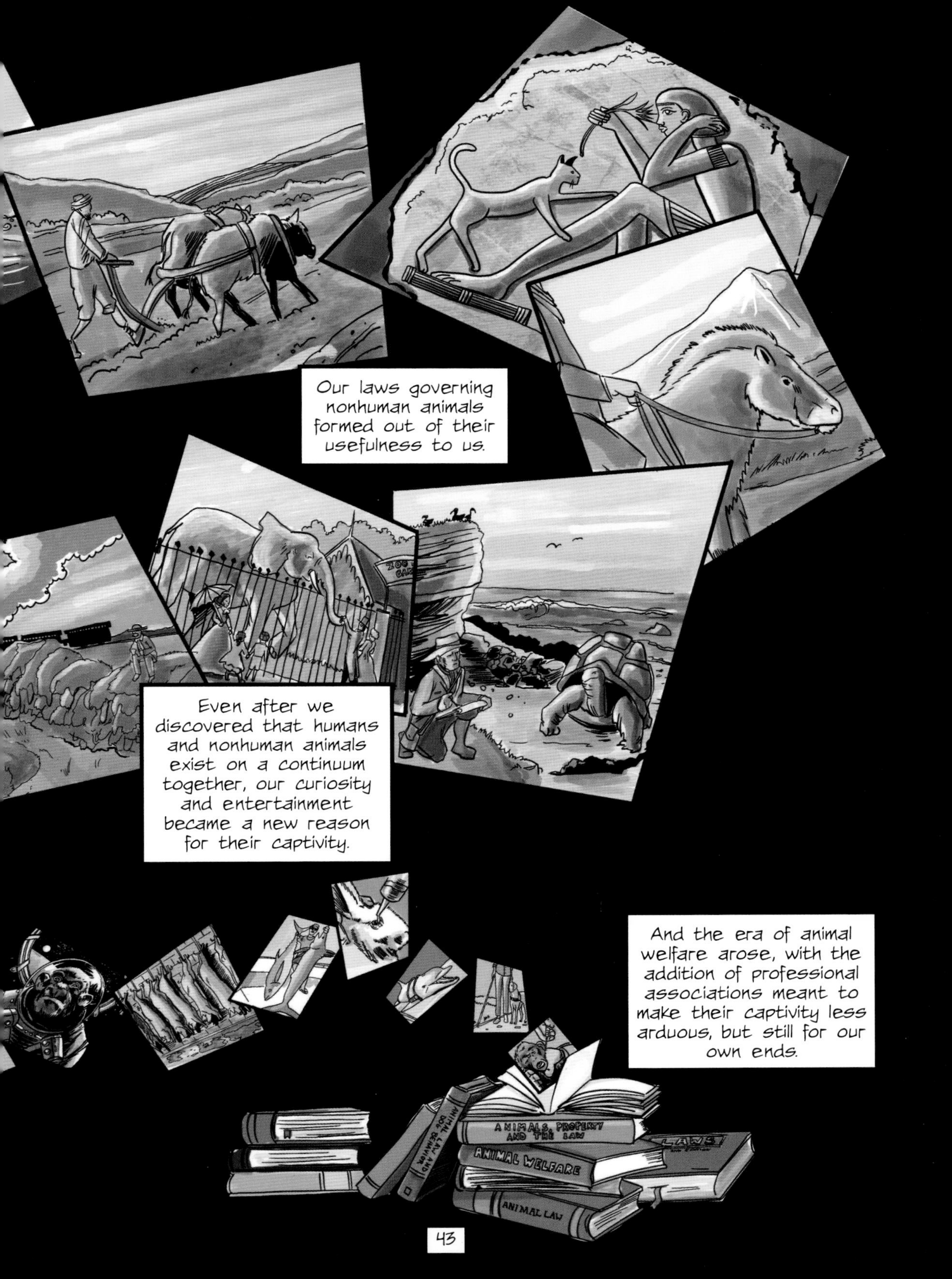
Our laws governing nonhuman animals formed out of their usefulness to us.
Even after we discovered that humans and nonhuman animals exist on a continuum together, our curiosity and entertainment became a new reason for their captivity.
And the era of animal welfare arose, with the addition of professional associations meant to make their captivity less arduous, but still for our own ends.
ANIMAL LAW AND DOG BEHAVIOR
ANIMALS, PROPERTY AND THE LAW
ANIMAL WELFARE
ANIMAL LAW

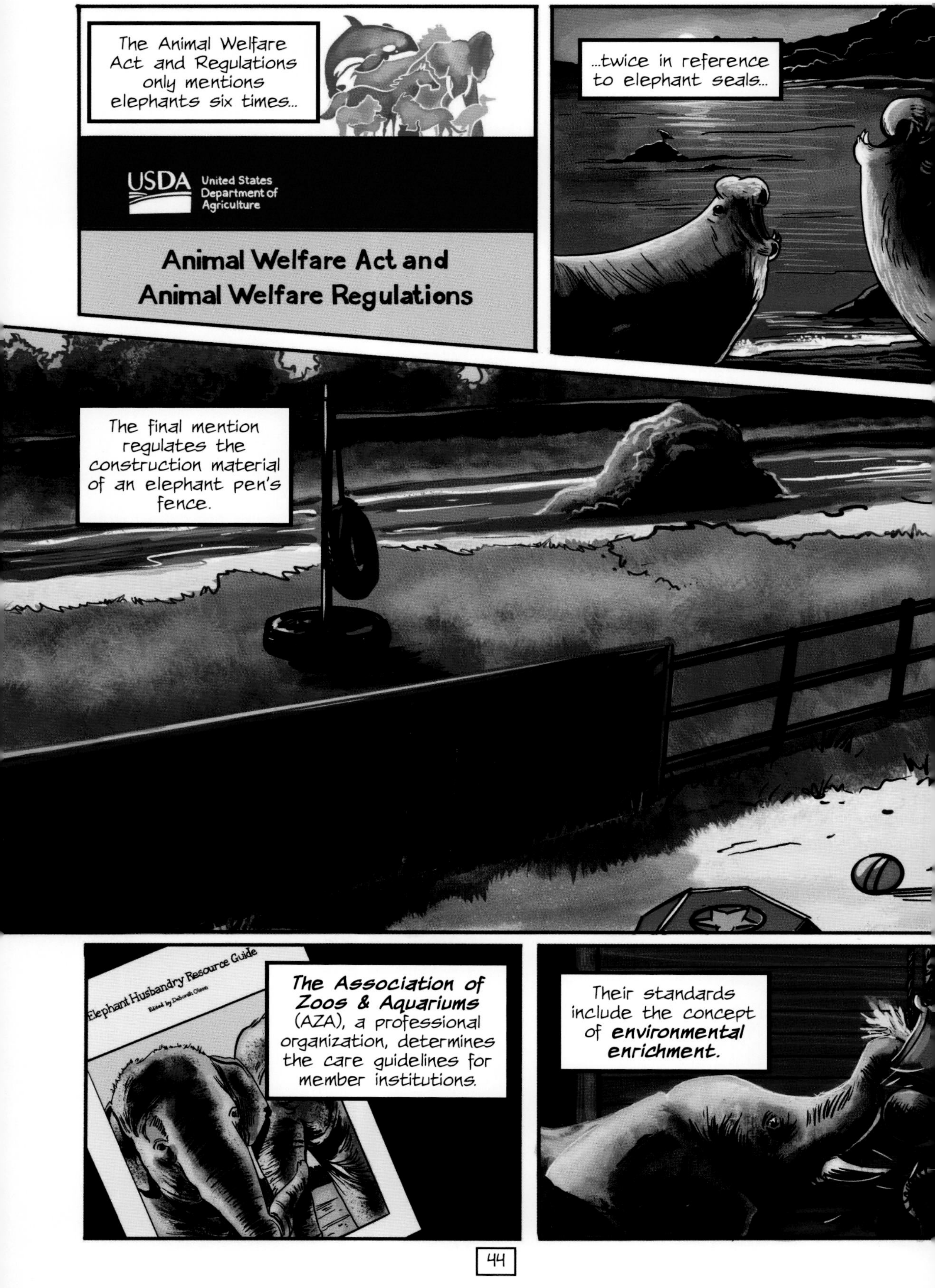

The Animal Welfare Act and Regulations only mentions elephants six times...
USDA United States Department of Agriculture
Animal Welfare Act and Animal Welfare Regulations
...twice in reference to elephant seals...
The final mention regulates the construction material of an elephant pen's fence.
Elephant Husbandry Resource Guide
Edited by Deborah Olson
The Association of Zoos & Aquariums (AZA), a professional organization, determines the care guidelines for member institutions.
Their standards include the concept of **environmental enrichment.**

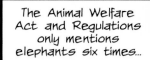

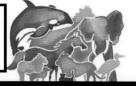

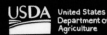

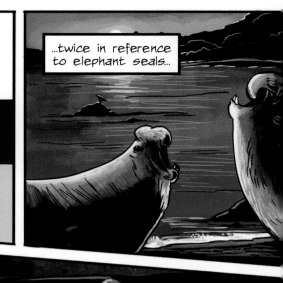

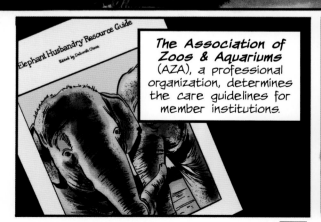

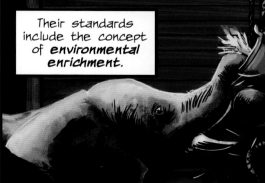

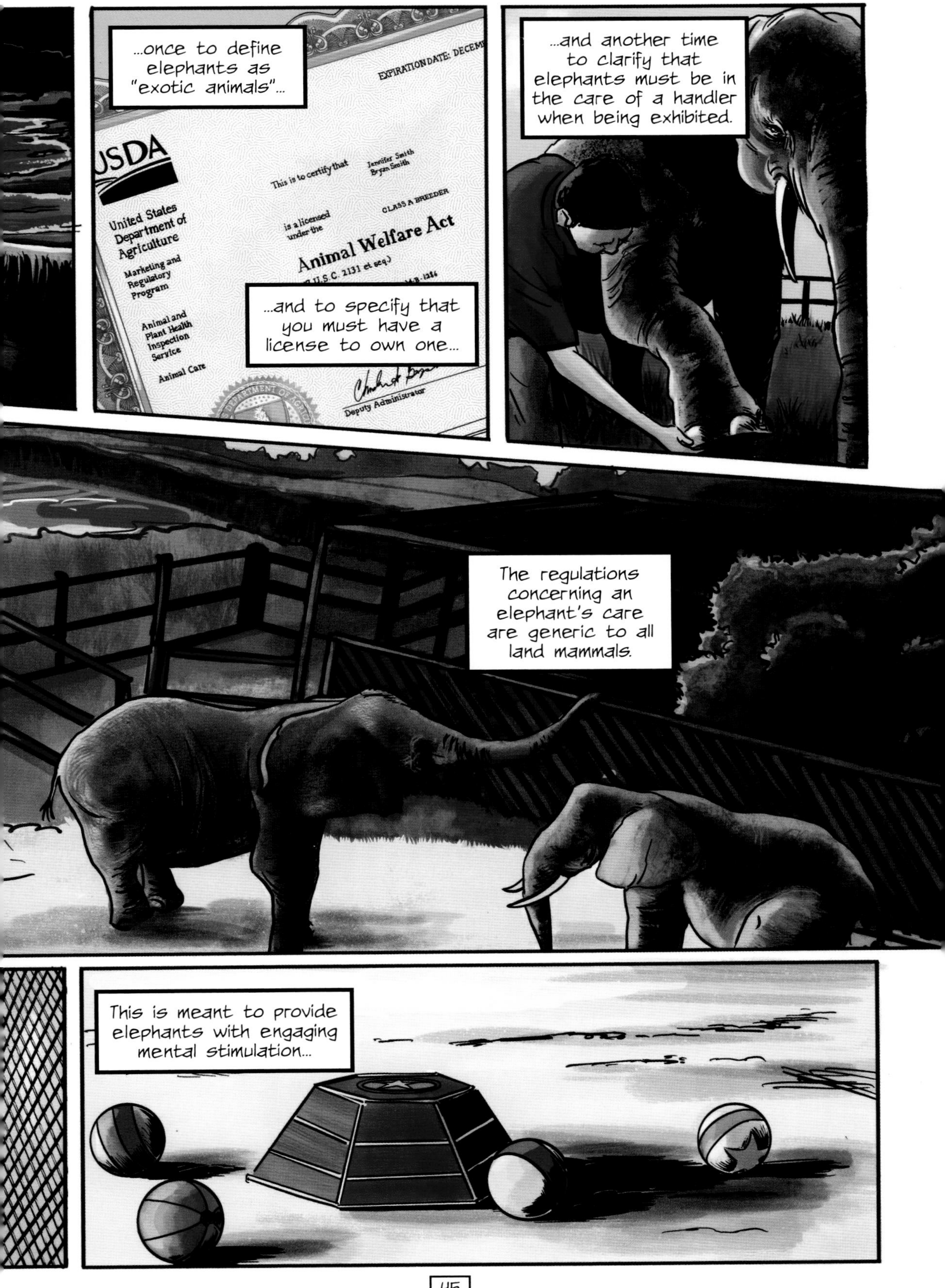
...once to define elephants as "exotic animals"...
EXPIRATION DATE: DECEM
USDA
United States Department of Agriculture
Marketing and Regulatory Program
Animal and Plant Health Inspection Service
Animal Care
This is to certify that
Jennifer Smith
Bryan Smith
CLASS A BREEDER
is a licensed under the
Animal Welfare Act
U.S.C. 2131 et seq.)
...and to specify that you must have a license to own one...
Deputy Administrator
...and another time to clarify that elephants must be in the care of a handler when being exhibited.
The regulations concerning an elephant's care are generic to all land mammals.
This is meant to provide elephants with engaging mental stimulation...

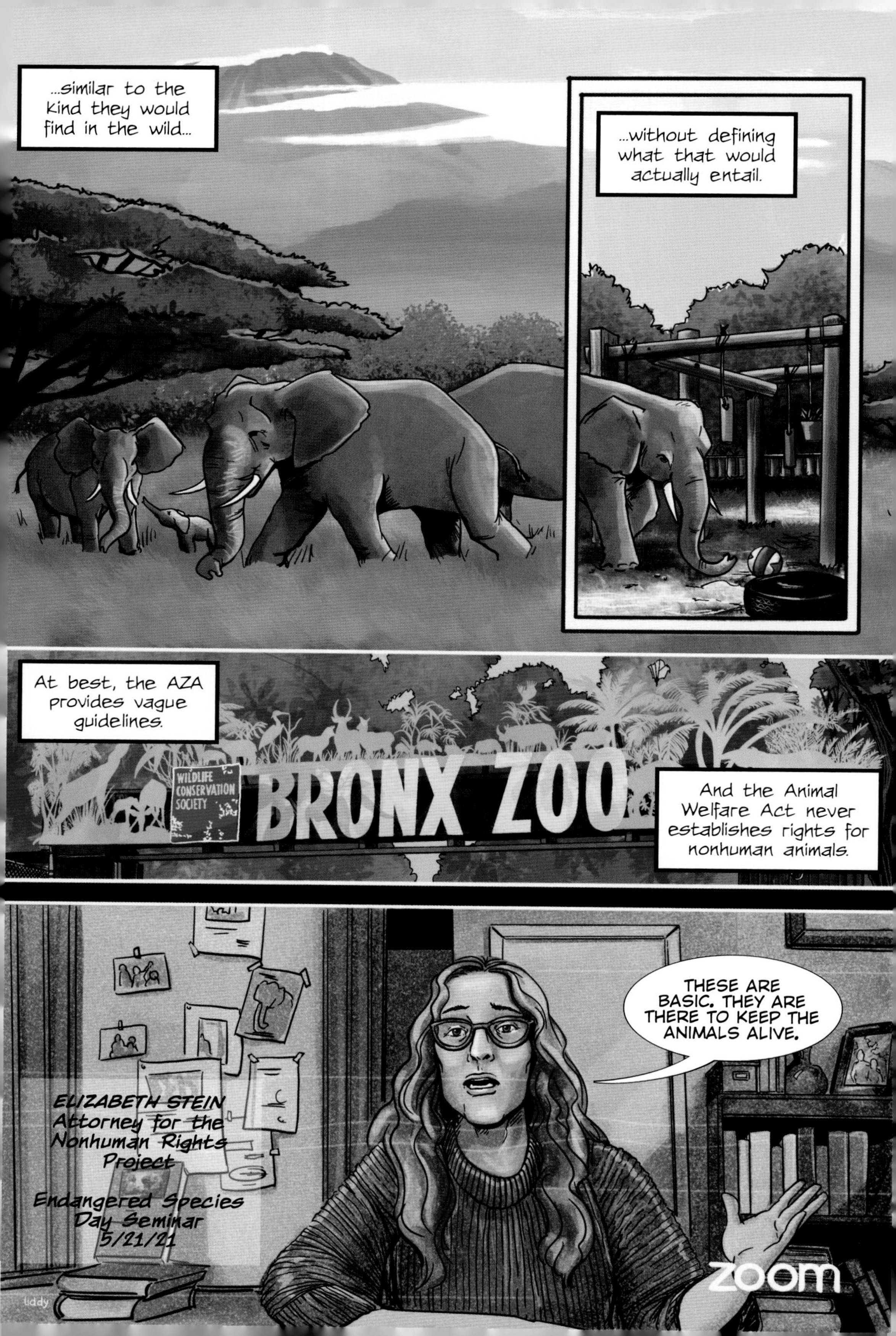
...similar to the kind they would find in the wild...
...without defining what that would actually entail.
At best, the AZA provides vague guidelines.
WILDLIFE CONSERVATION SOCIETY
BRONX ZOO
And the Animal Welfare Act never establishes rights for nonhuman animals.
ELIZABETH STEIN Attorney for the Nonhuman Rights Project
Endangered Species Day Seminar 5/21/21
THESE ARE BASIC. THEY ARE THERE TO KEEP THE ANIMALS ALIVE.
ZOOM
liddy

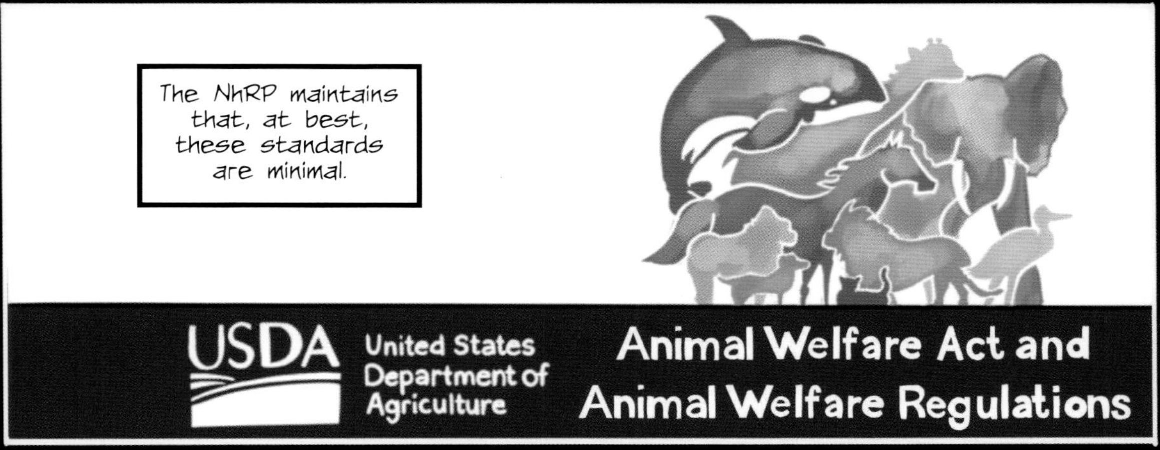
The NhRP maintains that, at best, these standards are minimal.
USDA
United States Department of Agriculture
Animal Welfare Act and Animal Welfare Regulations

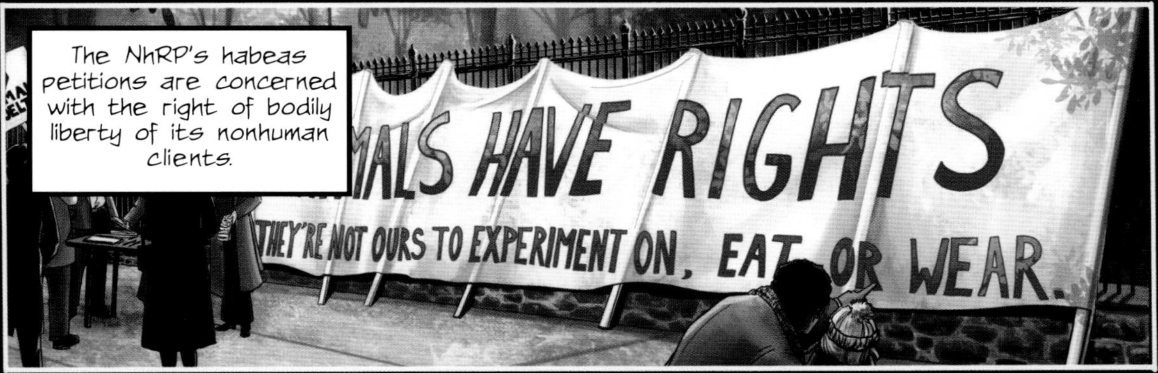
The NhRP's habeas petitions are concerned with the right of bodily liberty of its nonhuman clients.
MALS HAVE RIGHTS
THEY'RE NOT OURS TO EXPERIMENT ON, EAT OR WEAR.

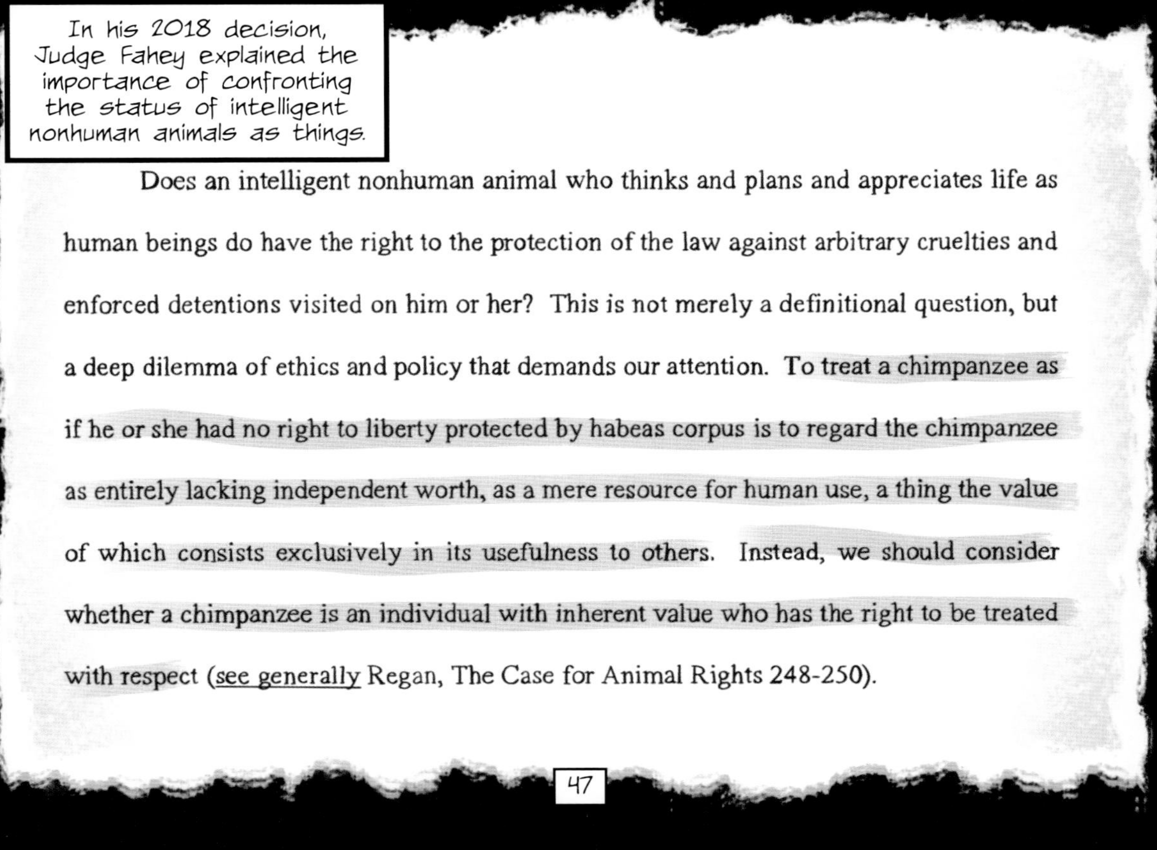
In his 2018 decision, Judge Fahey explained the importance of confronting the status of intelligent nonhuman animals as things.

Does an intelligent nonhuman animal who thinks and plans and appreciates life as human beings do have the right to the protection of the law against arbitrary cruelties and enforced detentions visited on him or her? This is not merely a definitional question, but a deep dilemma of ethics and policy that demands our attention. To treat a chimpanzee as if he or she had no right to liberty protected by habeas corpus is to regard the chimpanzee as entirely lacking independent worth, as a mere resource for human use, a thing the value of which consists exclusively in its usefulness to others. Instead, we should consider whether a chimpanzee is an individual with inherent value who has the right to be treated with respect (see generally Regan, The Case for Animal Rights 248-250).

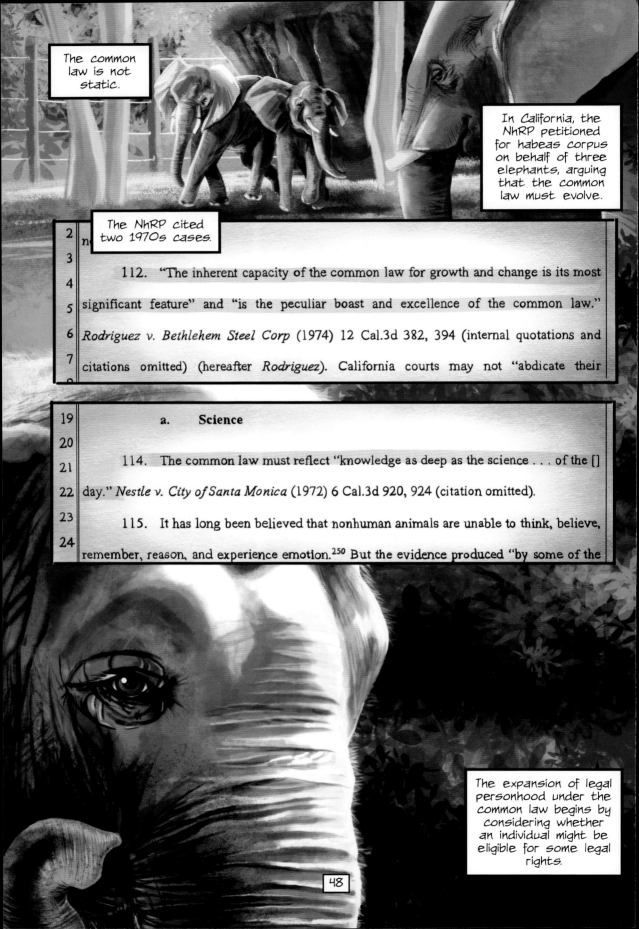

The common law is not static.

In California, the NhRP petitioned for habeas corpus on behalf of three elephants, arguing that the common law must evolve.

The NhRP cited two 1970s cases.

2
3
4
5
6
7

112. "The inherent capacity of the common law for growth and change is its most significant feature" and "is the peculiar boast and excellence of the common law." *Rodriguez v. Bethlehem Steel Corp* (1974) 12 Cal.3d 382, 394 (internal quotations and citations omitted) (hereafter *Rodriguez*). California courts may not "abdicate their

19
20
21
22
23
24

a.   **Science**

114.   The common law must reflect "knowledge as deep as the science . . . of the [] day." *Nestle v. City of Santa Monica* (1972) 6 Cal.3d 920, 924 (citation omitted).

115.   It has long been believed that nonhuman animals are unable to think, believe, remember, reason, and experience emotion.[250] But the evidence produced "by some of the

The expansion of legal personhood under the common law begins by considering whether an individual might be eligible for some legal rights.

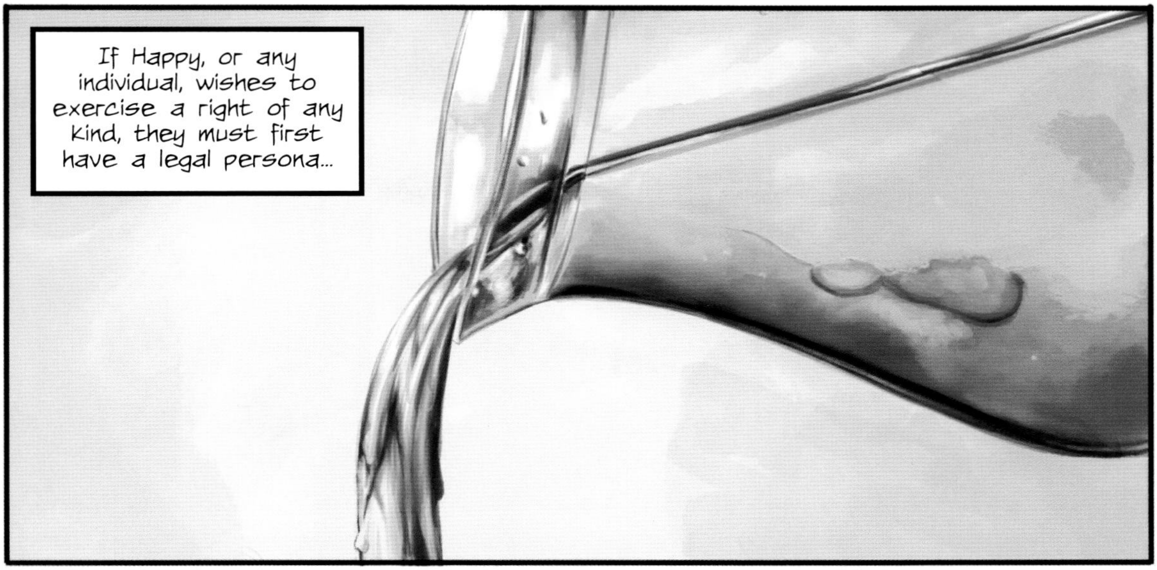
If Happy, or any individual, wishes to exercise a right of any kind, they must first have a legal persona...

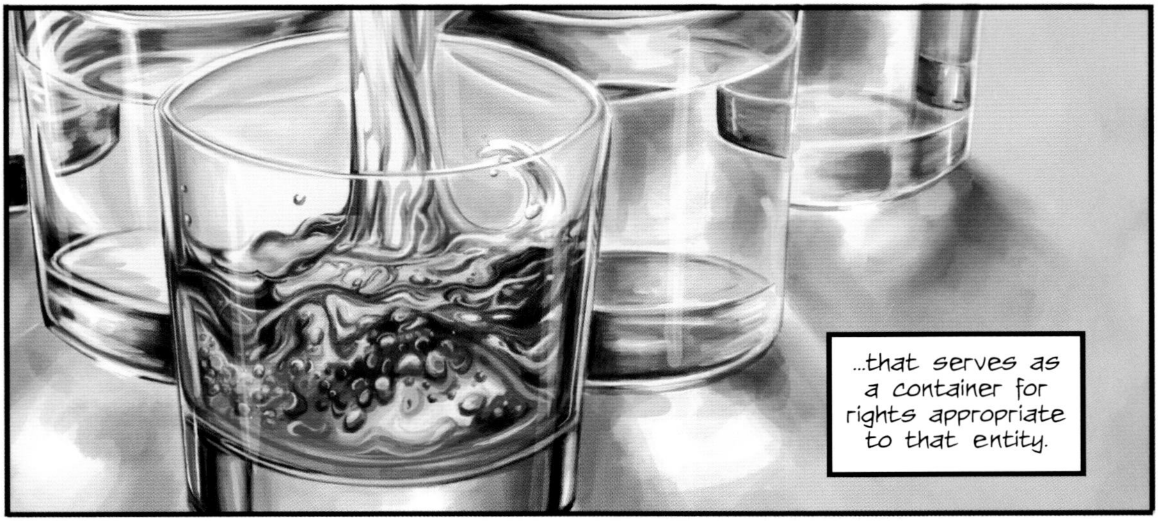
...that serves as a container for rights appropriate to that entity.

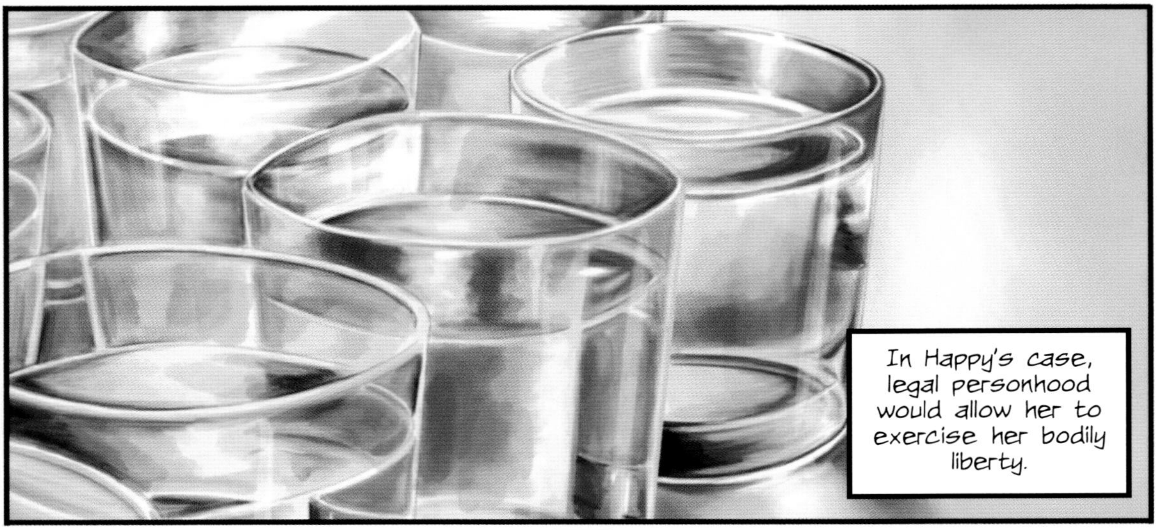
In Happy's case, legal personhood would allow her to exercise her bodily liberty.

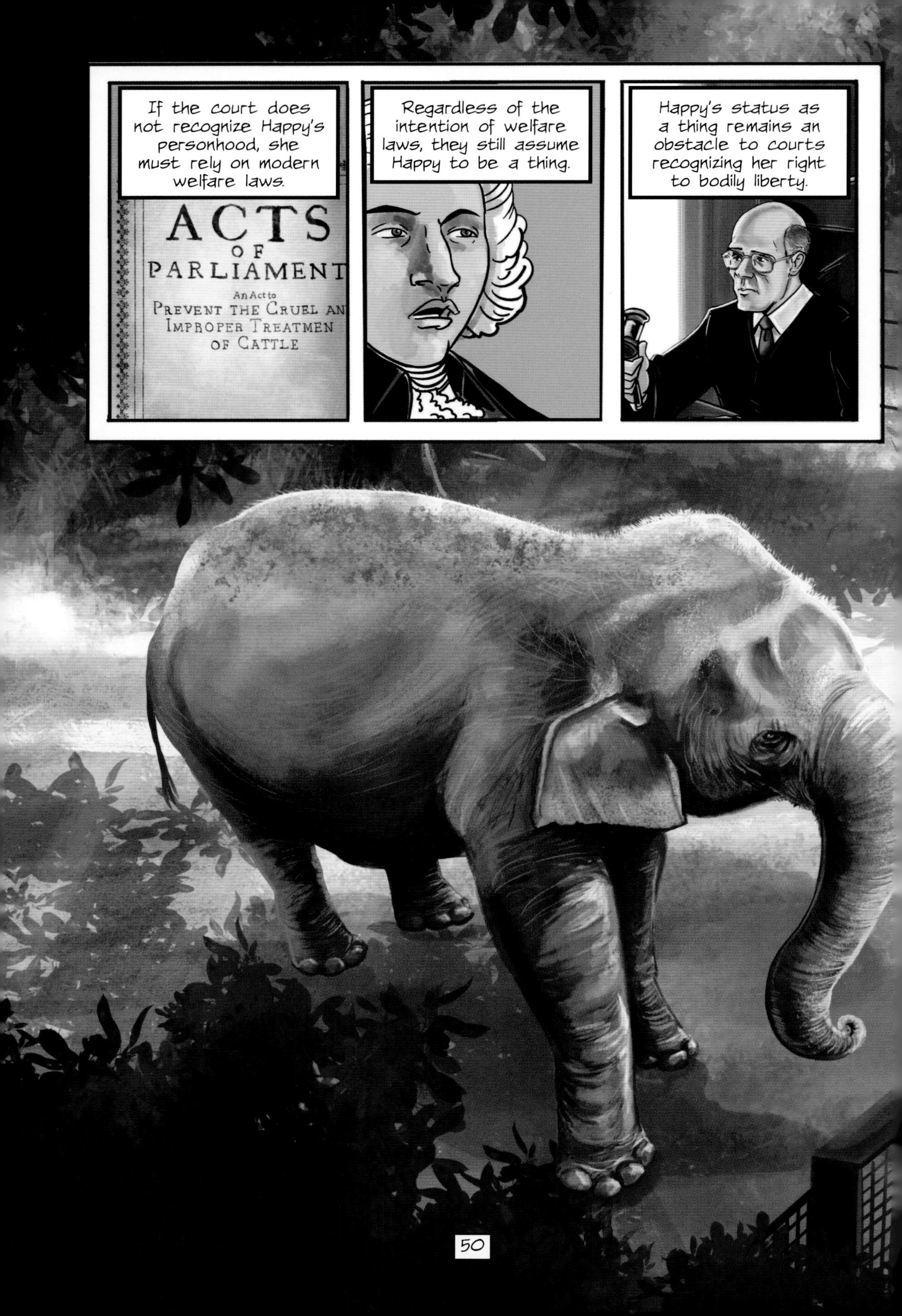
If the court does not recognize Happy's personhood, she must rely on modern welfare laws.
ACTS OF PARLIAMENT
An Act to Prevent the Cruel an Improper Treatmen of Cattle
Regardless of the intention of welfare laws, they still assume Happy to be a thing.
Happy's status as a thing remains an obstacle to courts recognizing her right to bodily liberty.

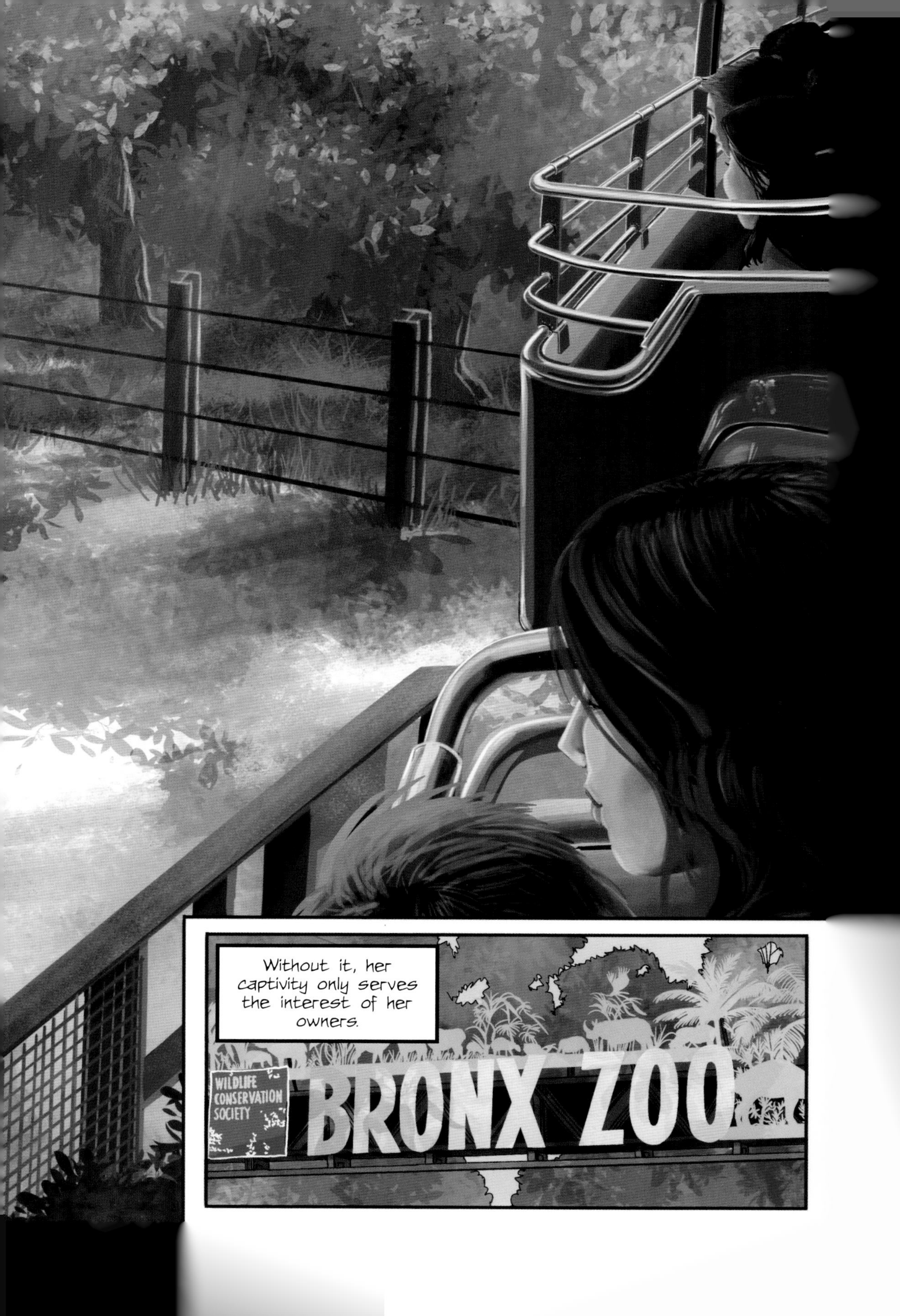
Without it, her captivity only serves the interest of her owners.
WILDLIFE CONSERVATION SOCIETY
BRONX ZOO

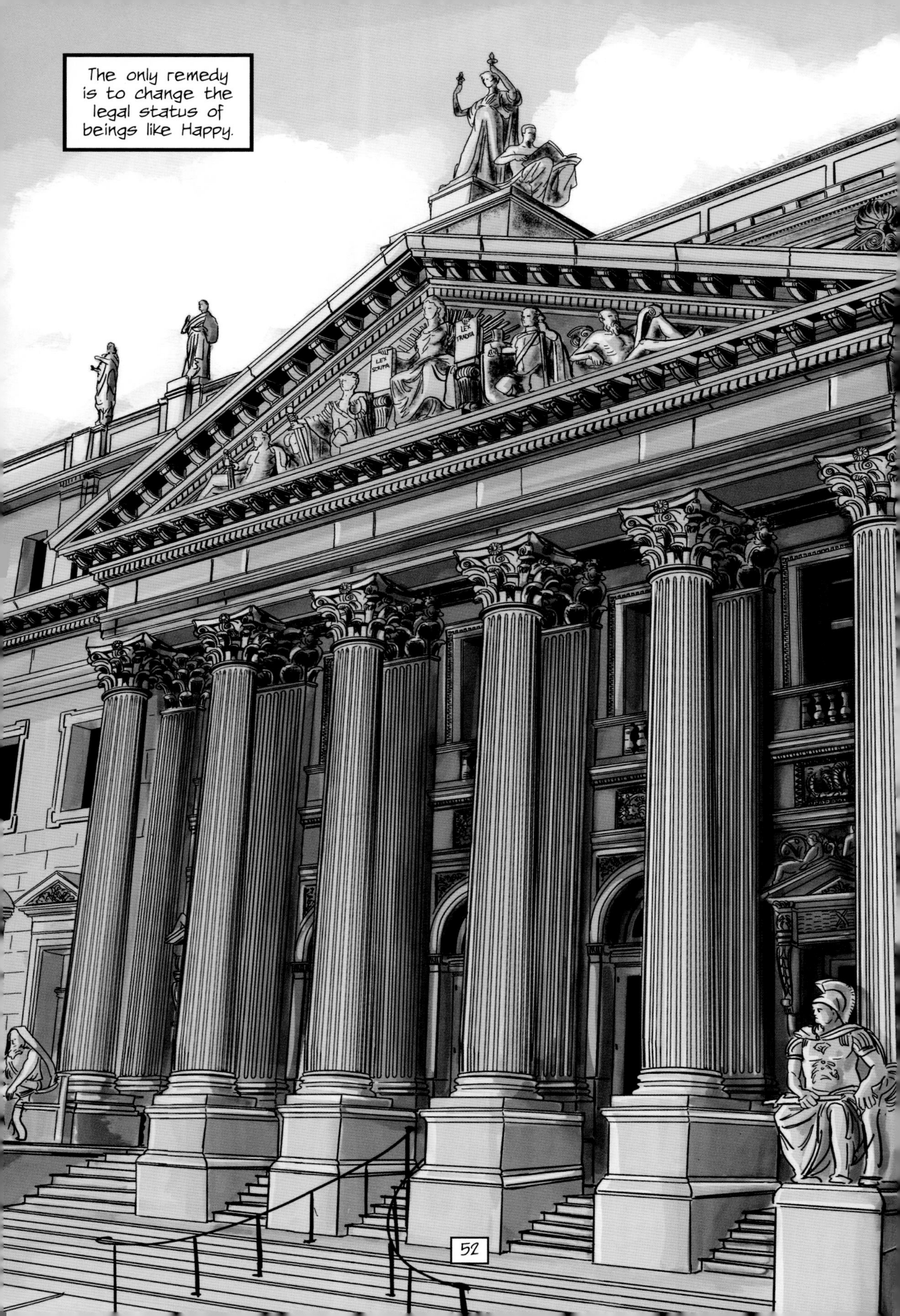
The only remedy is to change the legal status of beings like Happy.
LEX SCRIPTA
LEX TRADITA

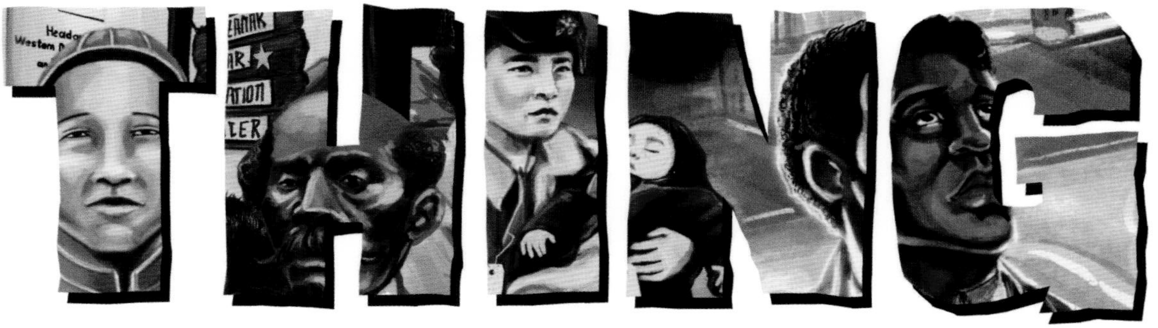
THING

CHAPTER 3: HABEAS CORPUS

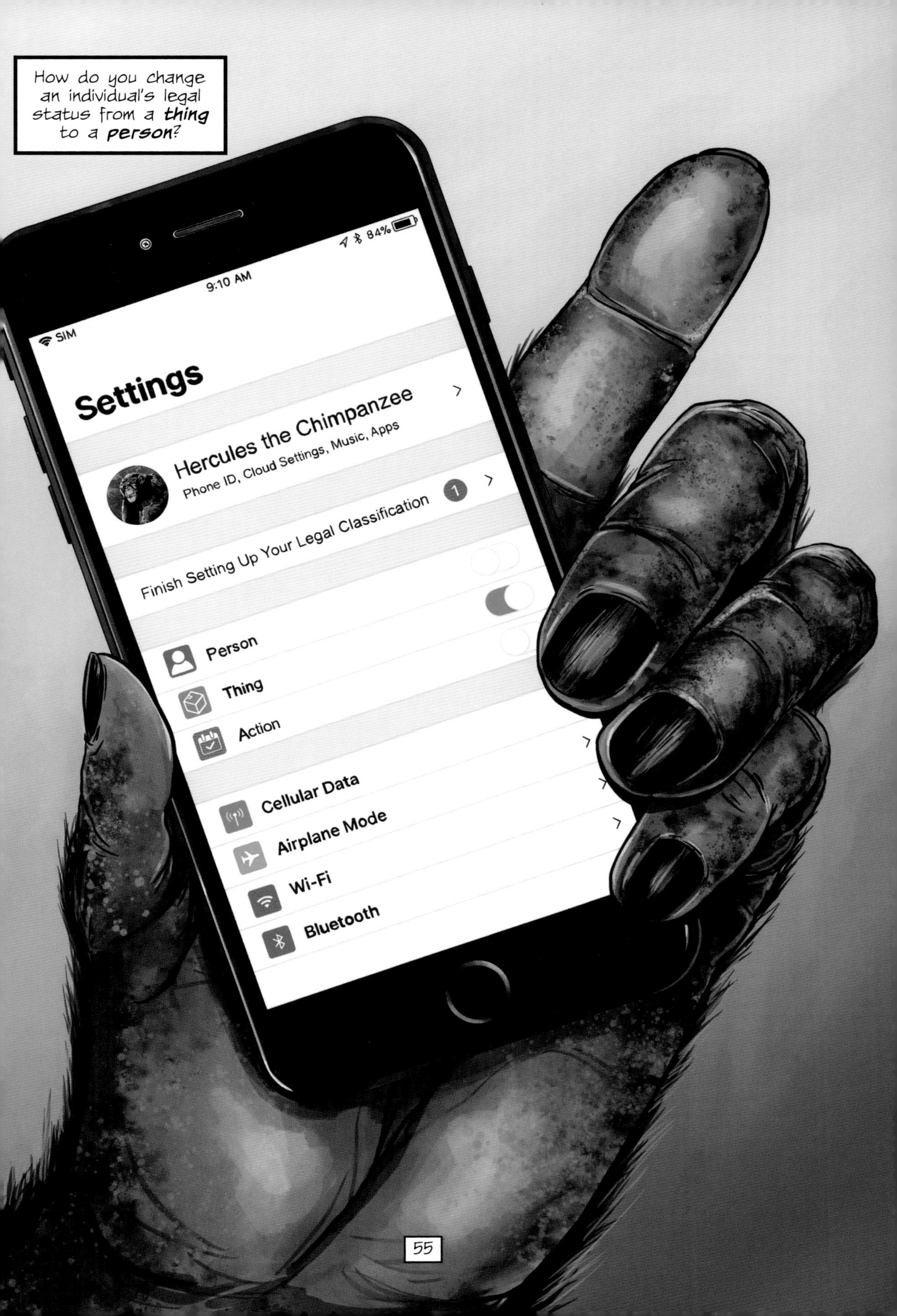

How do you change an individual's legal status from a ***thing*** to a ***person***?
9:10 AM
SIM
84%
Settings
Hercules the Chimpanzee
Phone ID, Cloud Settings, Music, Apps
Finish Setting Up Your Legal Classification
1
Person
Thing
Action
Cellular Data
Airplane Mode
Wi-Fi
Bluetooth

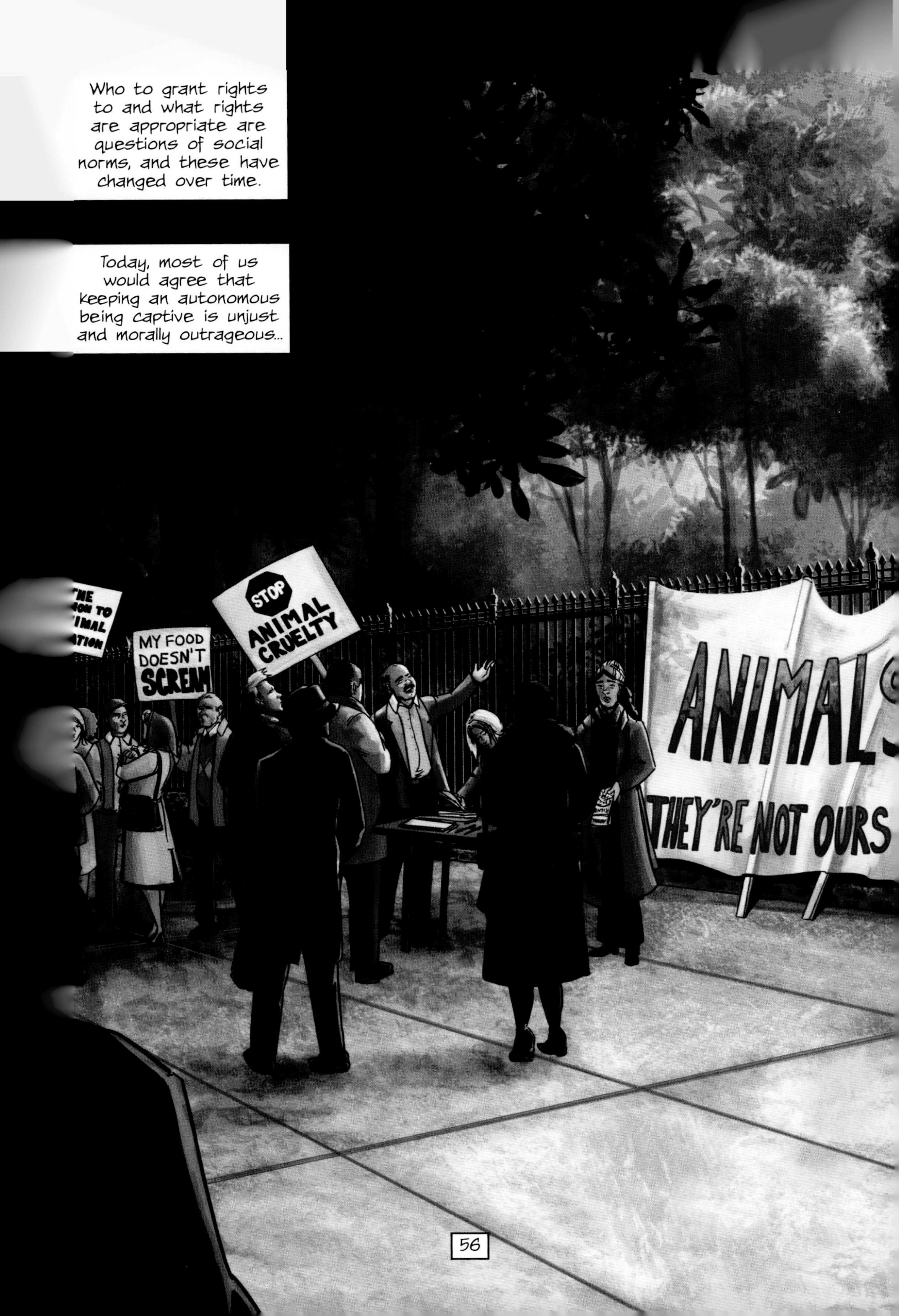
Who to grant rights to and what rights are appropriate are questions of social norms, and these have changed over time.
Today, most of us would agree that keeping an autonomous being captive is unjust and morally outrageous...
MY FOOD DOESN'T SCREAM
STOP
ANIMAL CRUELTY
ANIMALS
THEY'RE NOT OURS

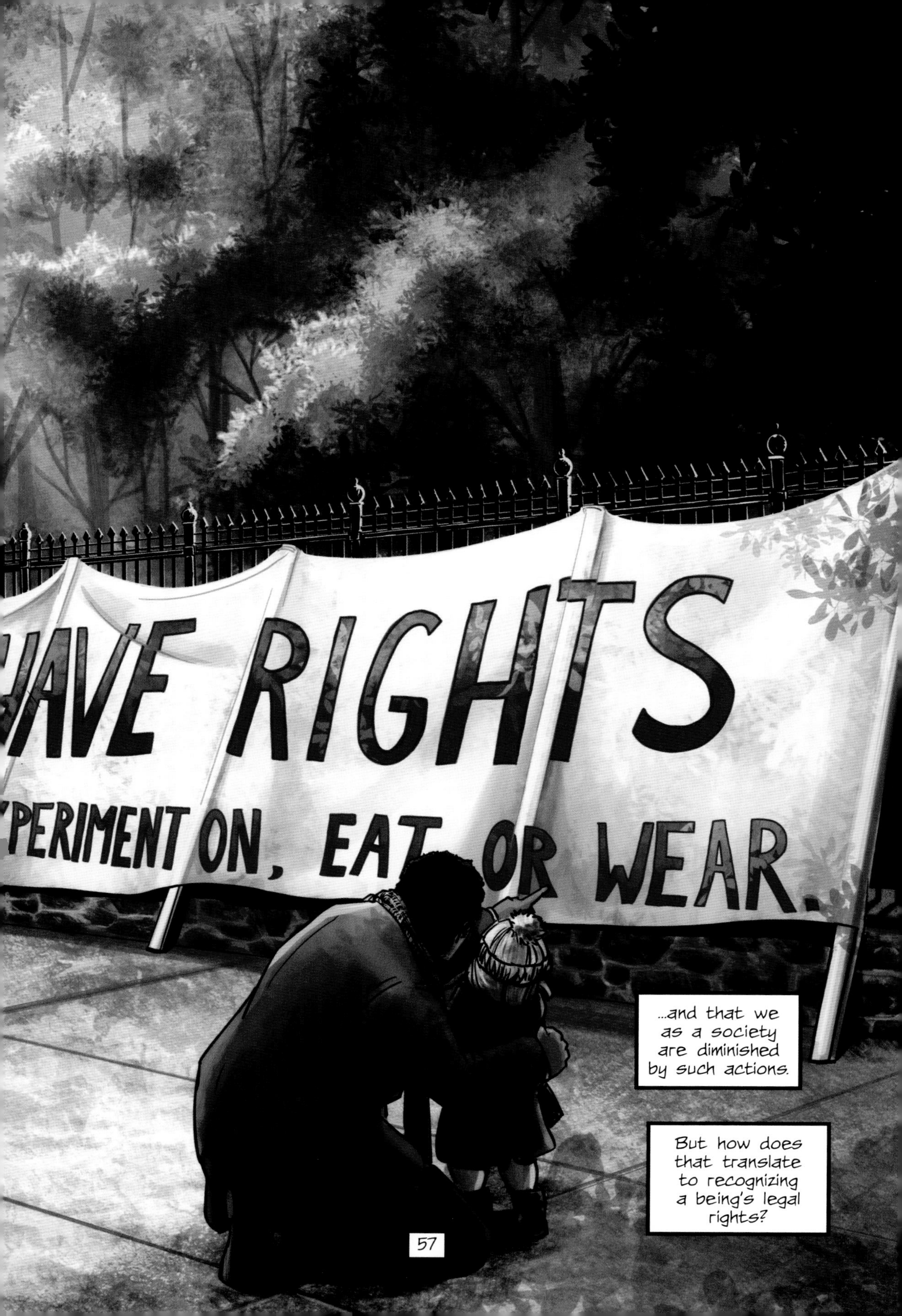

HAVE RIGHTS
PERIMENT ON, EAT OR WEAR.
...and that we as a society are diminished by such actions.
But how does that translate to recognizing a being's legal rights?

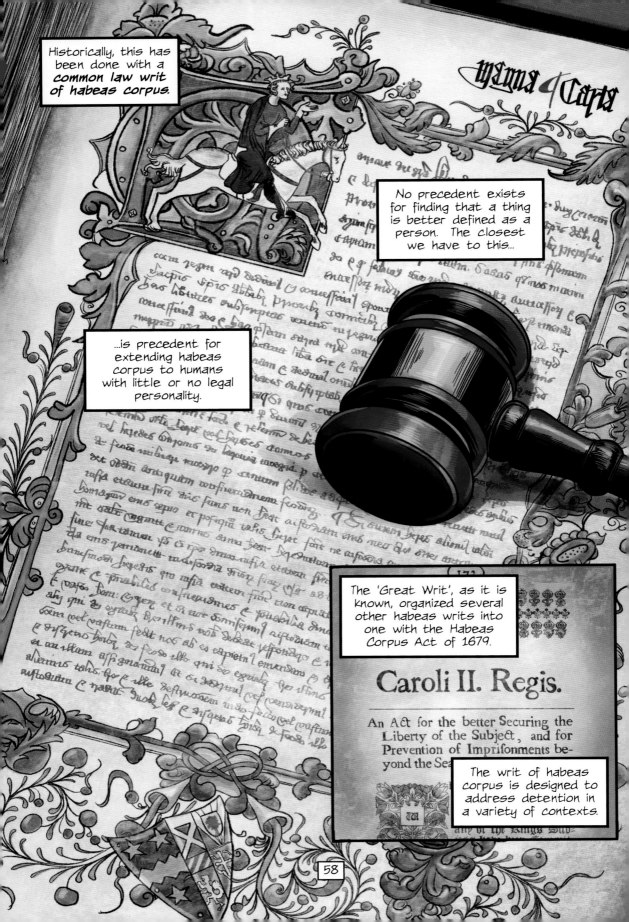

Historically, this has been done with a **common law writ of habeas corpus.**

No precedent exists for finding that a thing is better defined as a person. The closest we have to this...

...is precedent for extending habeas corpus to humans with little or no legal personality.

The 'Great Writ', as it is known, organized several other habeas writs into one with the Habeas Corpus Act of 1679.

Caroli II. Regis.

An Act for the better Securing the Liberty of the Subject, and for Prevention of Imprisonments beyond the Sea

The writ of habeas corpus is designed to address detention in a variety of contexts.

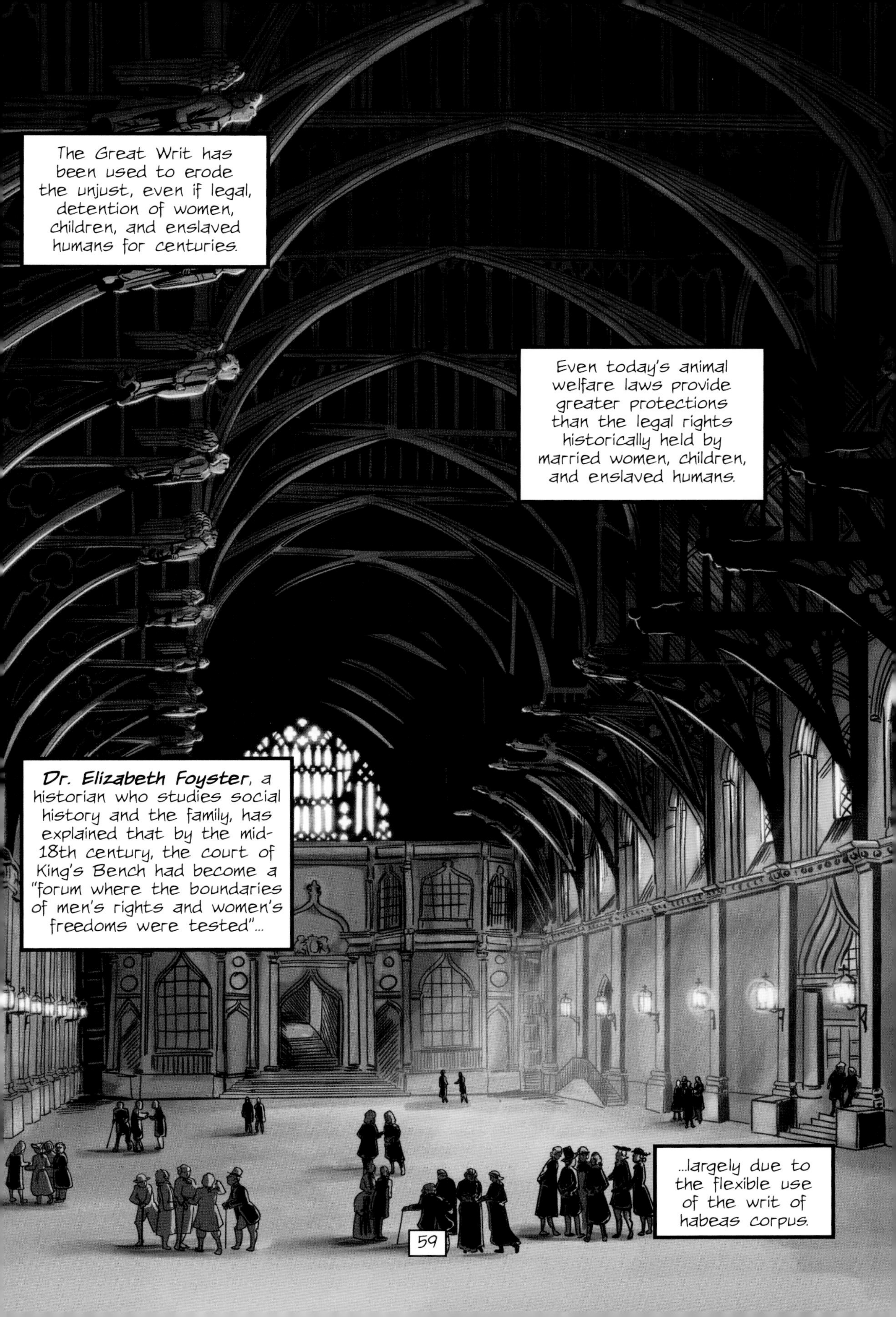
The Great Writ has been used to erode the unjust, even if legal, detention of women, children, and enslaved humans for centuries.
Even today's animal welfare laws provide greater protections than the legal rights historically held by married women, children, and enslaved humans.
Dr. Elizabeth Foyster, a historian who studies social history and the family, has explained that by the mid-18th century, the court of King's Bench had become a "forum where the boundaries of men's rights and women's freedoms were tested"...
...largely due to the flexible use of the writ of habeas corpus.

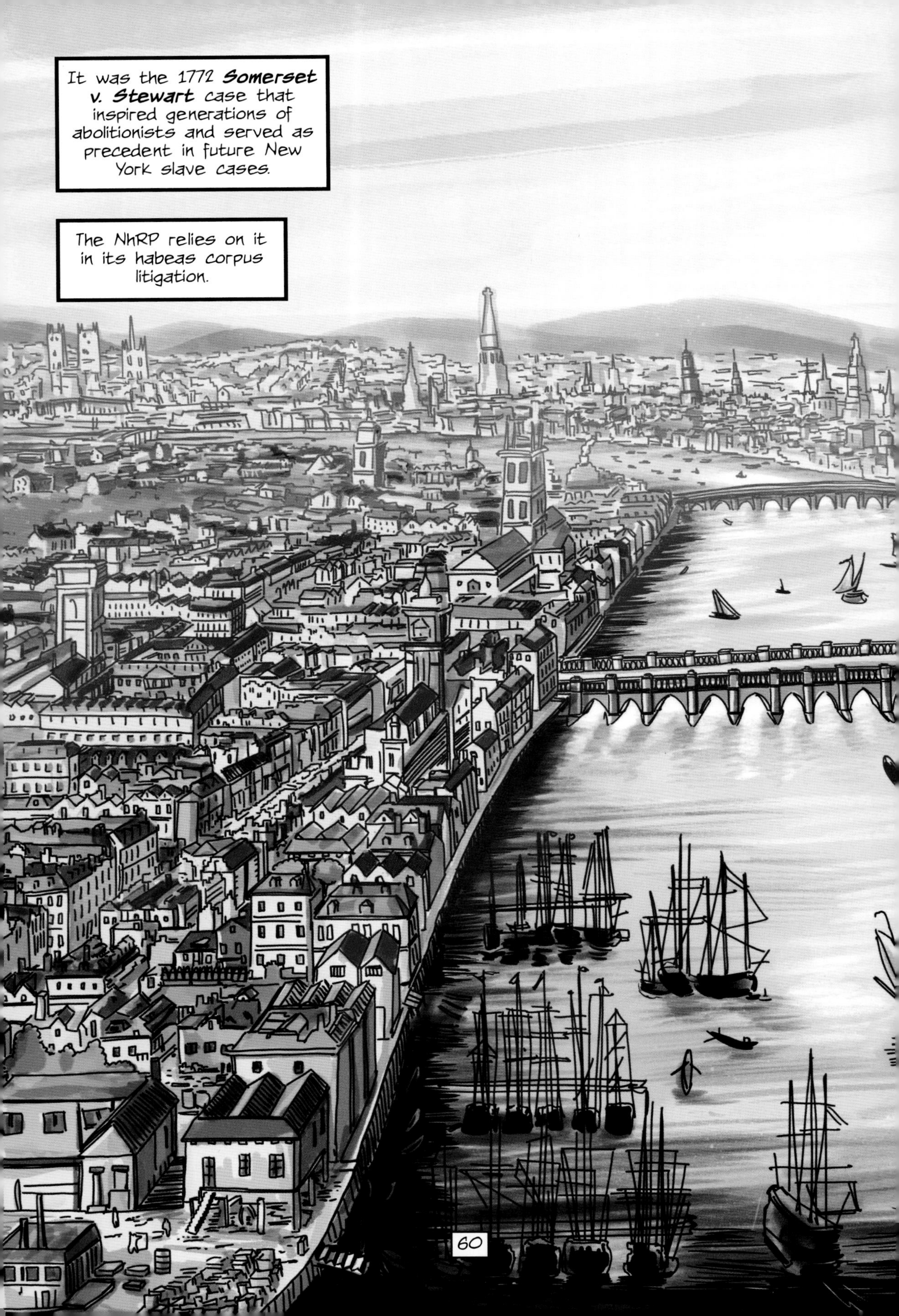
It was the 1772 **Somerset v. Stewart** case that inspired generations of abolitionists and served as precedent in future New York slave cases.
The NhRP relies on it in its habeas corpus litigation.

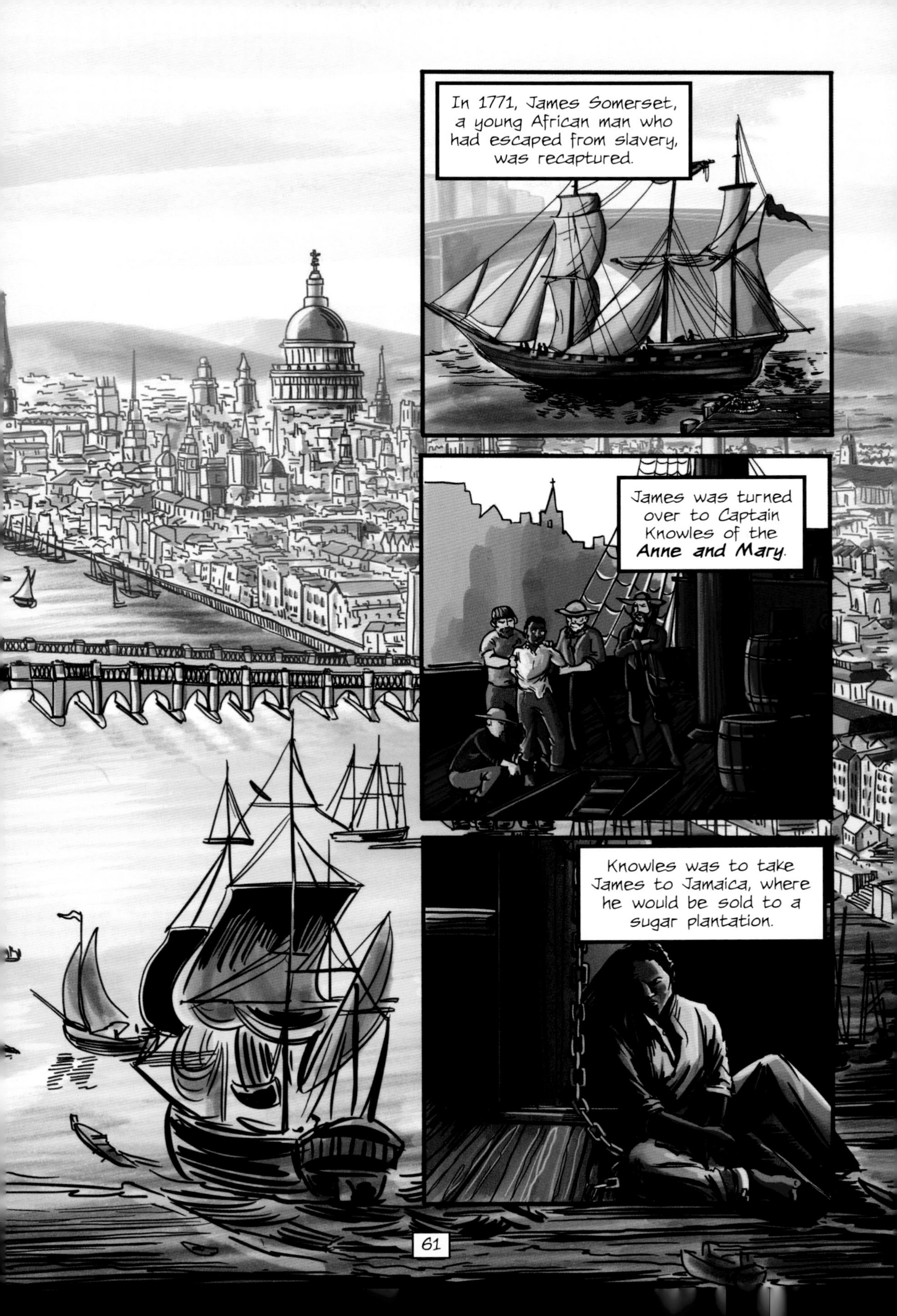
In 1771, James Somerset, a young African man who had escaped from slavery, was recaptured.
James was turned over to Captain Knowles of the **Anne and Mary**.
Knowles was to take James to Jamaica, where he would be sold to a sugar plantation.

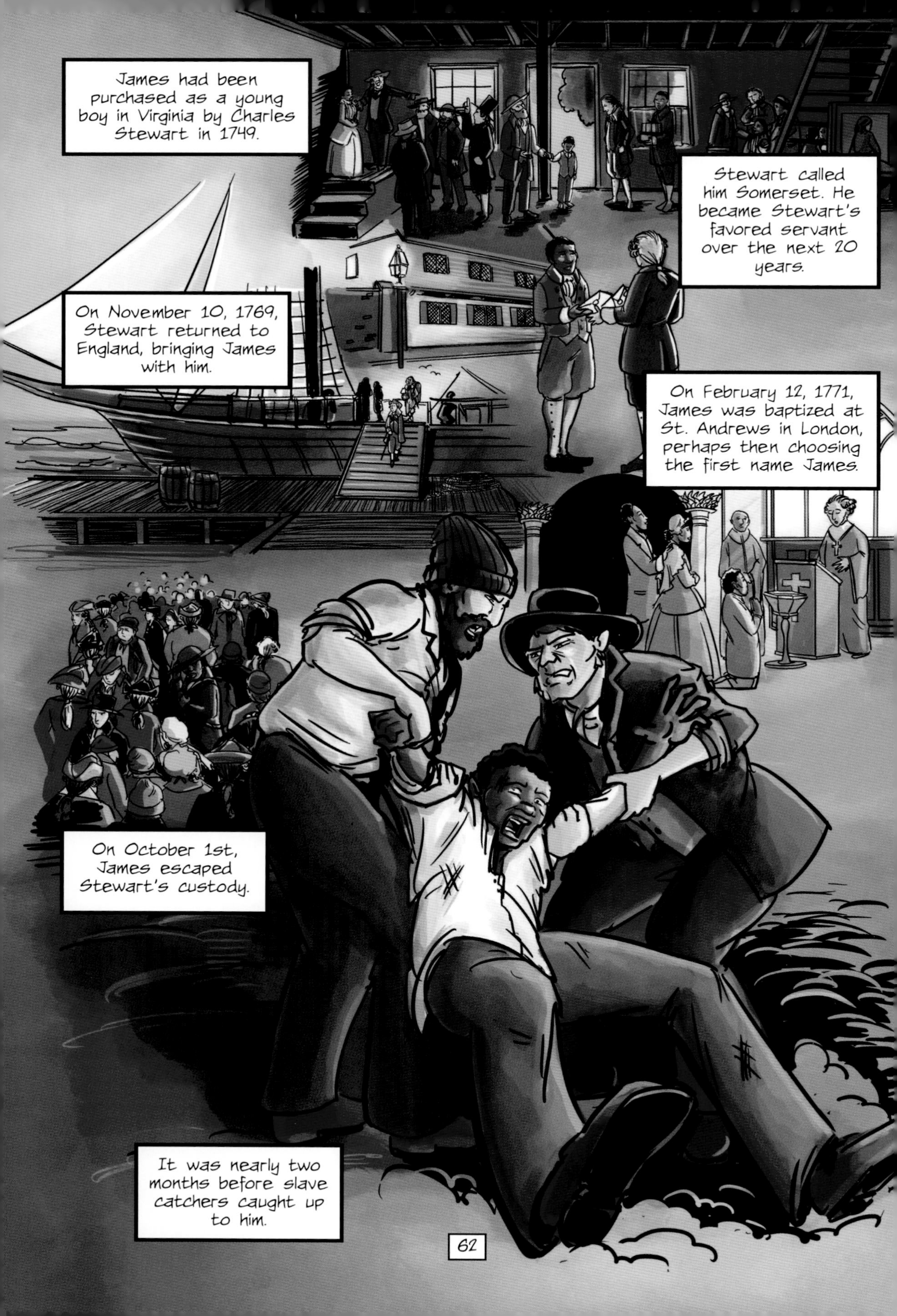
James had been purchased as a young boy in Virginia by Charles Stewart in 1749.
Stewart called him Somerset. He became Stewart's favored servant over the next 20 years.
On November 10, 1769, Stewart returned to England, bringing James with him.
On February 12, 1771, James was baptized at St. Andrews in London, perhaps then choosing the first name James.
On October 1st, James escaped Stewart's custody.
It was nearly two months before slave catchers caught up to him.

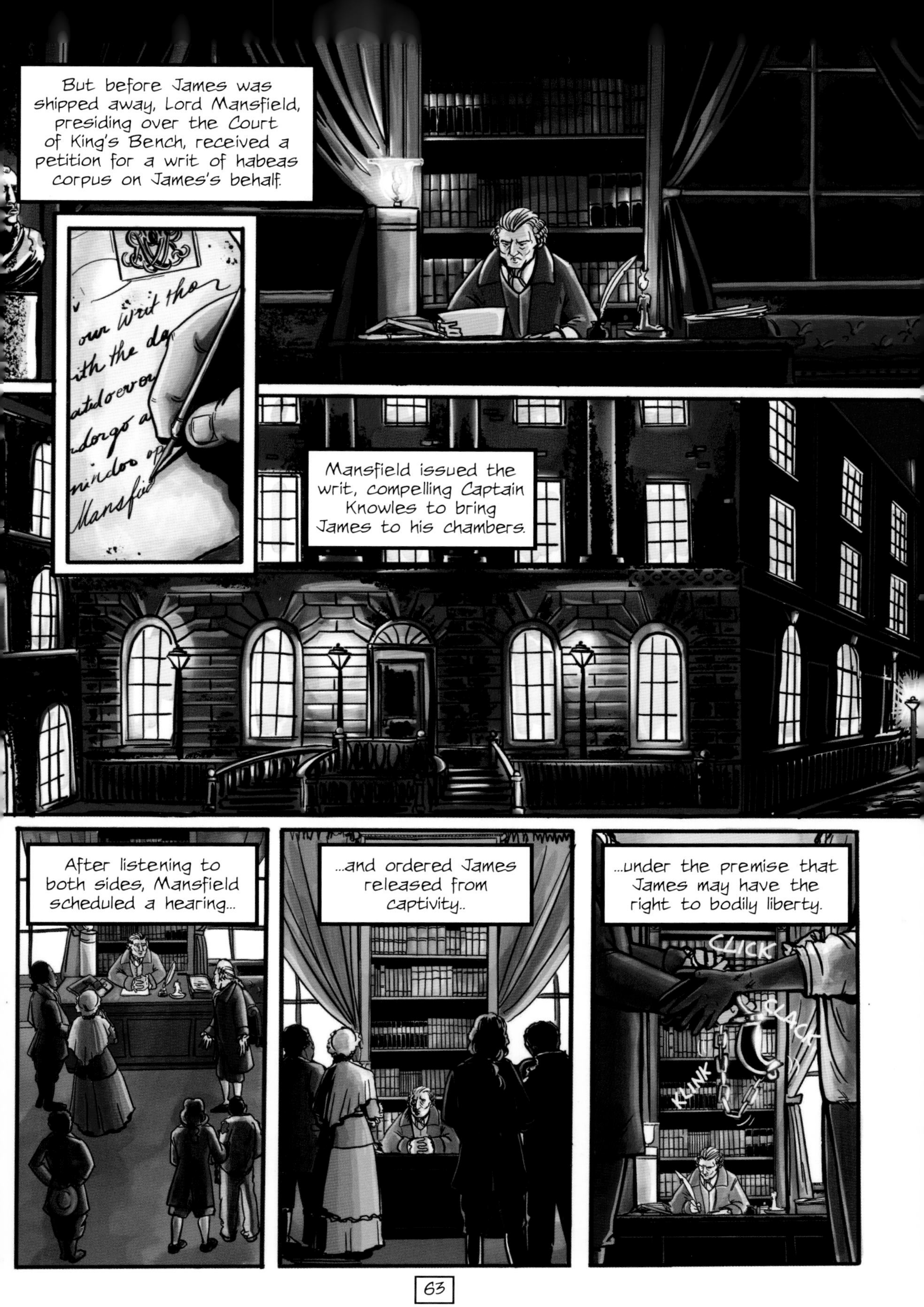
But before James was shipped away, Lord Mansfield, presiding over the Court of King's Bench, received a petition for a writ of habeas corpus on James's behalf.
Mansfield issued the writ, compelling Captain Knowles to bring James to his chambers.
After listening to both sides, Mansfield scheduled a hearing...
...and ordered James released from captivity..
...under the premise that James may have the right to bodily liberty.
CLICK
CLACK
KLINK

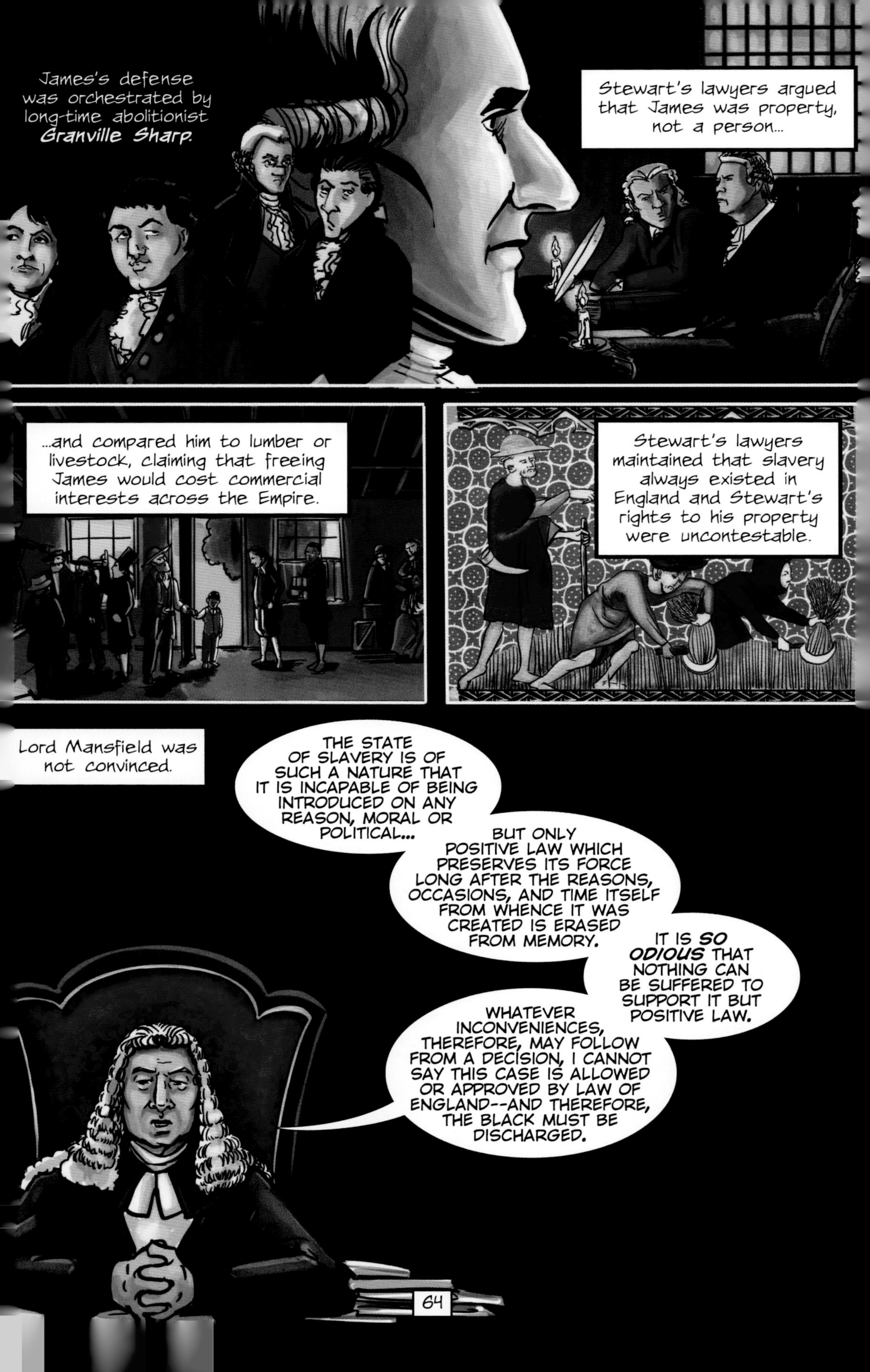
James's defense was orchestrated by long-time abolitionist Granville Sharp.
Stewart's lawyers argued that James was property, not a person...
...and compared him to lumber or livestock, claiming that freeing James would cost commercial interests across the Empire.
Stewart's lawyers maintained that slavery always existed in England and Stewart's rights to his property were uncontestable.
Lord Mansfield was not convinced.
THE STATE OF SLAVERY IS OF SUCH A NATURE THAT IT IS INCAPABLE OF BEING INTRODUCED ON ANY REASON, MORAL OR POLITICAL...
BUT ONLY POSITIVE LAW WHICH PRESERVES ITS FORCE LONG AFTER THE REASONS, OCCASIONS, AND TIME ITSELF FROM WHENCE IT WAS CREATED IS ERASED FROM MEMORY.
IT IS SO ODIOUS THAT NOTHING CAN BE SUFFERED TO SUPPORT IT BUT POSITIVE LAW.
WHATEVER INCONVENIENCES, THEREFORE, MAY FOLLOW FROM A DECISION, I CANNOT SAY THIS CASE IS ALLOWED OR APPROVED BY LAW OF ENGLAND--AND THEREFORE, THE BLACK MUST BE DISCHARGED.

With the ruling, James's right to liberty was recognized...

...and his questionable legal status immediately changed...

...to that of a **person**.

Finish Setting Up Your Legal Classification

Person

Thing

Action

Finish Setting Up Your Legal Classification

Person *SWIPE*

Thing

Action

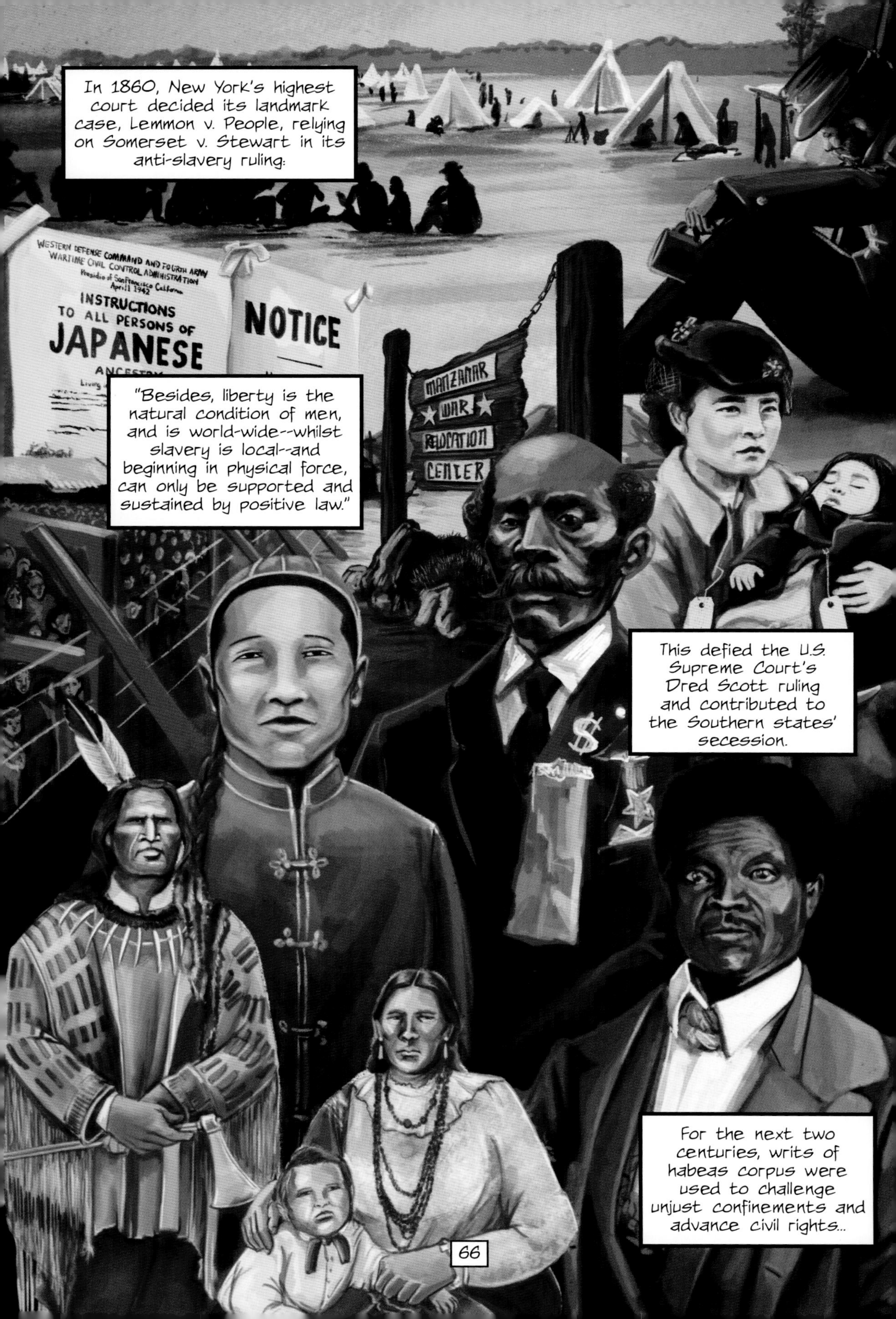
In 1860, New York's highest court decided its landmark case, Lemmon v. People, relying on Somerset v. Stewart in its anti-slavery ruling:
WESTERN DEFENSE COMMAND AND FOURTH ARMY WARTIME CIVIL CONTROL ADMINISTRATION
INSTRUCTIONS TO ALL PERSONS OF JAPANESE
NOTICE
MANZANAR WAR RELOCATION CENTER
"Besides, liberty is the natural condition of men, and is world-wide--whilst slavery is local--and beginning in physical force, can only be supported and sustained by positive law."
This defied the U.S. Supreme Court's Dred Scott ruling and contributed to the Southern states' secession.
For the next two centuries, writs of habeas corpus were used to challenge unjust confinements and advance civil rights...

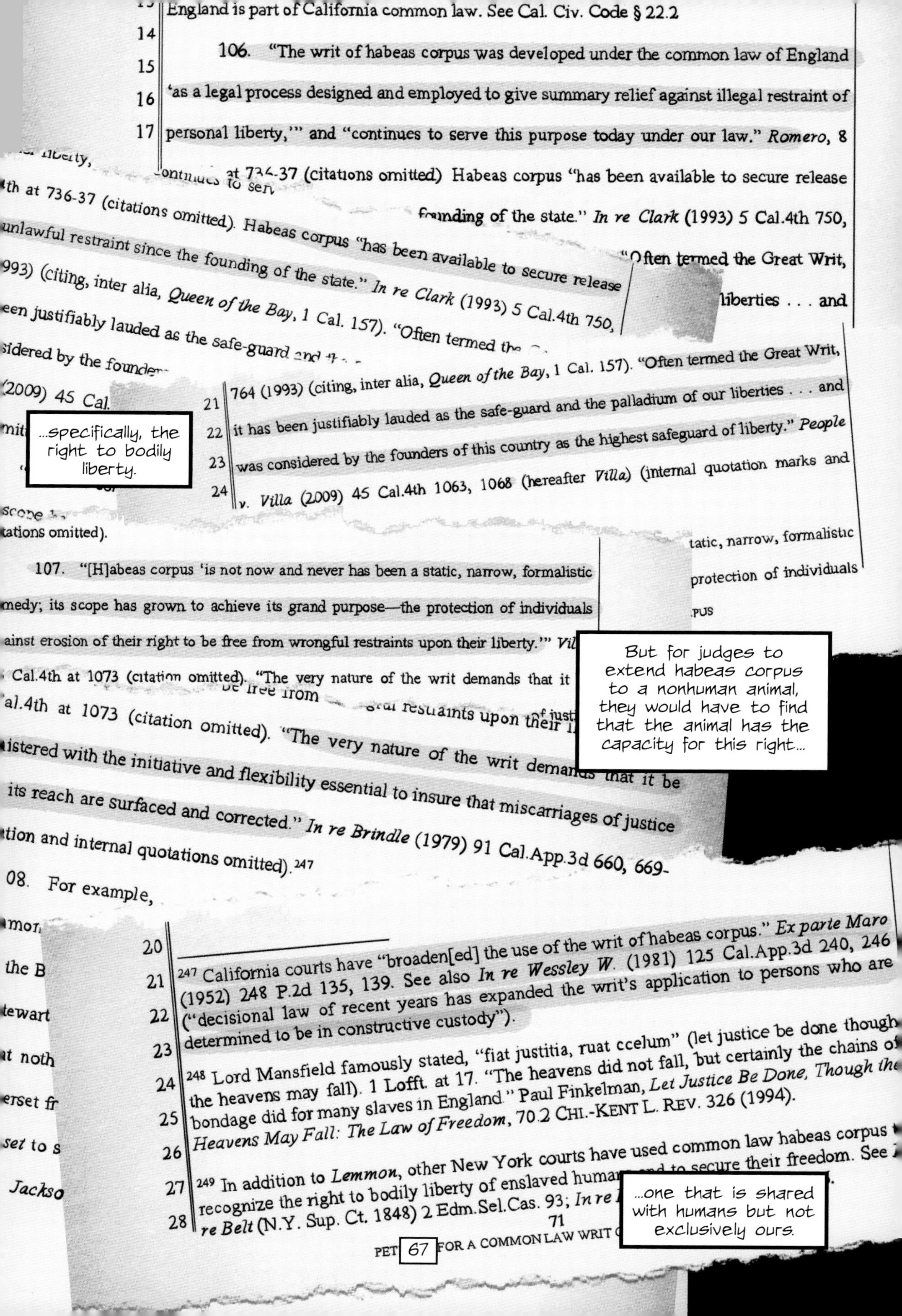
England is part of California common law. See Cal. Civ. Code § 22.2
106. "The writ of habeas corpus was developed under the common law of England 'as a legal process designed and employed to give summary relief against illegal restraint of personal liberty,'" and "continues to serve this purpose today under our law." Romero, 8
at 736-37 (citations omitted) Habeas corpus "has been available to secure release
founding of the state." In re Clark (1993) 5 Cal.4th 750,
"Often termed the Great Writ,
liberties . . . and
4th at 736-37 (citations omitted). Habeas corpus "has been available to secure release
unlawful restraint since the founding of the state." In re Clark (1993) 5 Cal.4th 750,
993) (citing, inter alia, Queen of the Bay, 1 Cal. 157). "Often termed the
een justifiably lauded as the safe-guard
sidered by the founder
(2009) 45 Cal.
...specifically, the right to bodily liberty.
764 (1993) (citing, inter alia, Queen of the Bay, 1 Cal. 157). "Often termed the Great Writ, it has been justifiably lauded as the safe-guard and the palladium of our liberties . . . and was considered by the founders of this country as the highest safeguard of liberty." People v. Villa (2009) 45 Cal.4th 1063, 1068 (hereafter Villa) (internal quotation marks and
tations omitted).
107. "[H]abeas corpus 'is not now and never has been a static, narrow, formalistic
medy; its scope has grown to achieve its grand purpose—the protection of individuals
ainst erosion of their right to be free from wrongful restraints upon their liberty.'" Vil
Cal.4th at 1073 (citation omitted). "The very nature of the writ demands that it
tatic, narrow, formalistic
protection of individuals
But for judges to extend habeas corpus to a nonhuman animal, they would have to find that the animal has the capacity for this right...
al.4th at 1073 (citation omitted). "The very nature of the writ demands that it be
istered with the initiative and flexibility essential to insure that miscarriages of justice
its reach are surfaced and corrected." In re Brindle (1979) 91 Cal.App.3d 660, 669-
tion and internal quotations omitted).247
08. For example,
247 California courts have "broaden[ed] the use of the writ of habeas corpus." Ex parte Maro (1952) 248 P.2d 135, 139. See also In re Wessley W. (1981) 125 Cal.App.3d 240, 246 ("decisional law of recent years has expanded the writ's application to persons who are determined to be in constructive custody").
248 Lord Mansfield famously stated, "fiat justitia, ruat ccelum" (let justice be done though the heavens may fall). 1 Lofft. at 17. "The heavens did not fall, but certainly the chains of bondage did for many slaves in England." Paul Finkelman, Let Justice Be Done, Though the Heavens May Fall: The Law of Freedom, 70.2 CHI.-KENT L. REV. 326 (1994).
249 In addition to Lemmon, other New York courts have used common law habeas corpus to secure their freedom. See
recognize the right to bodily liberty of enslaved human
re Belt (N.Y. Sup. Ct. 1848) 2 Edm.Sel.Cas. 93; In re
71
PET FOR A COMMON LAW WRIT
...one that is shared with humans but not exclusively ours.

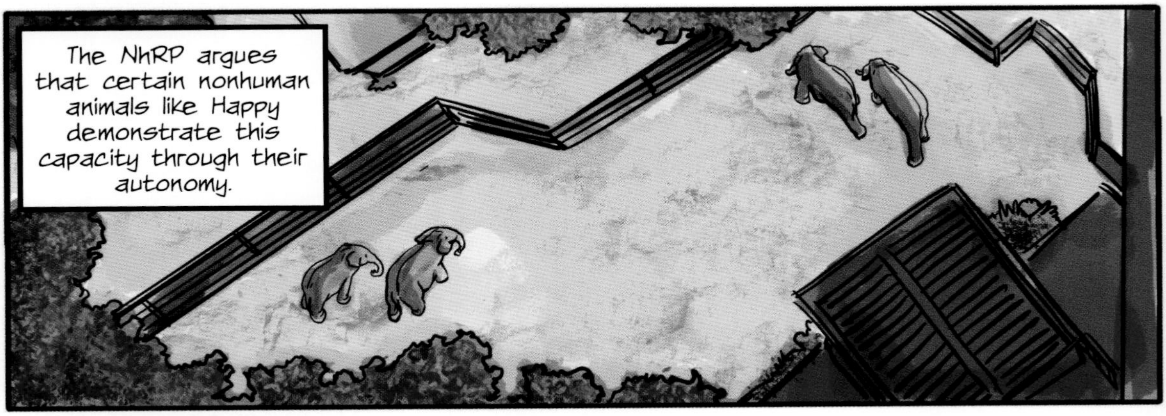
The NhRP argues that certain nonhuman animals like Happy demonstrate this capacity through their autonomy.

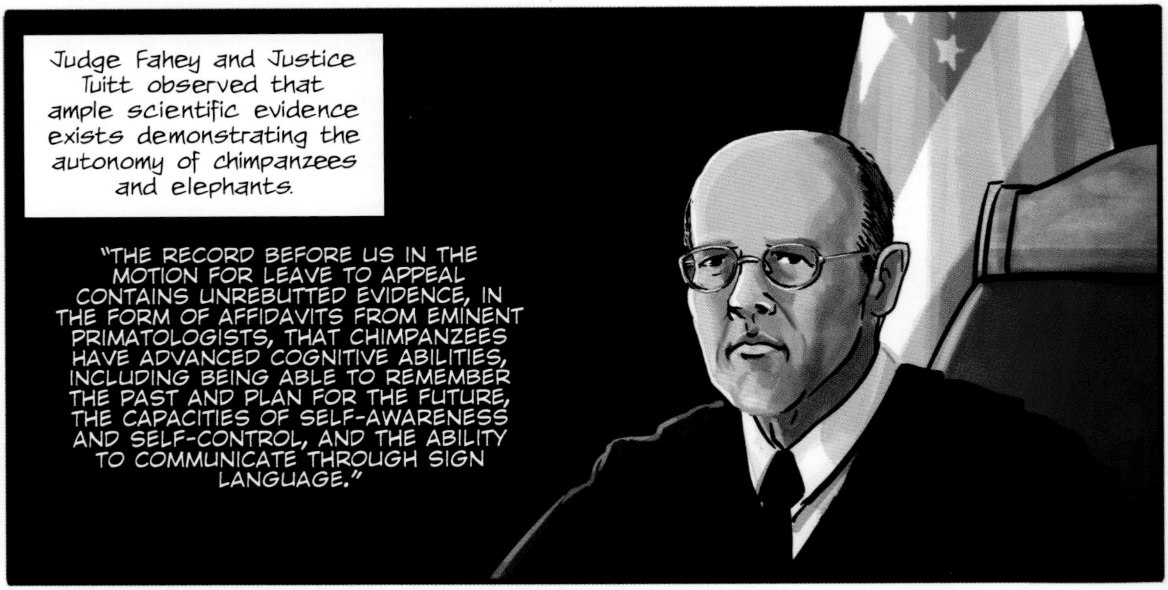
Judge Fahey and Justice Tuitt observed that ample scientific evidence exists demonstrating the autonomy of chimpanzees and elephants.
"THE RECORD BEFORE US IN THE MOTION FOR LEAVE TO APPEAL CONTAINS UNREBUTTED EVIDENCE, IN THE FORM OF AFFIDAVITS FROM EMINENT PRIMATOLOGISTS, THAT CHIMPANZEES HAVE ADVANCED COGNITIVE ABILITIES, INCLUDING BEING ABLE TO REMEMBER THE PAST AND PLAN FOR THE FUTURE, THE CAPACITIES OF SELF-AWARENESS AND SELF-CONTROL, AND THE ABILITY TO COMMUNICATE THROUGH SIGN LANGUAGE."

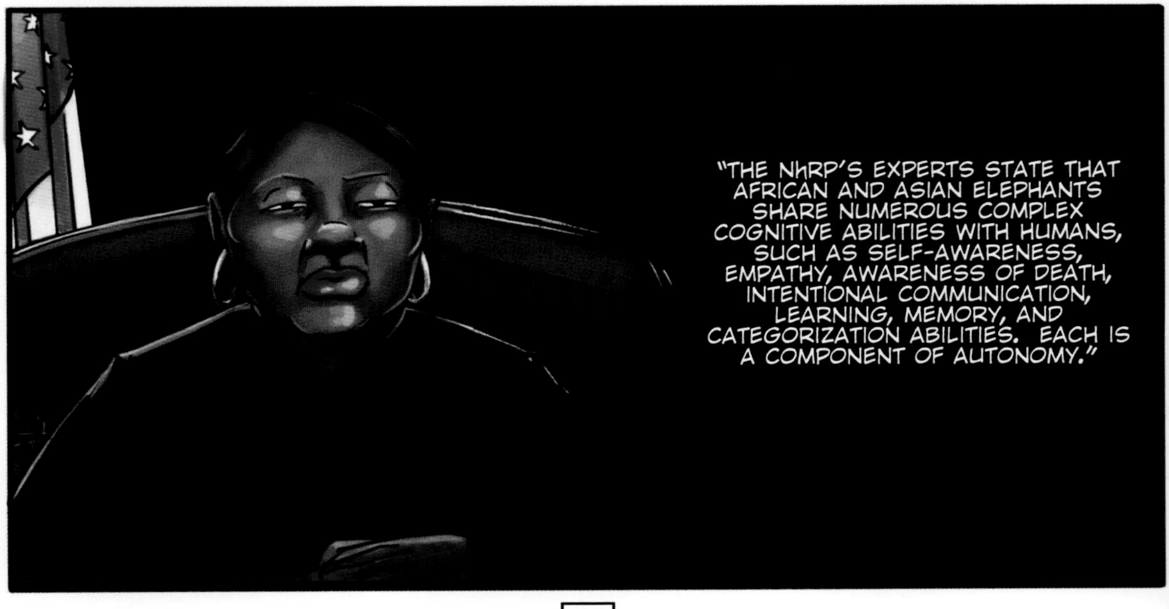
"THE NhRP'S EXPERTS STATE THAT AFRICAN AND ASIAN ELEPHANTS SHARE NUMEROUS COMPLEX COGNITIVE ABILITIES WITH HUMANS, SUCH AS SELF-AWARENESS, EMPATHY, AWARENESS OF DEATH, INTENTIONAL COMMUNICATION, LEARNING, MEMORY, AND CATEGORIZATION ABILITIES. EACH IS A COMPONENT OF AUTONOMY."

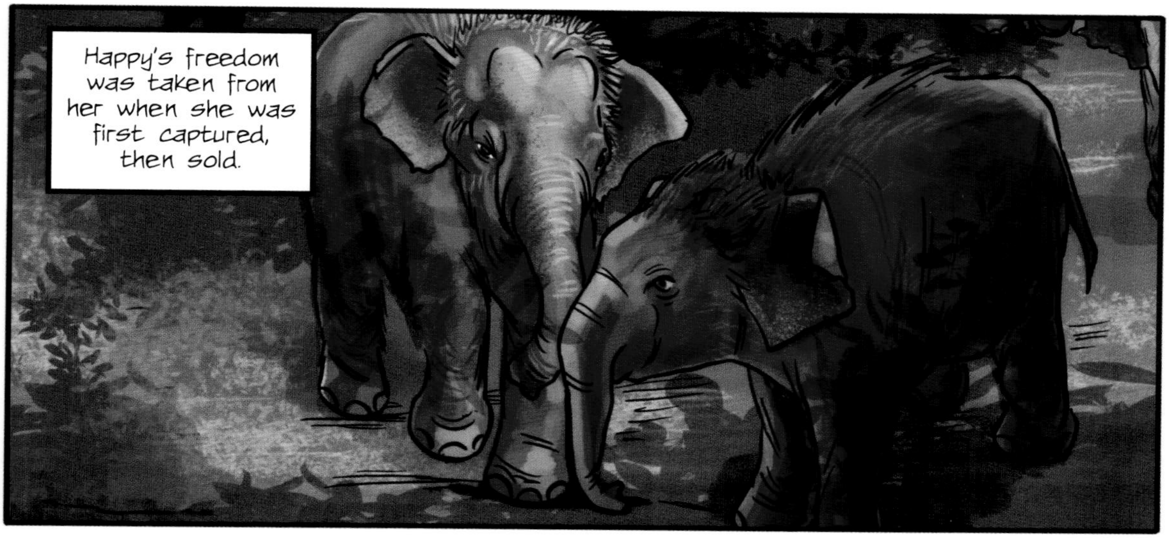
Happy's freedom was taken from her when she was first captured, then sold.

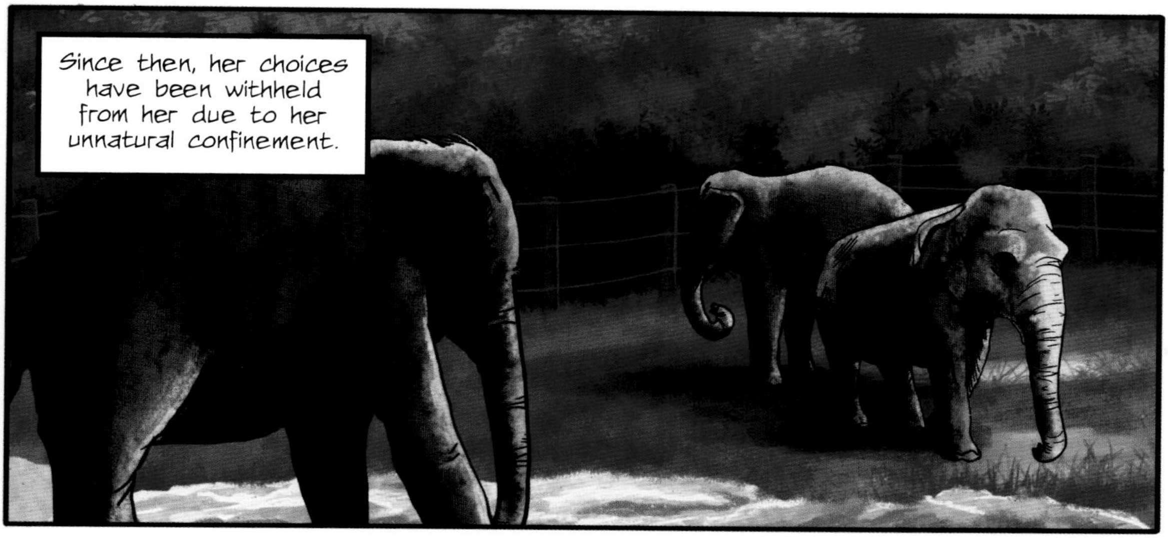
Since then, her choices have been withheld from her due to her unnatural confinement.

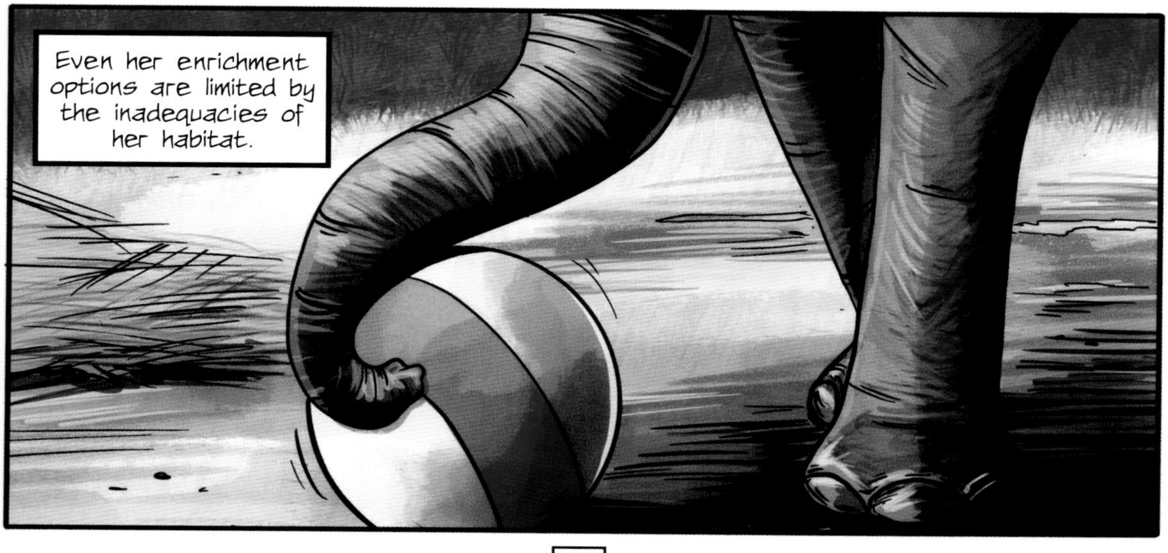
Even her enrichment options are limited by the inadequacies of her habitat.

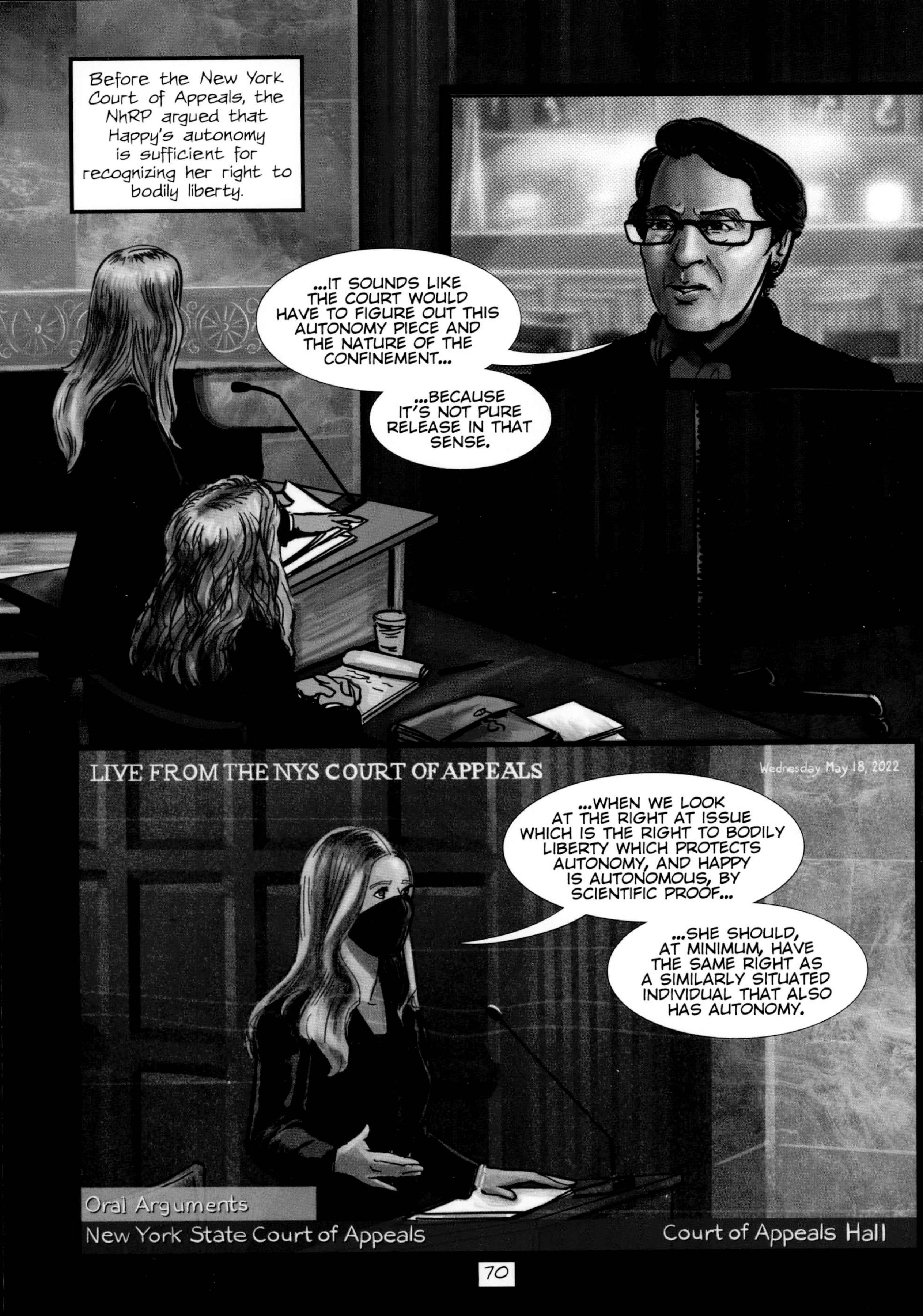
Before the New York Court of Appeals, the NhRP argued that Happy's autonomy is sufficient for recognizing her right to bodily liberty.
...IT SOUNDS LIKE THE COURT WOULD HAVE TO FIGURE OUT THIS AUTONOMY PIECE AND THE NATURE OF THE CONFINEMENT...
...BECAUSE IT'S NOT PURE RELEASE IN THAT SENSE.
LIVE FROM THE NYS COURT OF APPEALS
Wednesday May 18, 2022
...WHEN WE LOOK AT THE RIGHT AT ISSUE WHICH IS THE RIGHT TO BODILY LIBERTY WHICH PROTECTS AUTONOMY, AND HAPPY IS AUTONOMOUS, BY SCIENTIFIC PROOF...
...SHE SHOULD, AT MINIMUM, HAVE THE SAME RIGHT AS A SIMILARLY SITUATED INDIVIDUAL THAT ALSO HAS AUTONOMY.
Oral Arguments
New York State Court of Appeals
Court of Appeals Hall

New York courts have recognized the importance of protecting autonomy in humans.

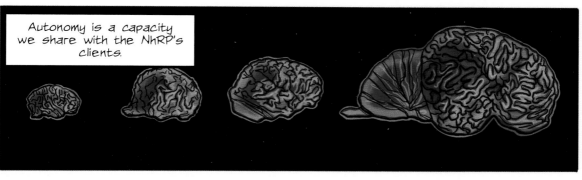

Autonomy is a capacity we share with the NhRP's clients.

In its filing, the NhRP cites *Byrn v. New York City Health and Hospital Corp.*, a precursor to Roe v. Wade.

In Byrn, the court made clear that a legal person is not synonymous with being a human.

Why should an autonomous being, like Happy, be denied the right to bodily liberty just because she isn't a member of the species Homo sapiens?

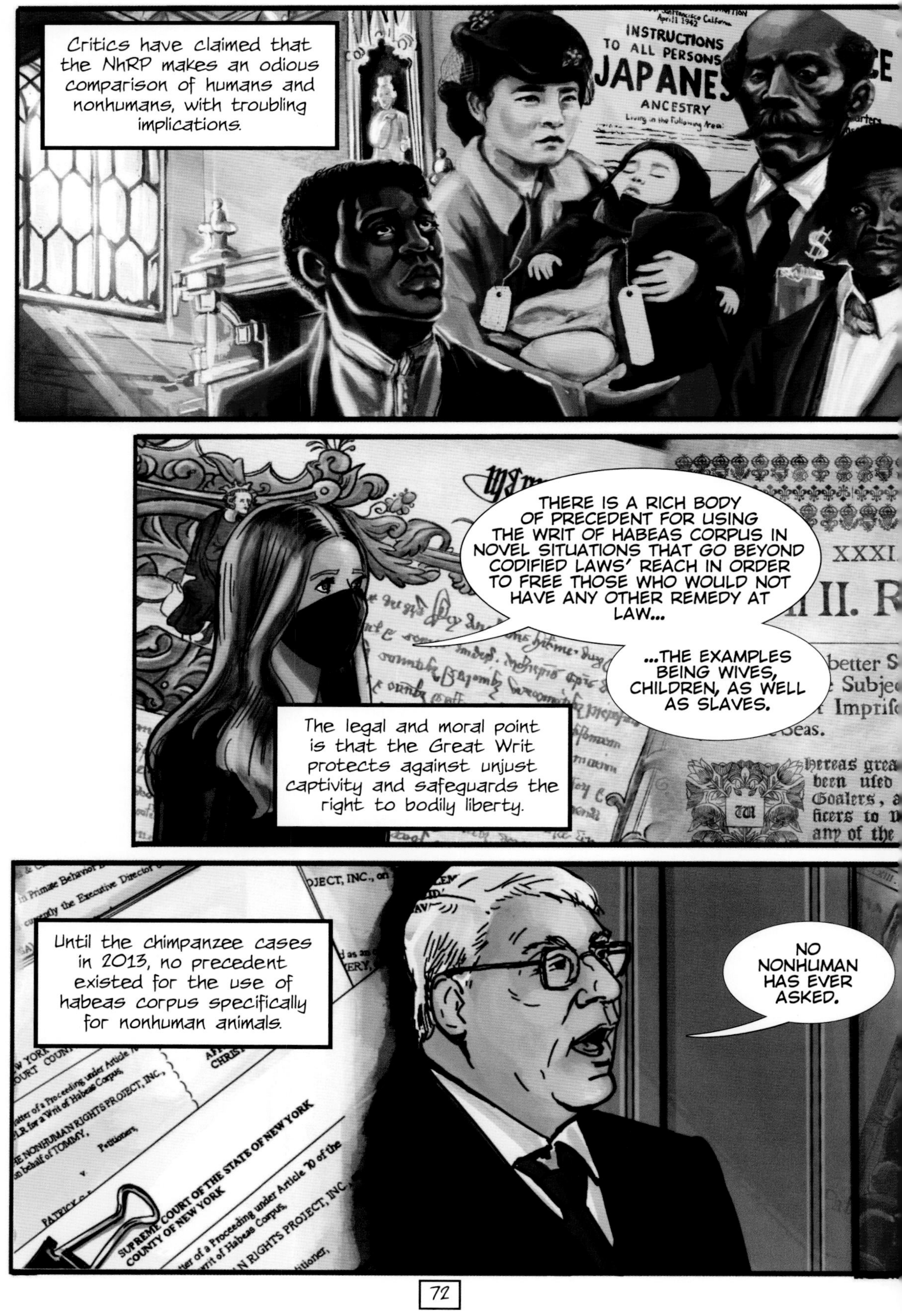

Critics have claimed that the NhRP makes an odious comparison of humans and nonhumans, with troubling implications.
INSTRUCTIONS TO ALL PERSONS
ANCESTRY
THERE IS A RICH BODY OF PRECEDENT FOR USING THE WRIT OF HABEAS CORPUS IN NOVEL SITUATIONS THAT GO BEYOND CODIFIED LAWS' REACH IN ORDER TO FREE THOSE WHO WOULD NOT HAVE ANY OTHER REMEDY AT LAW...
...THE EXAMPLES BEING WIVES, CHILDREN, AS WELL AS SLAVES.
The legal and moral point is that the Great Writ protects against unjust captivity and safeguards the right to bodily liberty.
Until the chimpanzee cases in 2013, no precedent existed for the use of habeas corpus specifically for nonhuman animals.
NO NONHUMAN HAS EVER ASKED.
SUPREME COURT OF THE STATE OF NEW YORK COUNTY OF NEW YORK

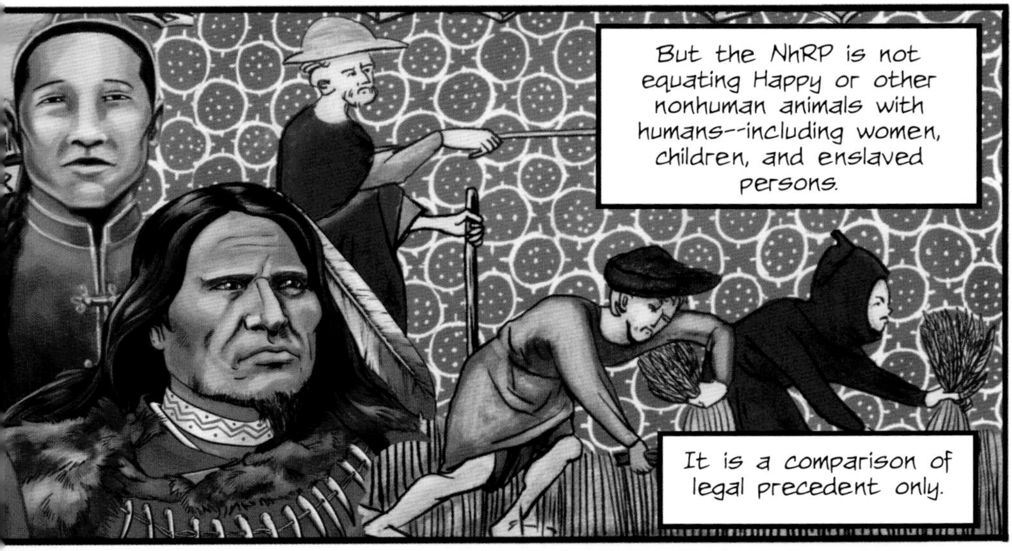
But the NhRP is not equating Happy or other nonhuman animals with humans--including women, children, and enslaved persons.
It is a comparison of legal precedent only.

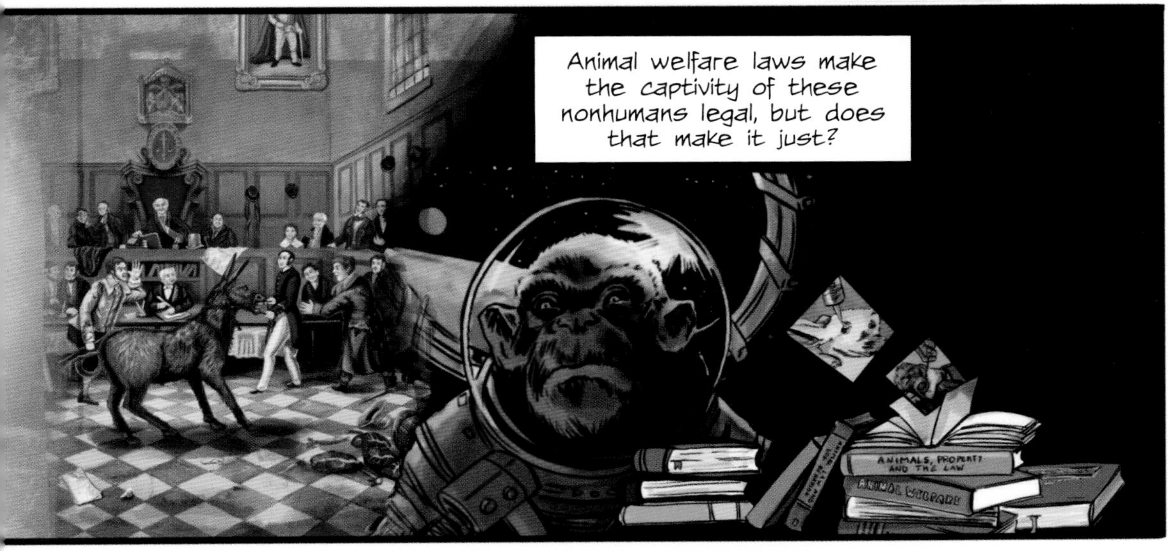
Animal welfare laws make the captivity of these nonhumans legal, but does that make it just?
ANIMALS, PROPERTY AND THE LAW
ANIMAL WELFARE

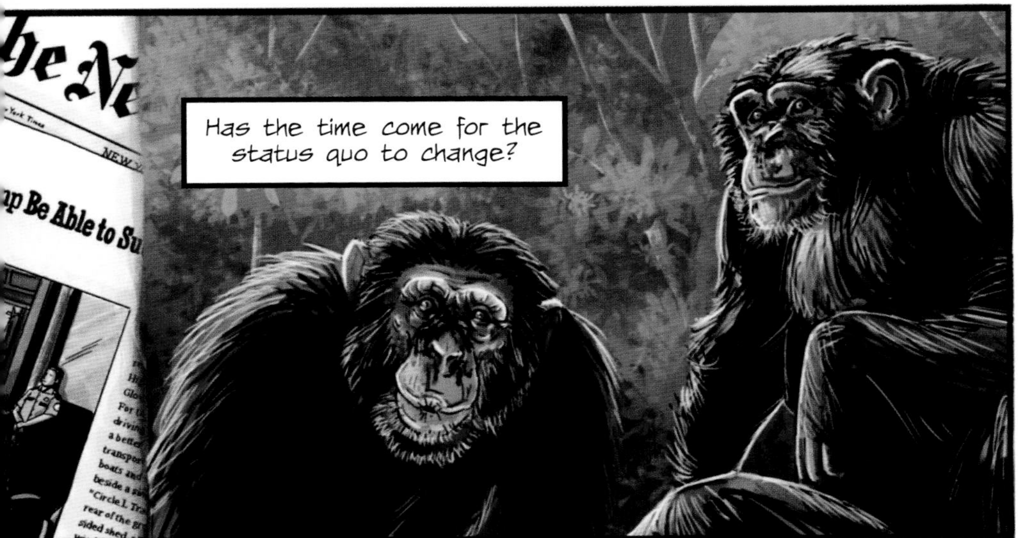
Has the time come for the status quo to change?

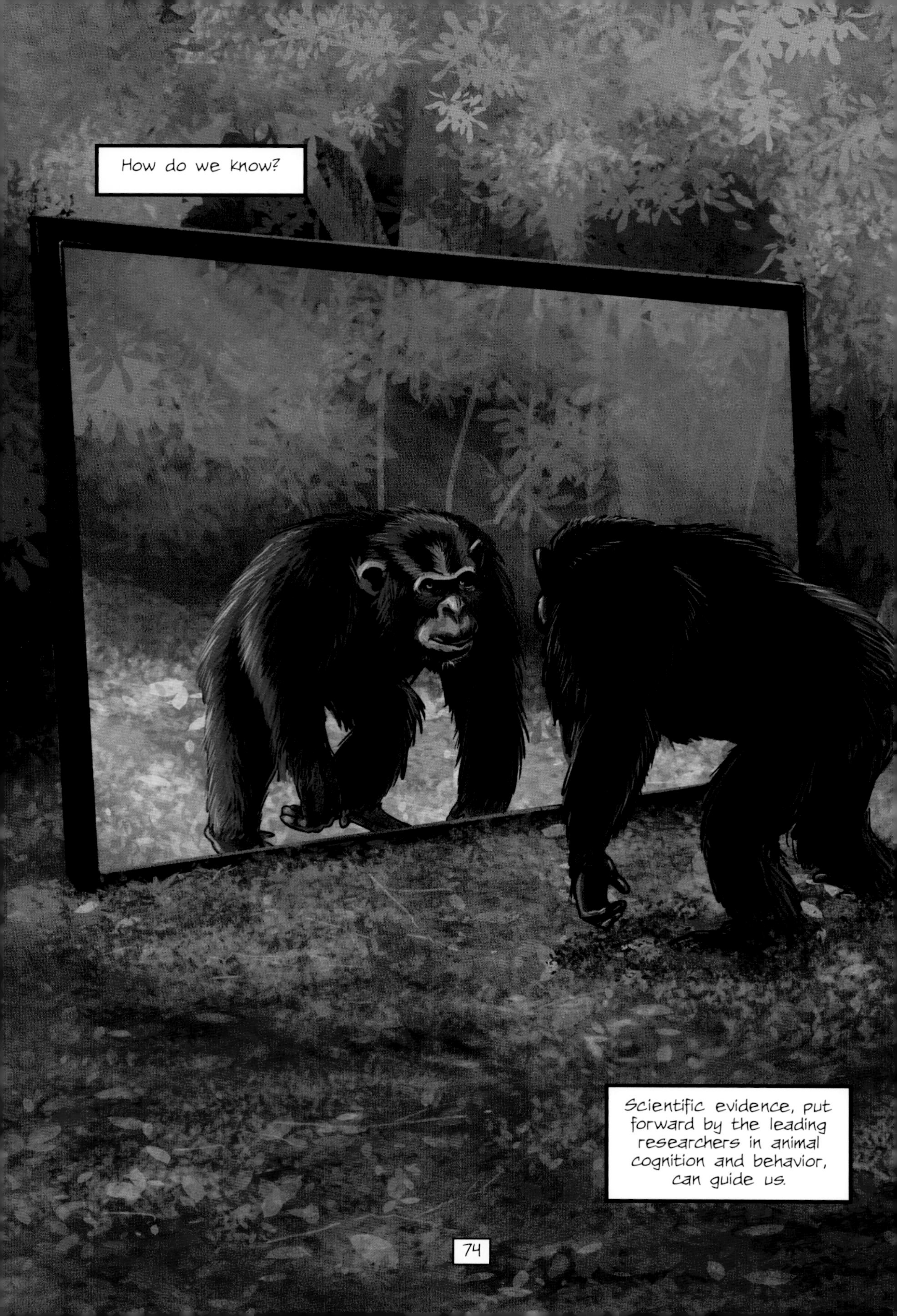
How do we know?
Scientific evidence, put forward by the leading researchers in animal cognition and behavior, can guide us.

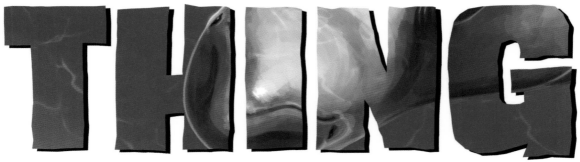

THING

CHAPTER 4: SELF-AWARENESS

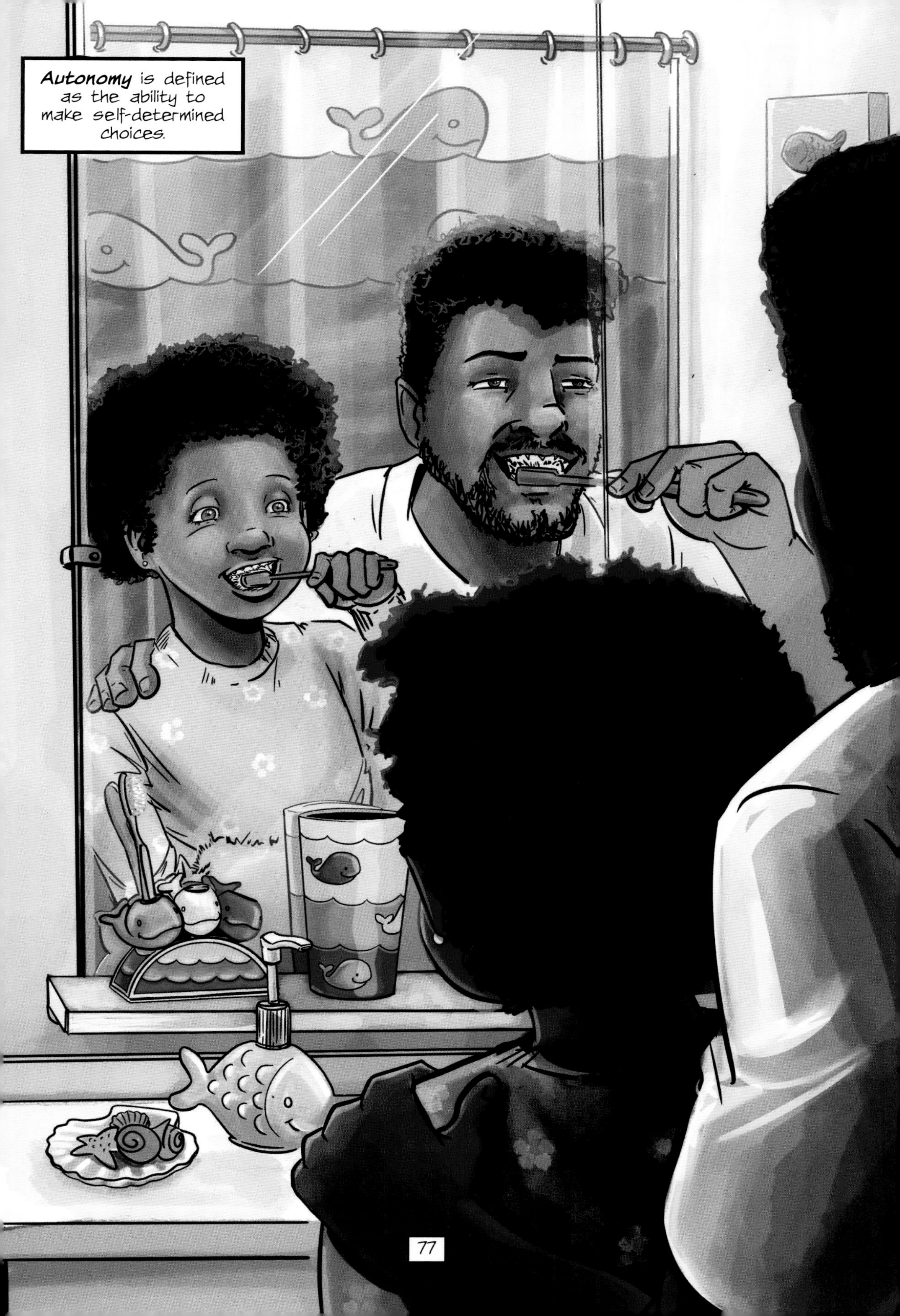
Autonomy is defined as the ability to make self-determined choices.

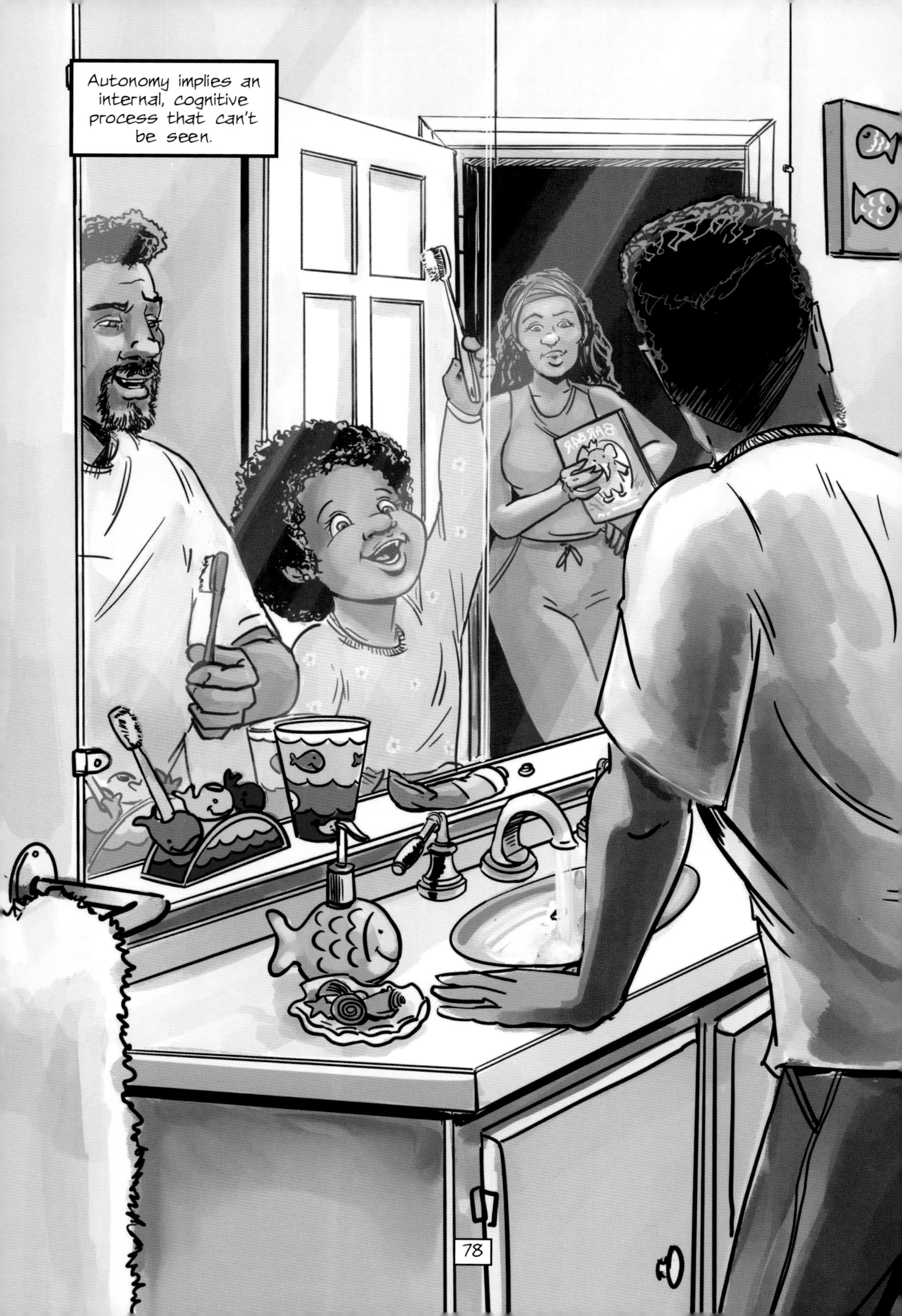
Autonomy implies an internal, cognitive process that can't be seen.

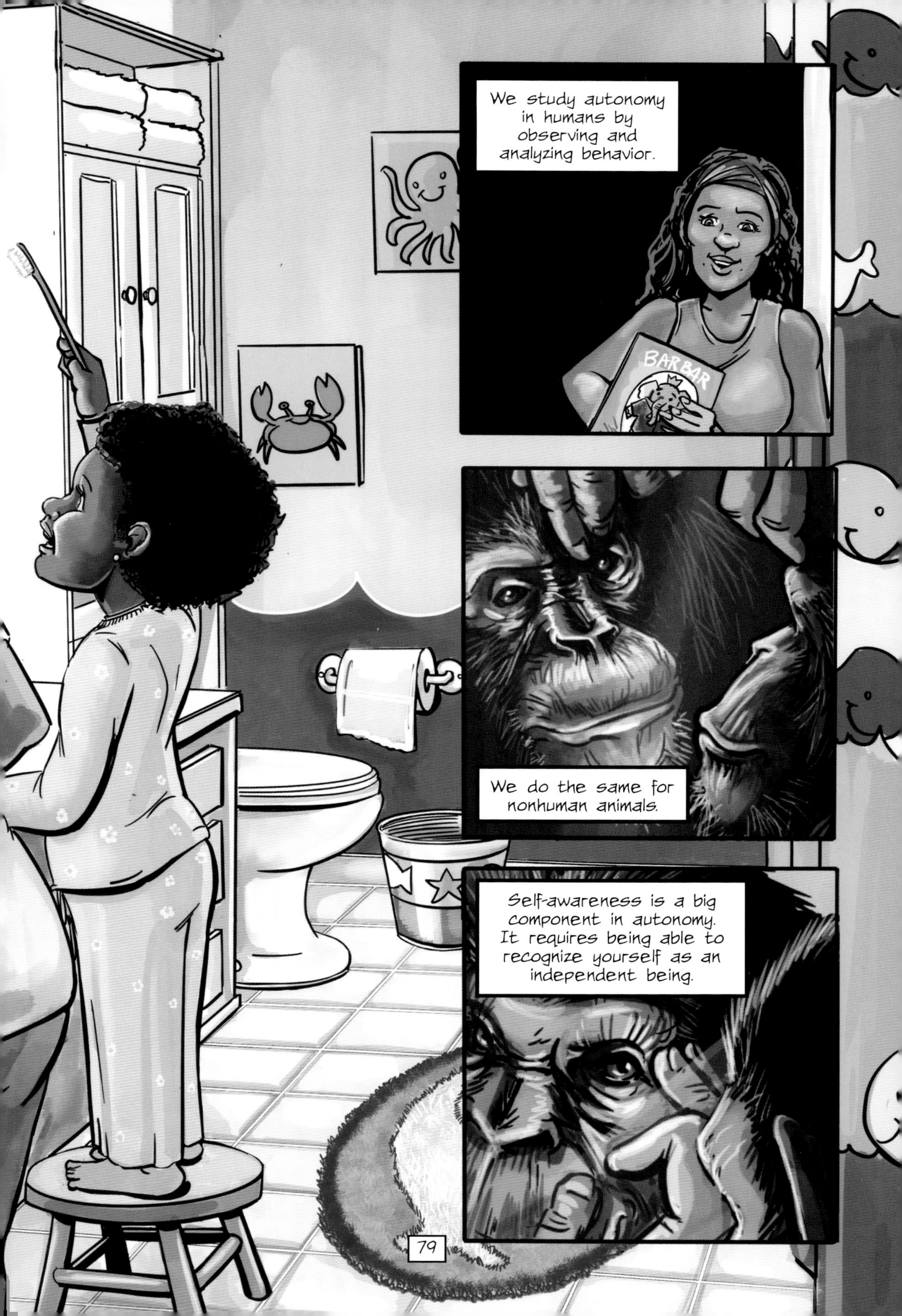
We study autonomy in humans by observing and analyzing behavior.
BARBAR
We do the same for nonhuman animals.
Self-awareness is a big component in autonomy. It requires being able to recognize yourself as an independent being.

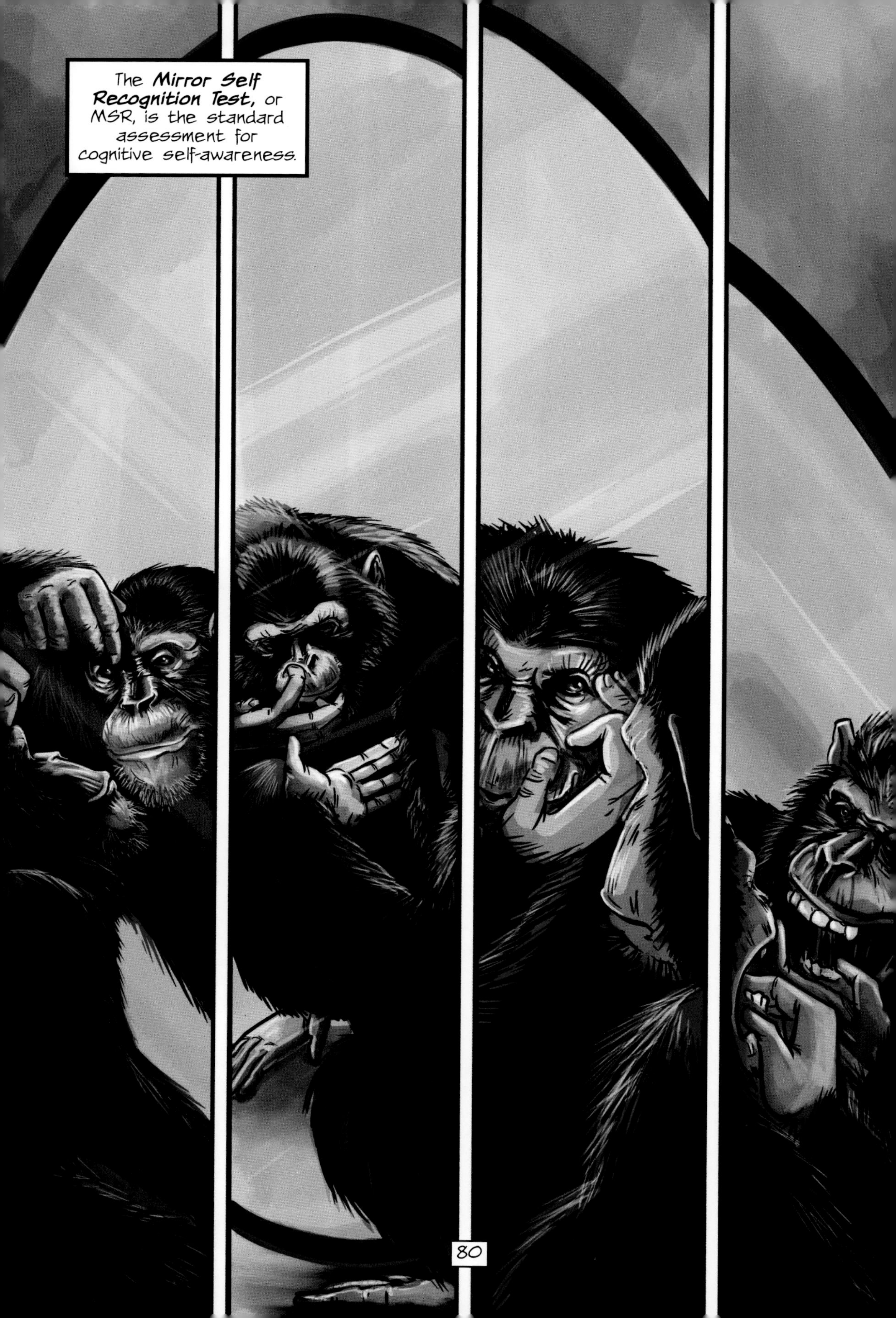
The **Mirror Self Recognition Test,** or MSR, is the standard assessment for cognitive self-awareness.

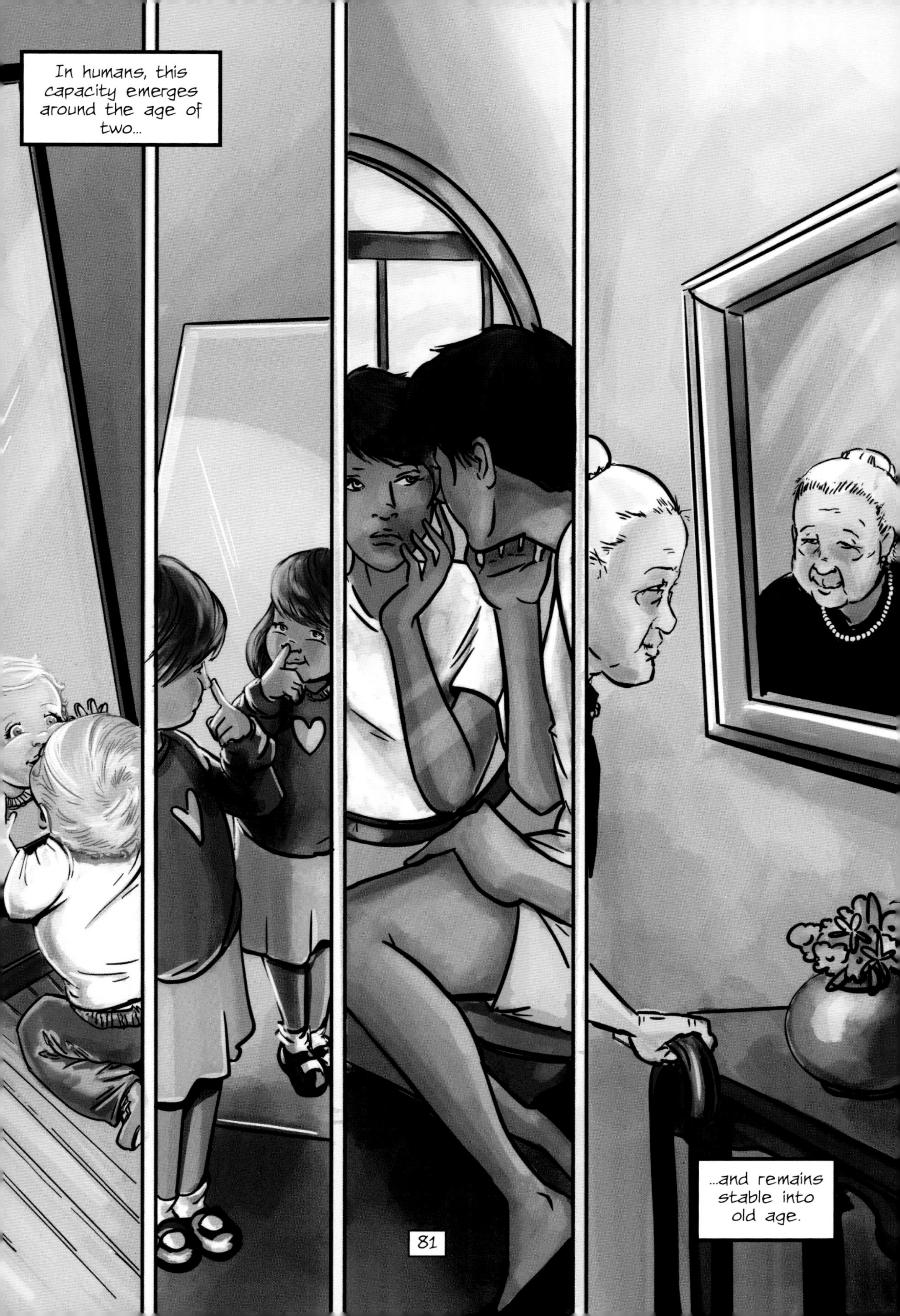
In humans, this capacity emerges around the age of two...
...and remains stable into old age.

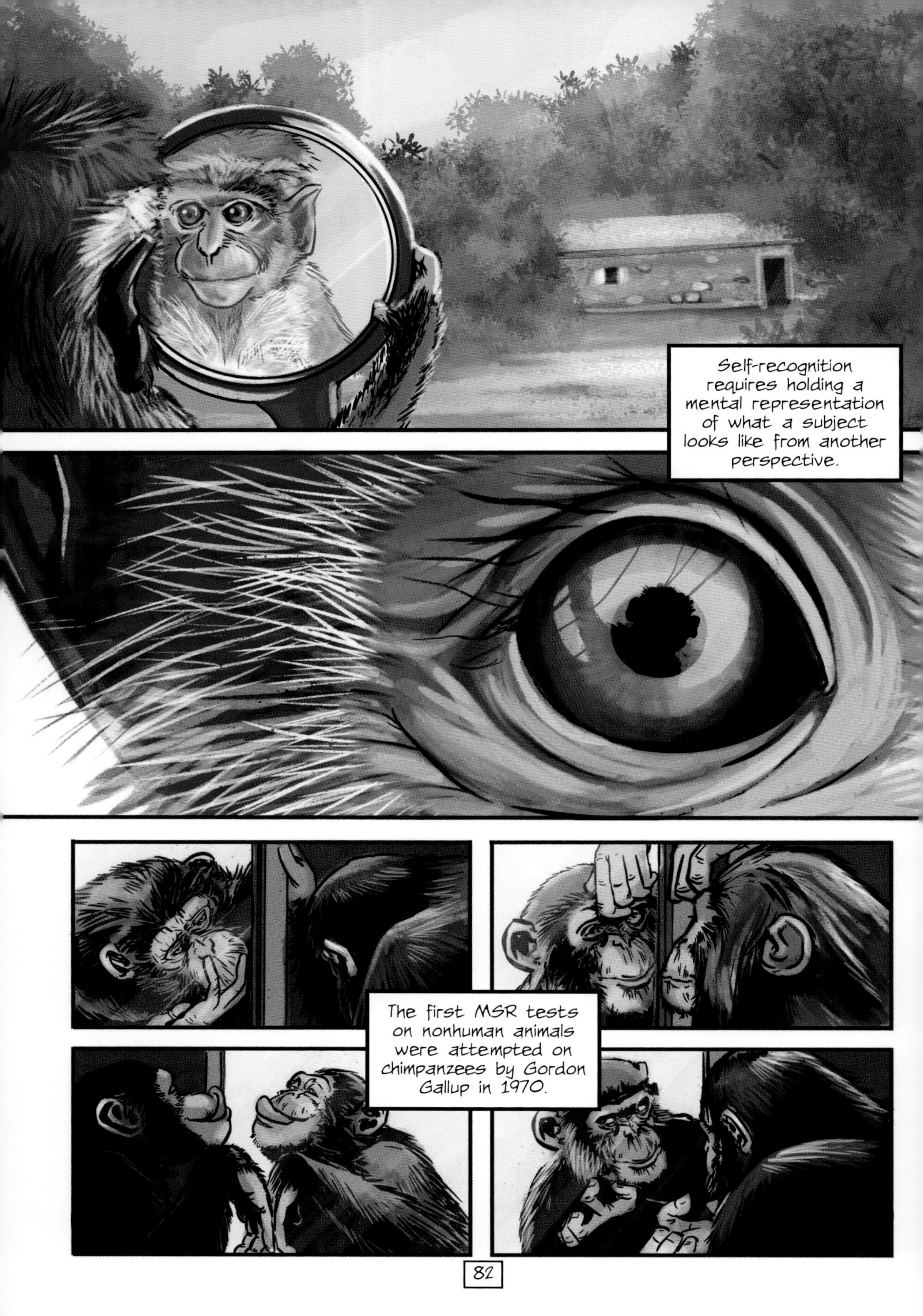
Self-recognition requires holding a mental representation of what a subject looks like from another perspective.
The first MSR tests on nonhuman animals were attempted on chimpanzees by Gordon Gallup in 1970.

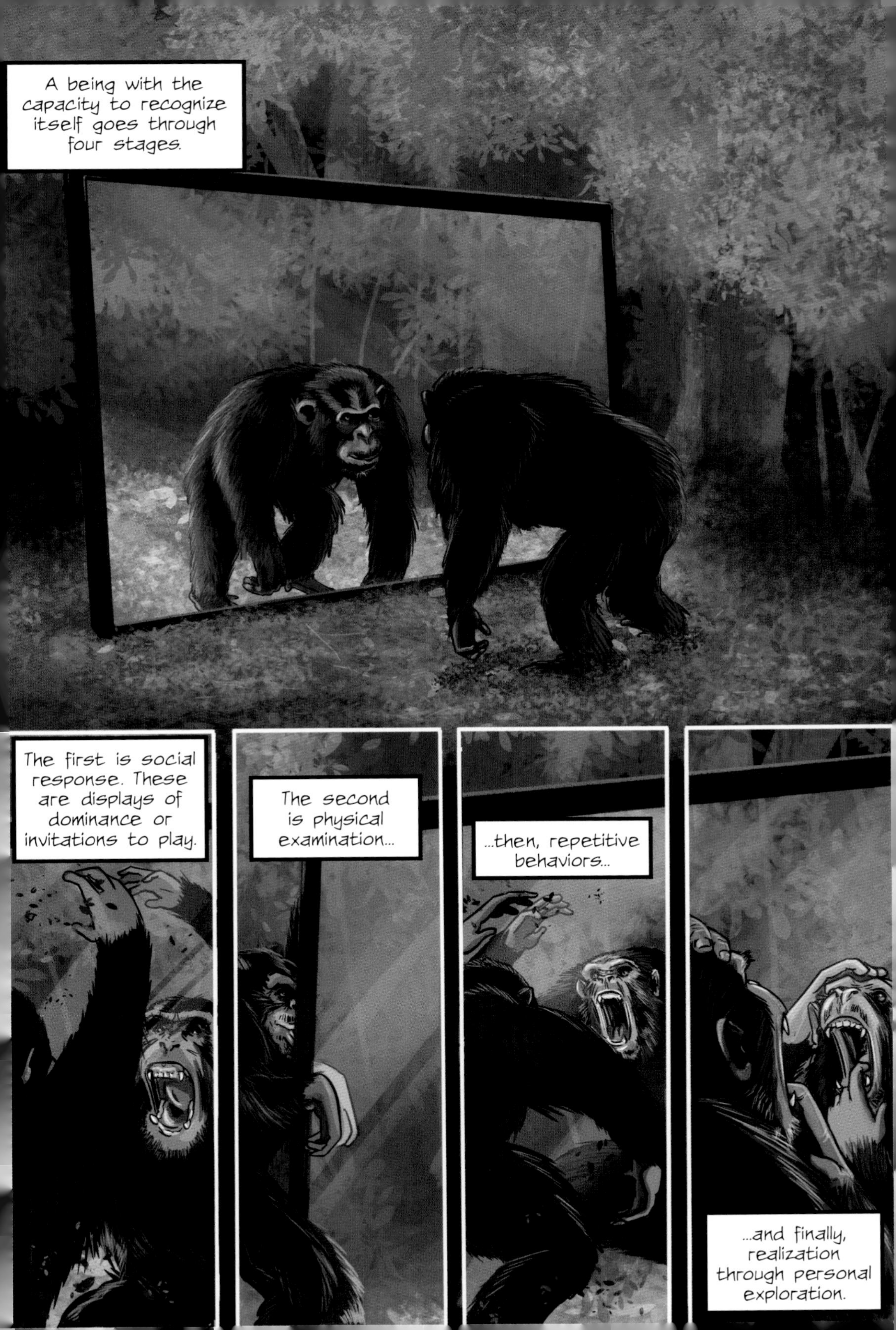
A being with the capacity to recognize itself goes through four stages.
The first is social response. These are displays of dominance or invitations to play.
The second is physical examination...
...then, repetitive behaviors...
...and finally, realization through personal exploration.

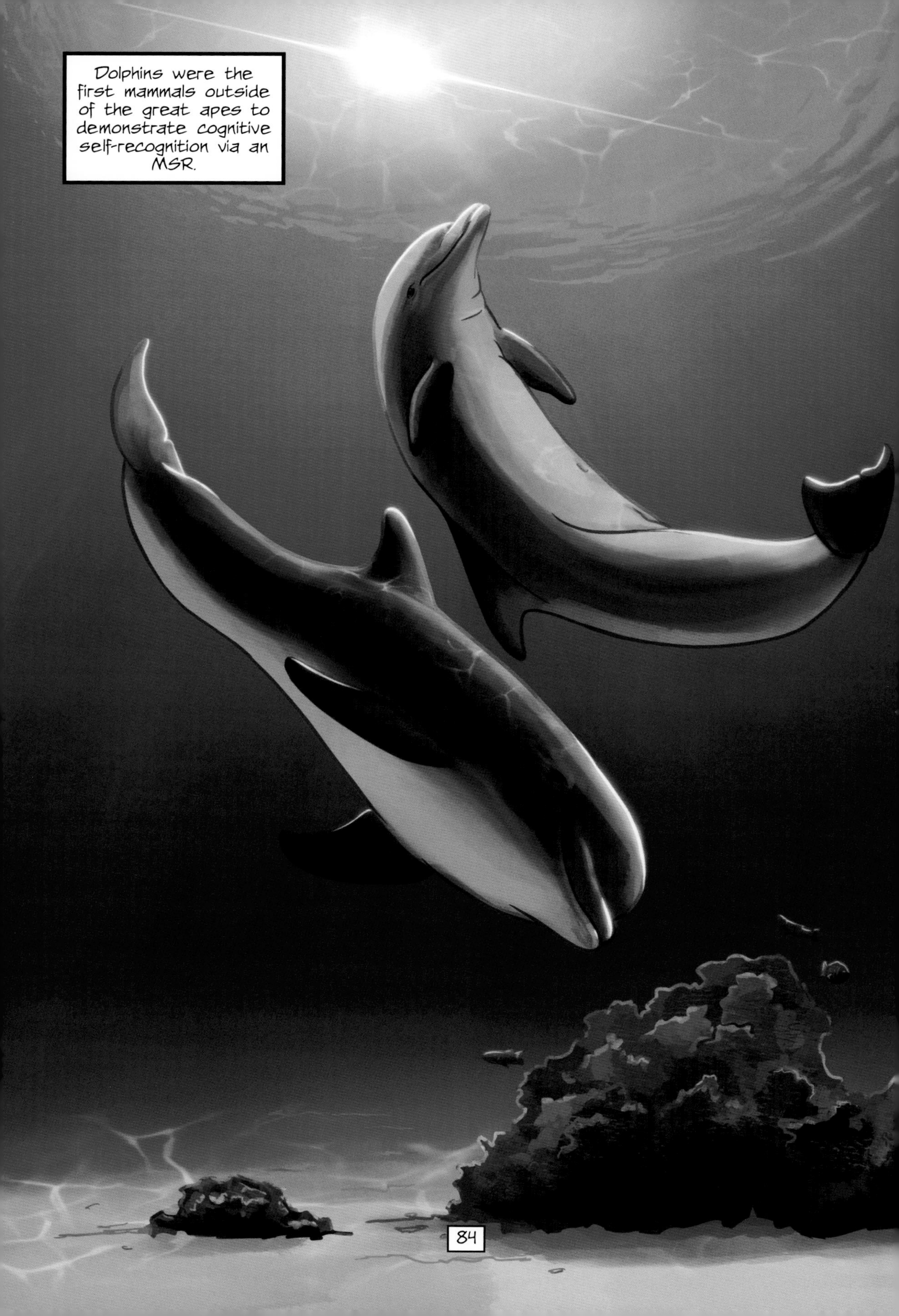
Dolphins were the first mammals outside of the great apes to demonstrate cognitive self-recognition via an MSR.

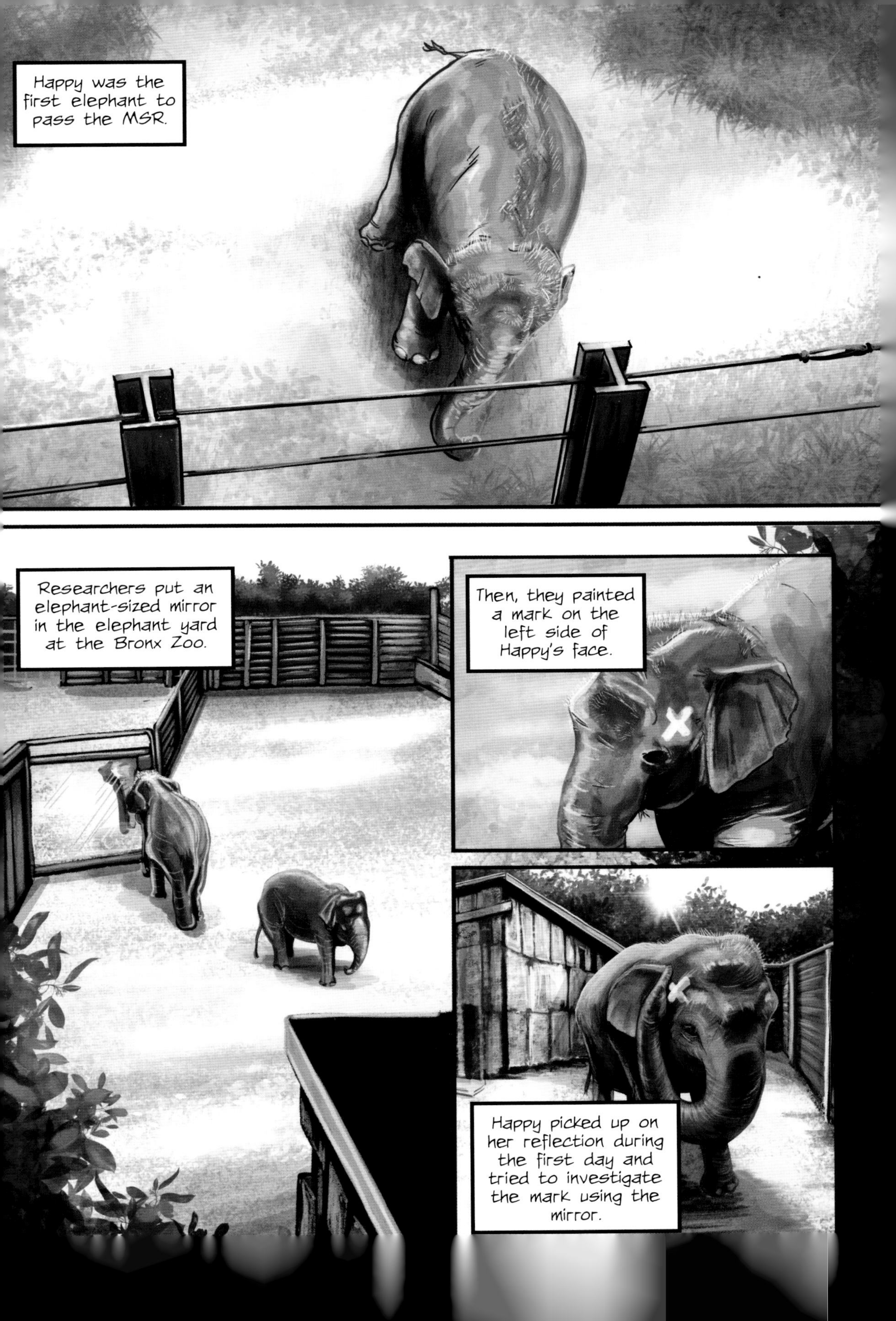
Happy was the first elephant to pass the MSR.
Researchers put an elephant-sized mirror in the elephant yard at the Bronx Zoo.
Then, they painted a mark on the left side of Happy's face.
Happy picked up on her reflection during the first day and tried to investigate the mark using the mirror.

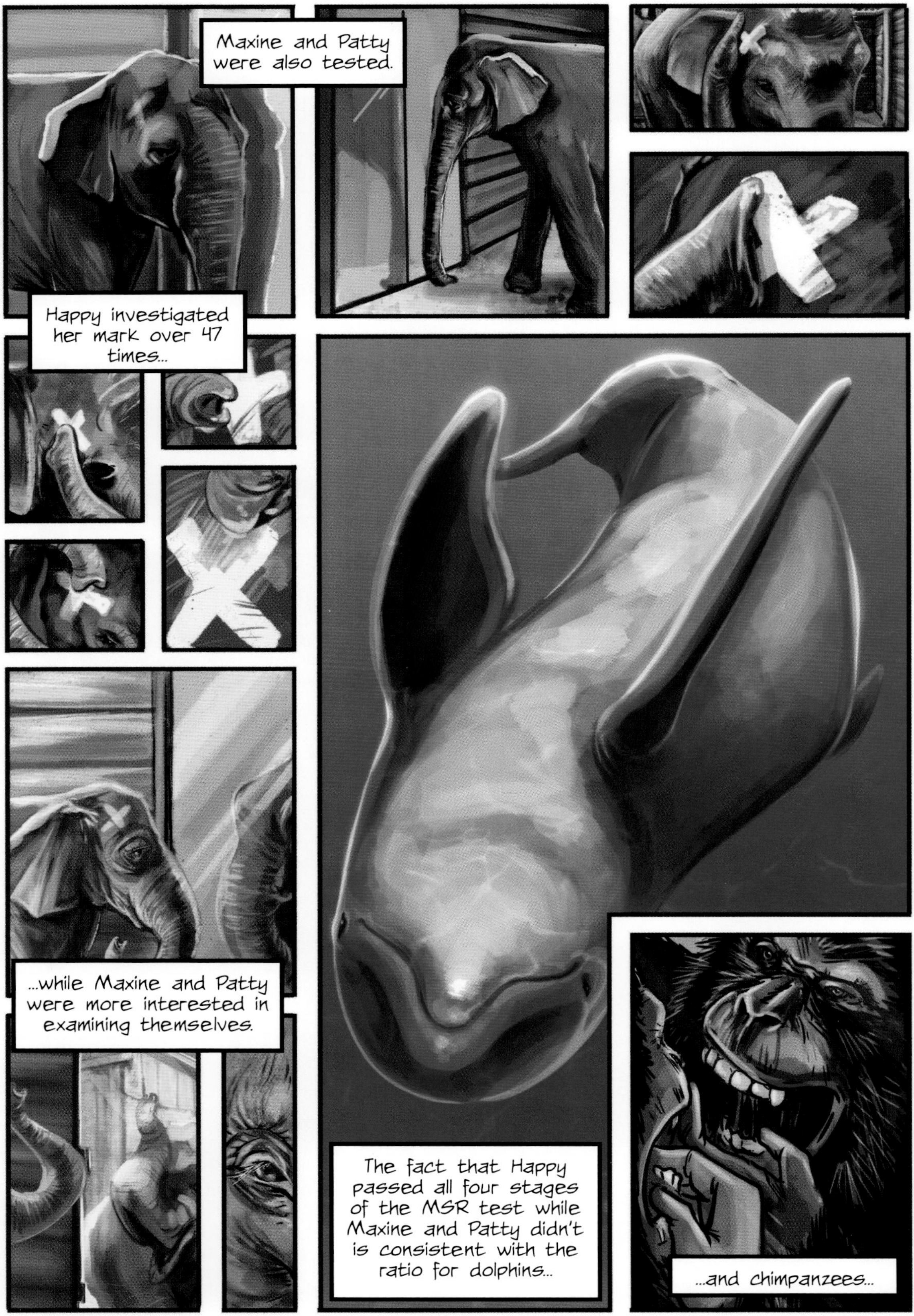
Maxine and Patty were also tested.
Happy investigated her mark over 47 times...
...while Maxine and Patty were more interested in examining themselves.
The fact that Happy passed all four stages of the MSR test while Maxine and Patty didn't is consistent with the ratio for dolphins...
...and chimpanzees...

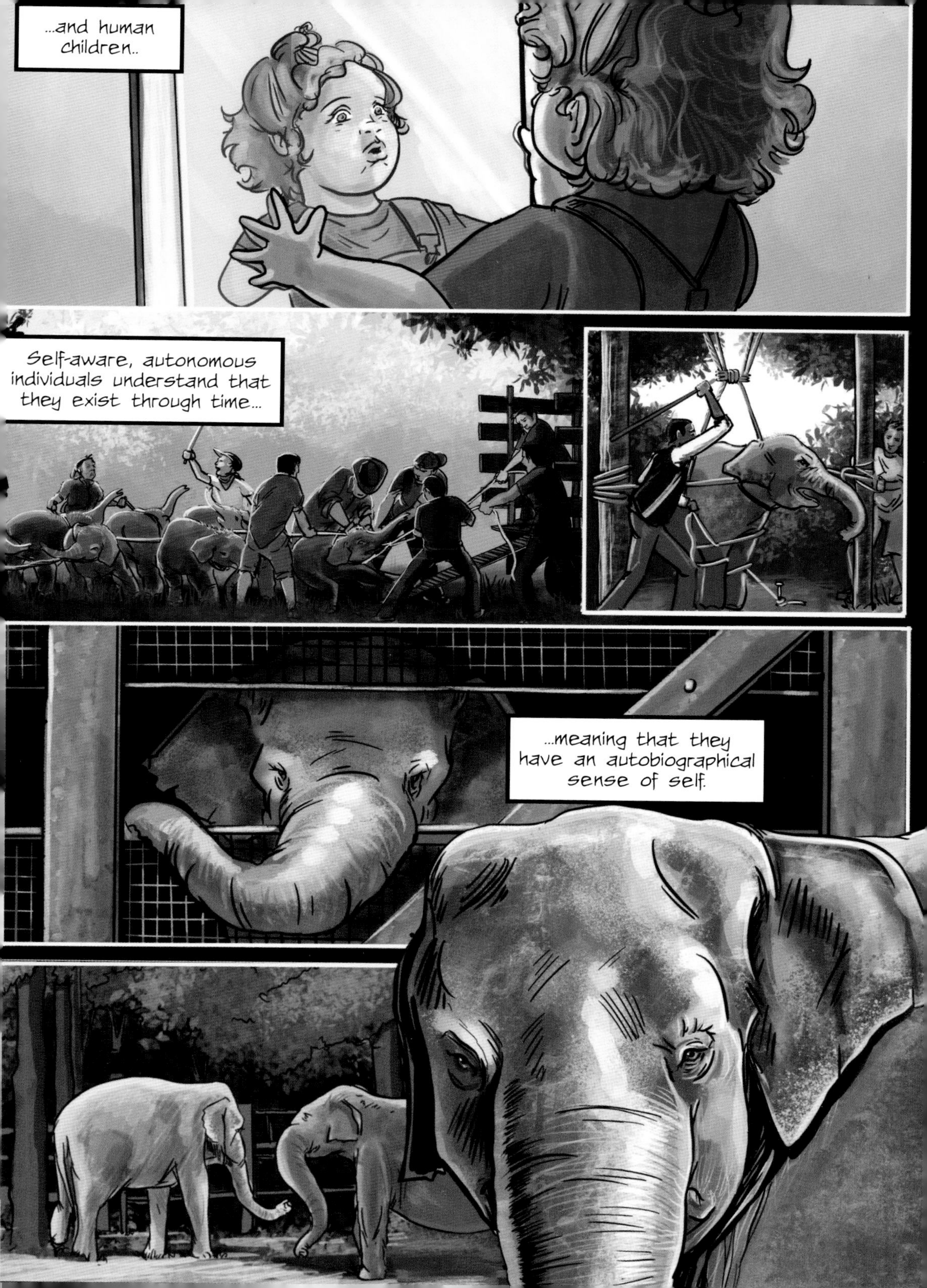
...and human children..
Self-aware, autonomous individuals understand that they exist through time...
...meaning that they have an autobiographical sense of self.

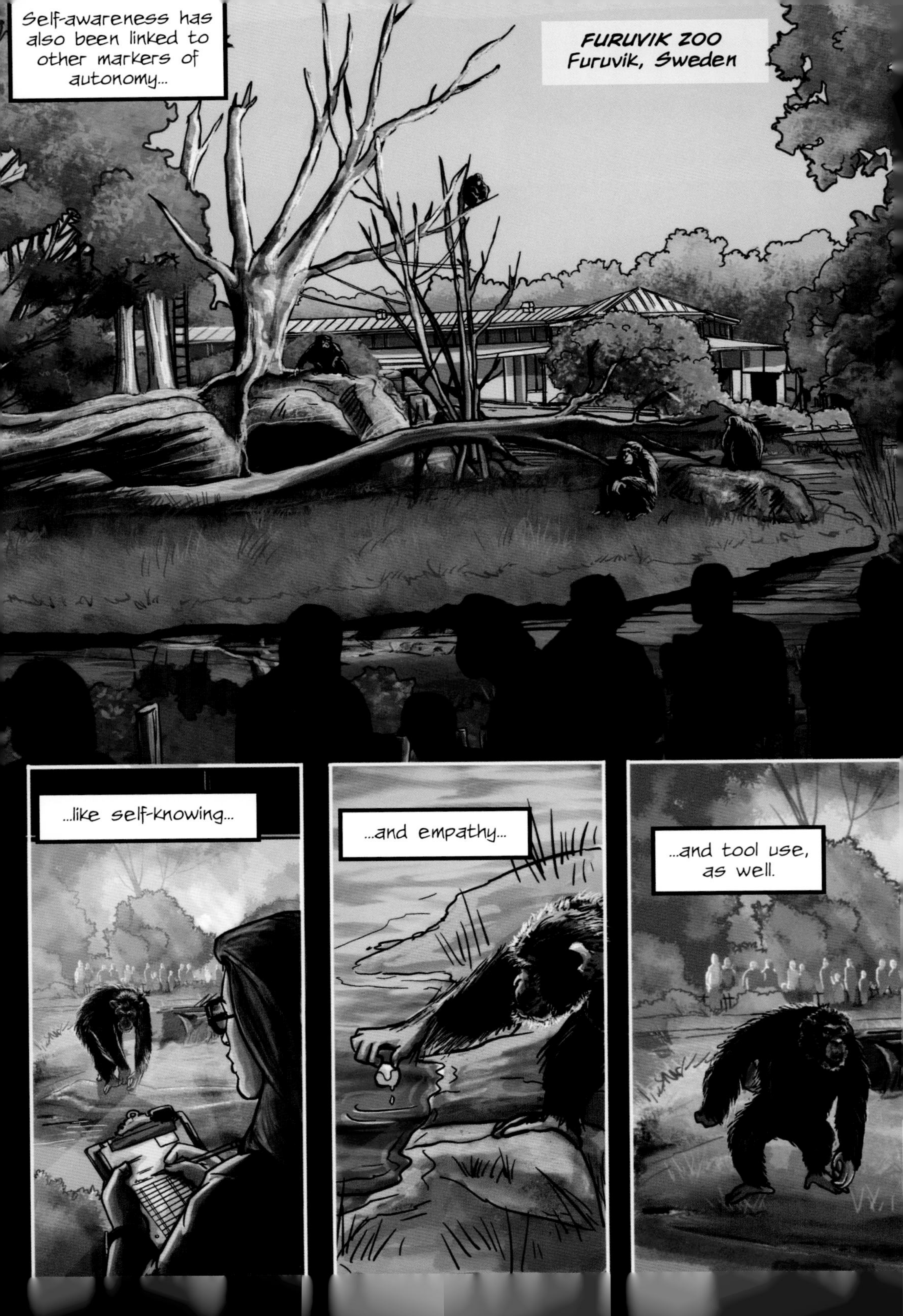
Self-awareness has also been linked to other markers of autonomy...
FURUVIK ZOO
Furuvik, Sweden
...like self-knowing...
...and empathy...
...and tool use, as well.

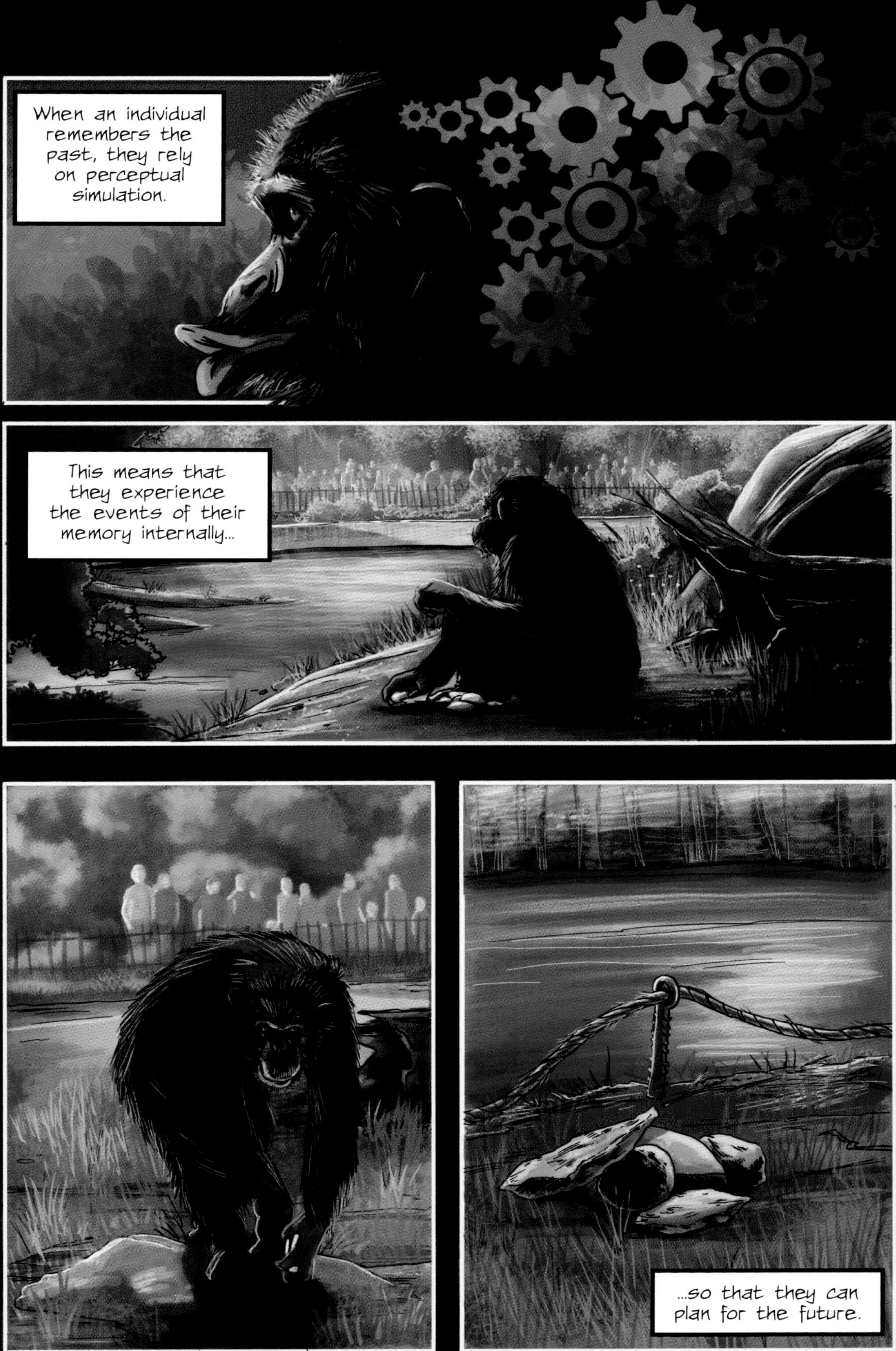
When an individual remembers the past, they rely on perceptual simulation.
This means that they experience the events of their memory internally...
...so that they can plan for the future.

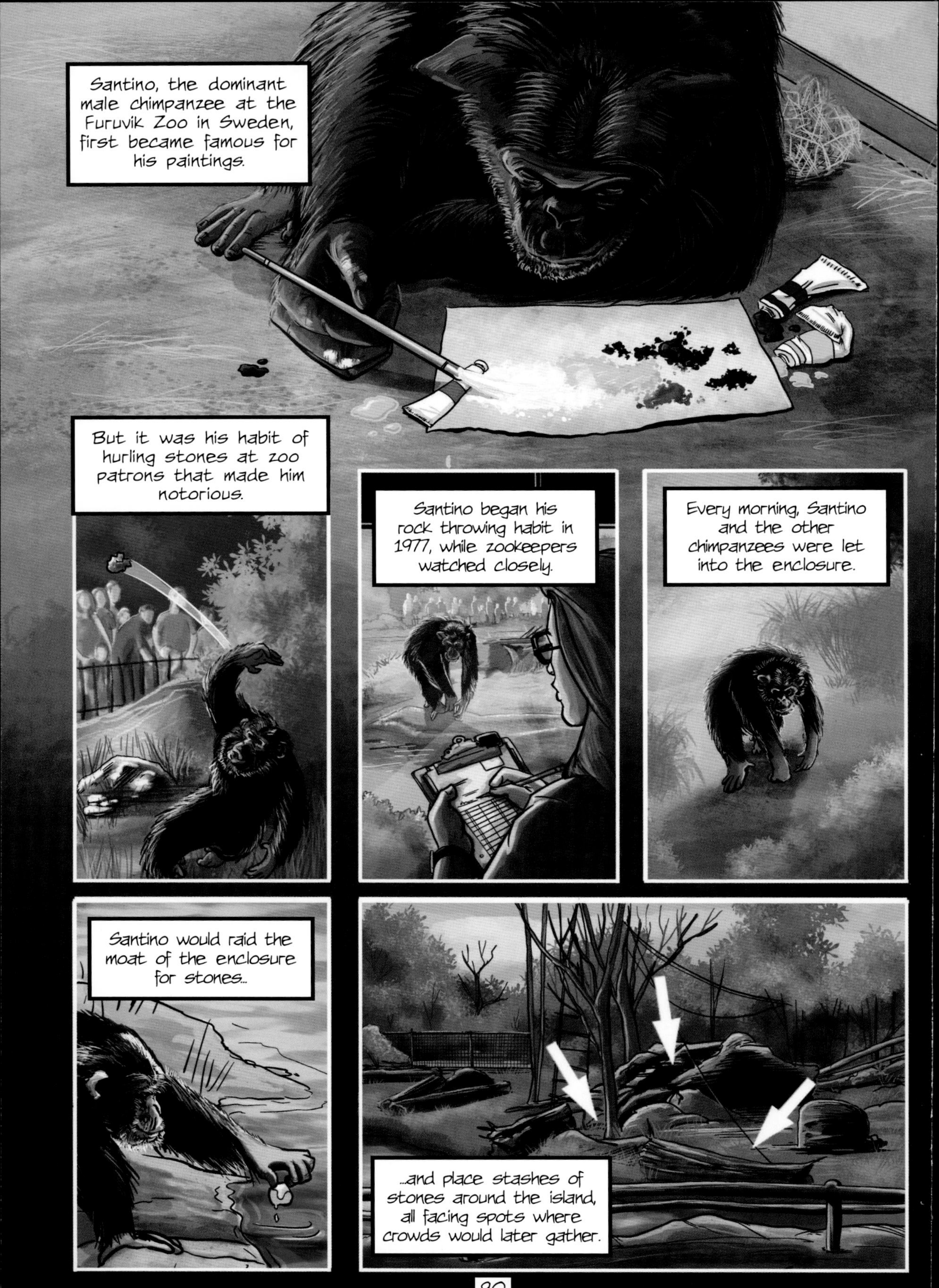
Santino, the dominant male chimpanzee at the Furuvik Zoo in Sweden, first became famous for his paintings.
But it was his habit of hurling stones at zoo patrons that made him notorious.
Santino began his rock throwing habit in 1977, while zookeepers watched closely.
Every morning, Santino and the other chimpanzees were let into the enclosure.
Santino would raid the moat of the enclosure for stones...
...and place stashes of stones around the island, all facing spots where crowds would later gather.

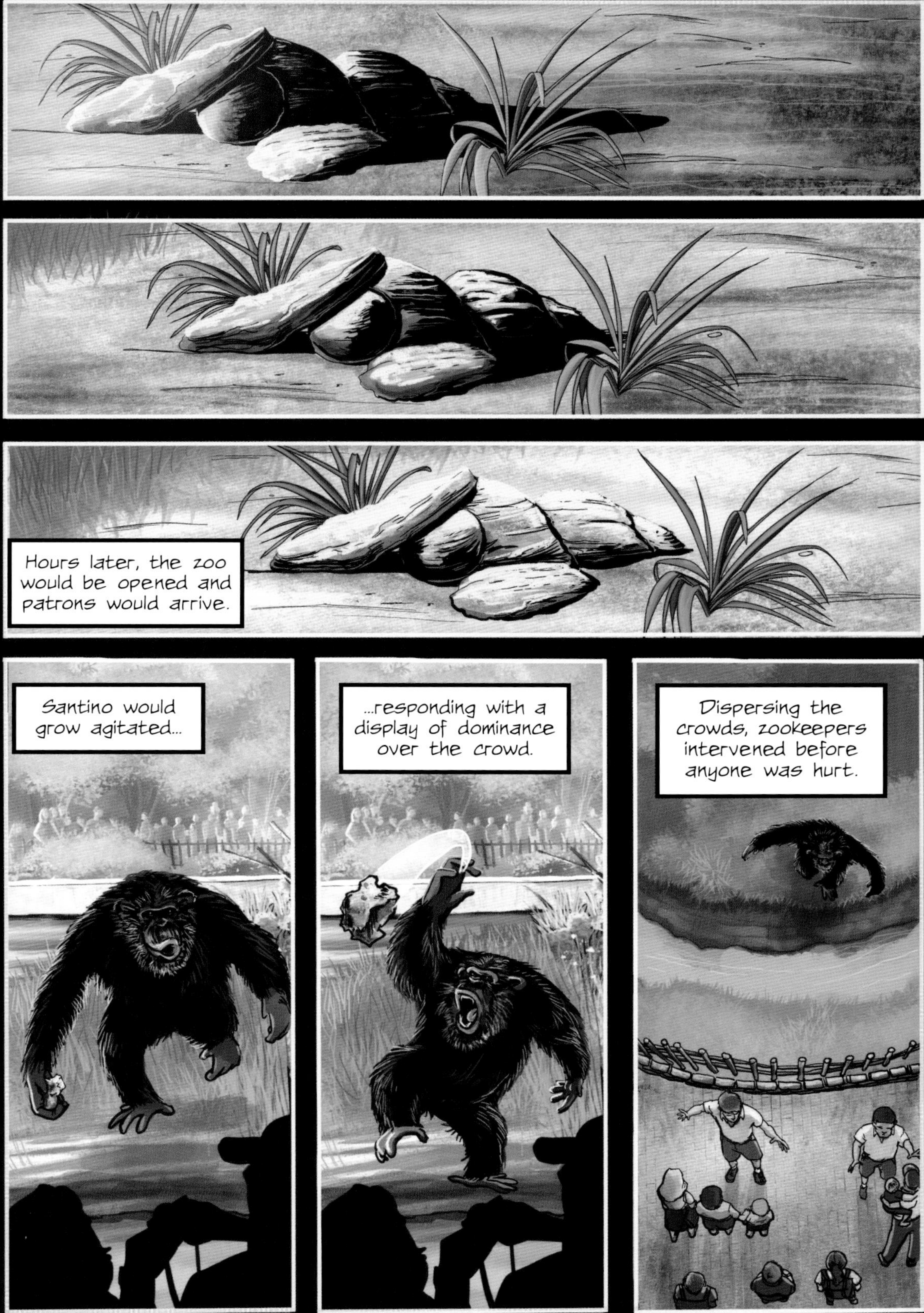
Hours later, the zoo would be opened and patrons would arrive.
Santino would grow agitated...
...responding with a display of dominance over the crowd.
Dispersing the crowds, zookeepers intervened before anyone was hurt.

In the wild, Santino's behavior would illicit compliance...

...and interlopers would submit and leave.

But zoo patrons were enthralled...

...which resulted in more aggression from Santino.

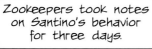

Zookeepers took notes on Santino's behavior for three days.

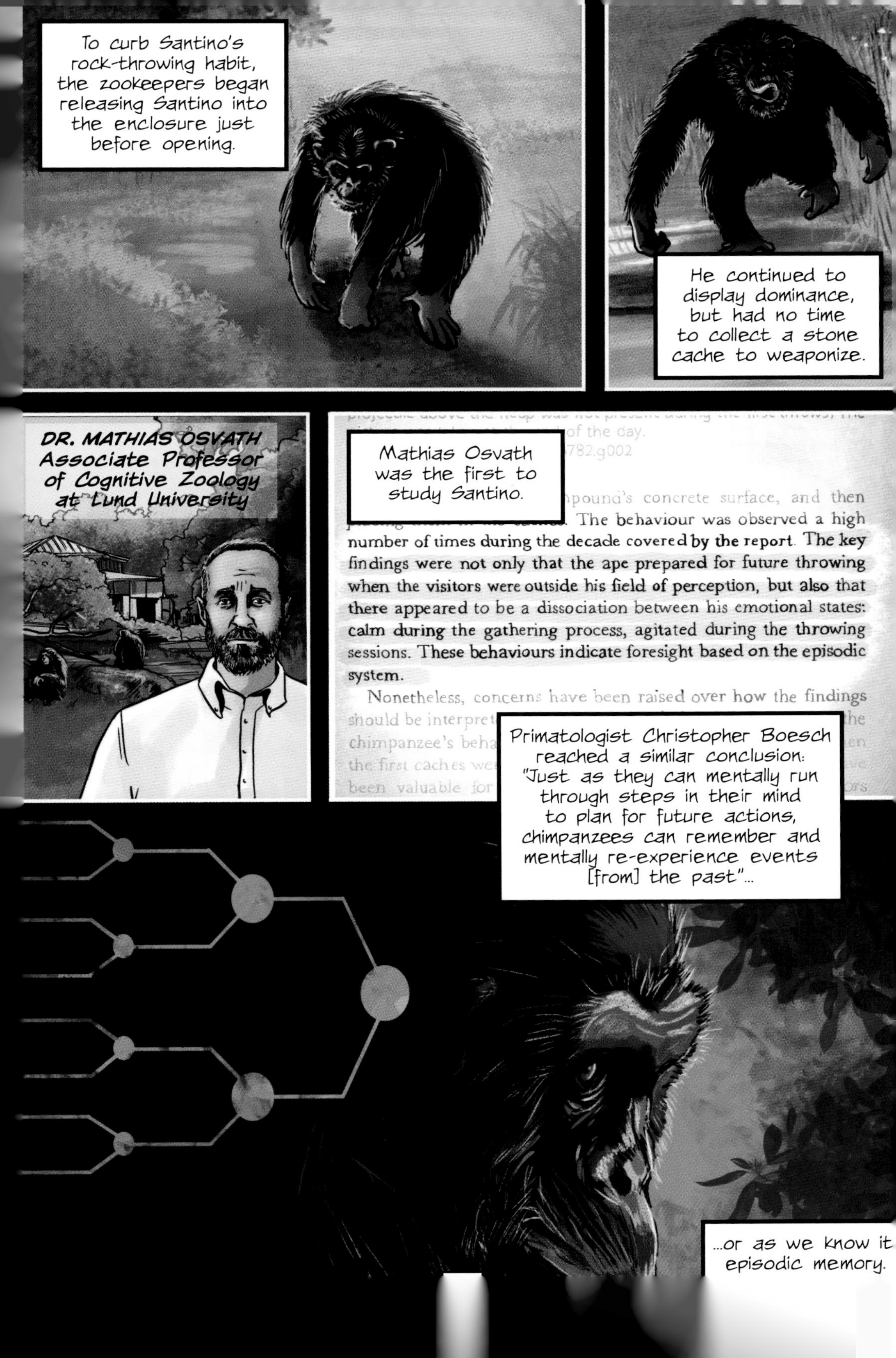

To curb Santino's rock-throwing habit, the zookeepers began releasing Santino into the enclosure just before opening.

He continued to display dominance, but had no time to collect a stone cache to weaponize.

DR. MATHIAS OSVATH
Associate Professor of Cognitive Zoology at Lund University

Mathias Osvath was the first to study Santino.

projectile above the heap was not present during the first throws; the ... of the day.
...782.g002

...pound's concrete surface, and then ... The behaviour was observed a high number of times during the decade covered by the report. The key findings were not only that the ape prepared for future throwing when the visitors were outside his field of perception, but also that there appeared to be a dissociation between his emotional states: calm during the gathering process, agitated during the throwing sessions. These behaviours indicate foresight based on the episodic system.

Nonetheless, concerns have been raised over how the findings should be interpret... chimpanzee's beha... the first caches we... been valuable for...

Primatologist Christopher Boesch reached a similar conclusion: "Just as they can mentally run through steps in their mind to plan for future actions, chimpanzees can remember and mentally re-experience events [from] the past"...

...or as we know it episodic memory.

himp makes elaborate plots to attack
ans

ers report that Santino, a male chimpanzee at Furuvik Zoo in Sweden,
ngly complex attacks against zoo visitors.

The Big Question: What does
forward-planning reveal about
chimps' relationship to humans

Steve Connor | Wednesday 11 March 2009 01:00 comments

Santino's story has a
second chapter, with
profound implications for our
understanding of episodic
memory and cognition.

SFGATE

News

Chimp collects rocks to throw at people later

Rising Associated Press

News Podcasts Video Technology Space Physics Health More

NewScientist

Missile-throwing chimp plots
attacks on tourists

LL NEWS > STONE-

ANIMALS

e-Throwing Chimp Is Back -- A
onal

r is further evidence of advanced planning in nonhum

RED   BACKCHANNEL BUSINESS CULTURE GEAR IDEAS SCIENCE SECURITY

CLIMATE   LATEST NEWS AND SOLUTIONS   ENVIRONMENT   EMISSIONS   GEOENGI

one-Throwing Chimp Thinks Ahead

p famous for throwing stones at visitors now conceals his w
r aim at his targets, providing further putative evidence that

The
Guardian
For 200 years

| | Opinion | Sport | Culture | Li |
| News | | | |

World Europe US Americas Asia Australia Middle East Africa Inequality Global devel

Chimp who threw stones at zoo visitors
showed human trait, says scientist

Assembling ammunition in advance reveals ape's unsuspected
ability to plan for future

LIVE SCIENCE

🏠   News   Space & Physics   Health   Planet Earth

ceptive Chimp Hides Ammo, Blasts
suspecting Zoo Visitors

arles Q. Choi May 17, 2012

Because eleven years
later, Santino changed
his strategy.

He figured out that
his keepers were
disassembling his caches...

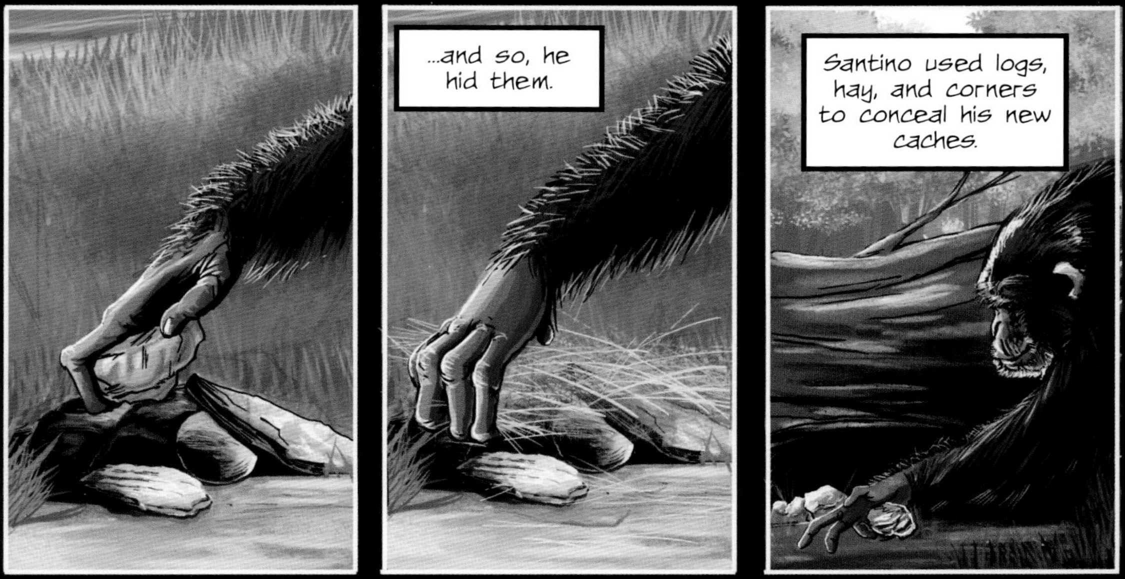
...and so, he hid them.
Santino used logs, hay, and corners to conceal his new caches.

of the planning behaviour of this chimpanzee (Osvath and Karvonen, 2012). We found very complex behaviours not documented before. The chimpanzee engaged in deception for the future by constructing hides for his stone caches and by inhibiting his aggressive displays (which are tell-tale signs of upcoming throws). The key finding was that chimpanzees are not only able to prepare for an upcoming event, but are also able to mentally construct a new situation which will alter the future (in this case the behaviours of human zoo visitors).

14. Part of being an autonomous individual is self-control. Chimpanzees, like

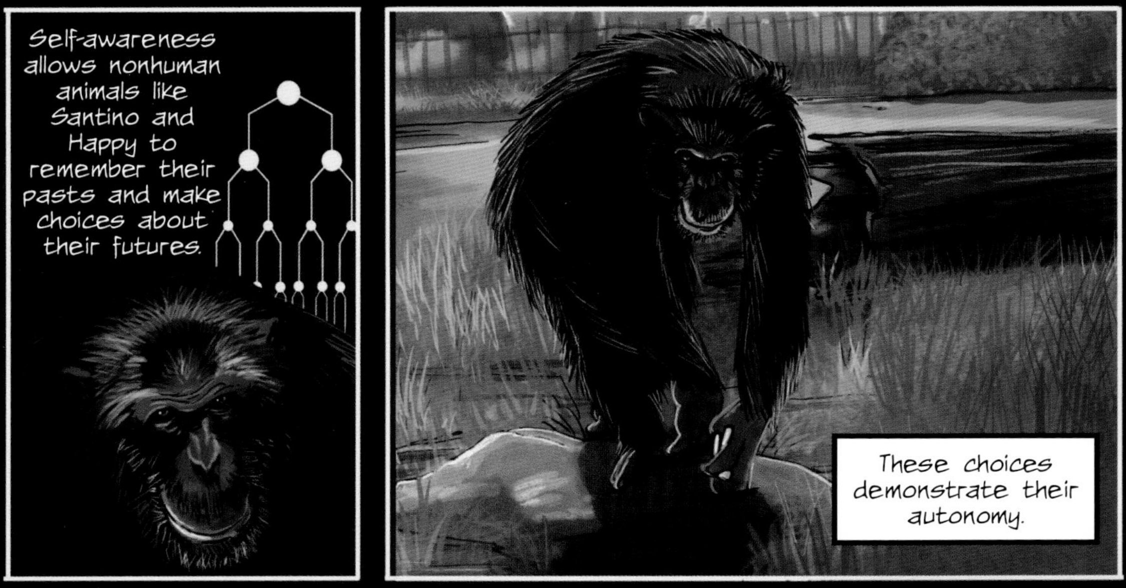
Self-awareness allows nonhuman animals like Santino and Happy to remember their pasts and make choices about their futures.
These choices demonstrate their autonomy.

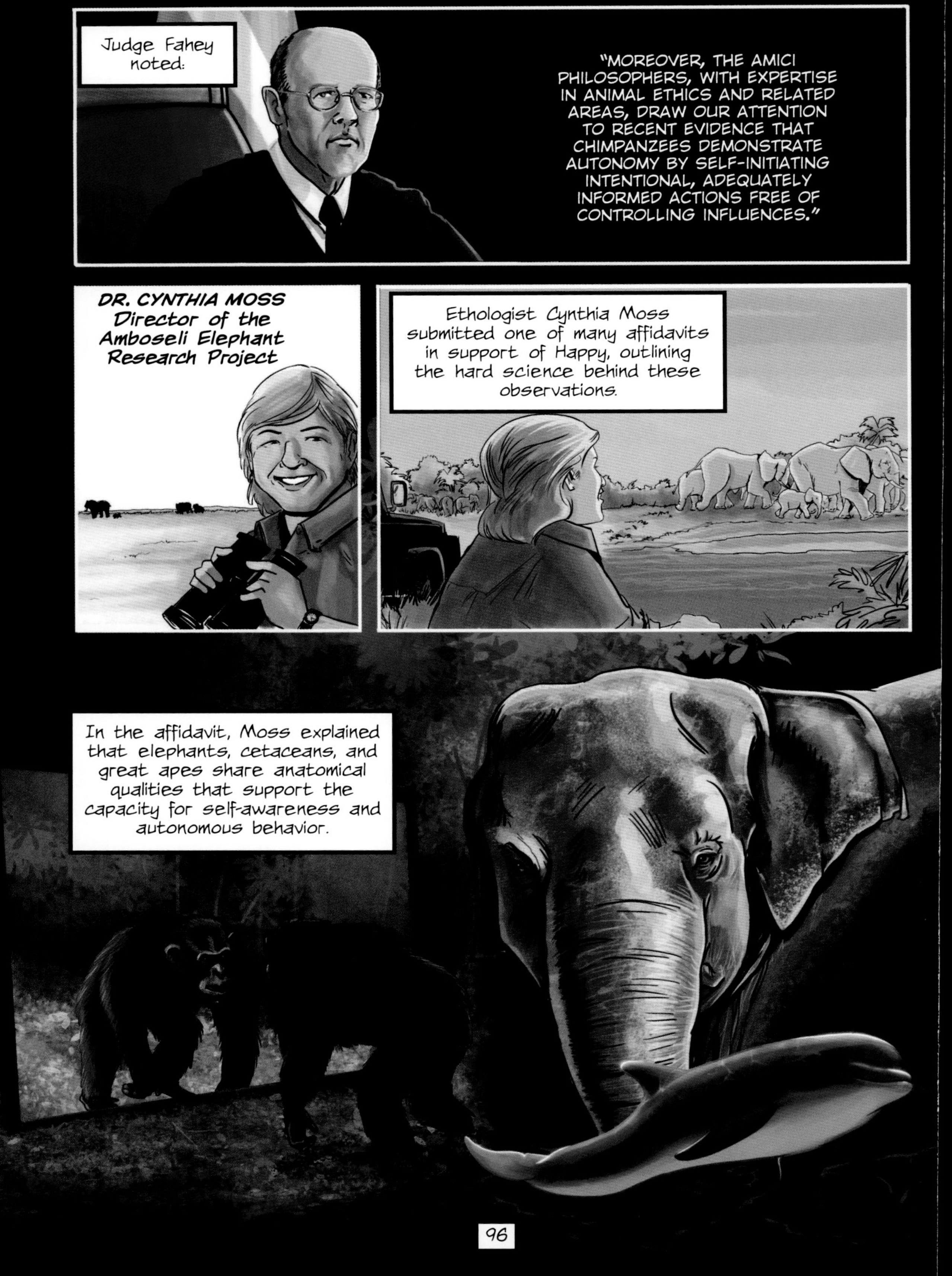
Judge Fahey noted:
"MOREOVER, THE AMICI PHILOSOPHERS, WITH EXPERTISE IN ANIMAL ETHICS AND RELATED AREAS, DRAW OUR ATTENTION TO RECENT EVIDENCE THAT CHIMPANZEES DEMONSTRATE AUTONOMY BY SELF-INITIATING INTENTIONAL, ADEQUATELY INFORMED ACTIONS FREE OF CONTROLLING INFLUENCES."
DR. CYNTHIA MOSS Director of the Amboseli Elephant Research Project
Ethologist Cynthia Moss submitted one of many affidavits in support of Happy, outlining the hard science behind these observations.
In the affidavit, Moss explained that elephants, cetaceans, and great apes share anatomical qualities that support the capacity for self-awareness and autonomous behavior.

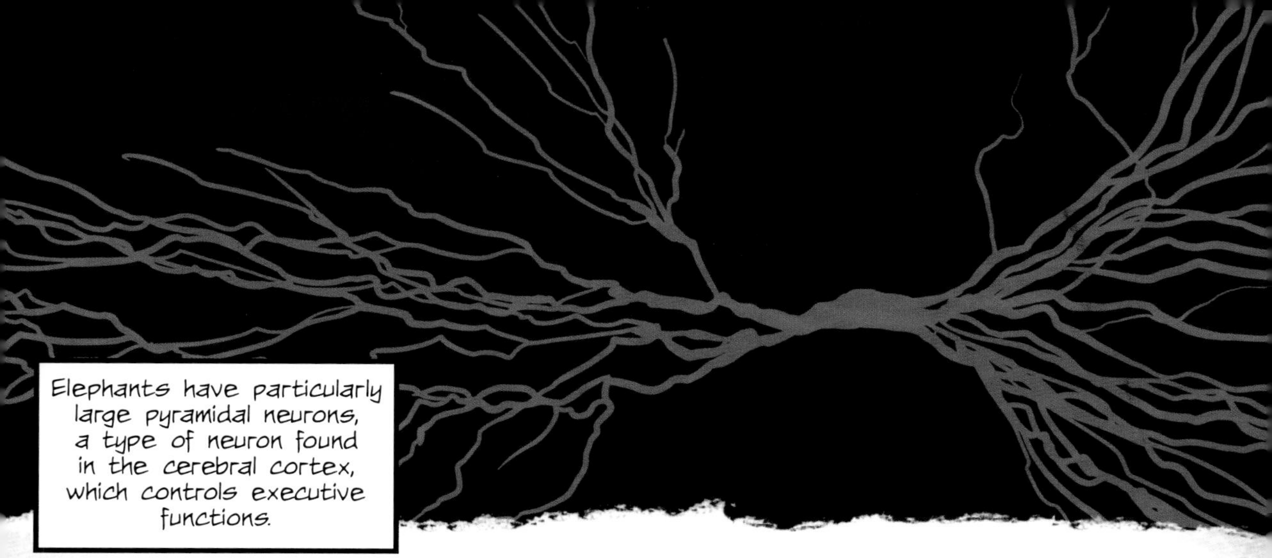
Elephants have particularly large pyramidal neurons, a type of neuron found in the cerebral cortex, which controls executive functions.

elephants: 1 1 x … 6.2 x 10 ; dolphins: 5.8 x 10^9, Roth & Dicke 2005). Elephants' pyramidal neurons (a class of neuron that is found in the cerebral cortex, particularly the pre-frontal cortex - the brain area that controls executive functions) are larger than in humans and most other species (Cozzi et al 2001). The degree of complexity of pyramidal neurons is linked to cognitive ability, with more (and more complex) connections between pyramidal neurons being associated with increased cognitive capabilities (Elston 2003). Elephant pyramidal neurons have a large dendritic tree, i.e. a large number of connections with other neurons for receiving and sending signals (Cozzi et al 2001).

24. Elephants, like humans, great apes and some cetaceans, possess *von Economo*

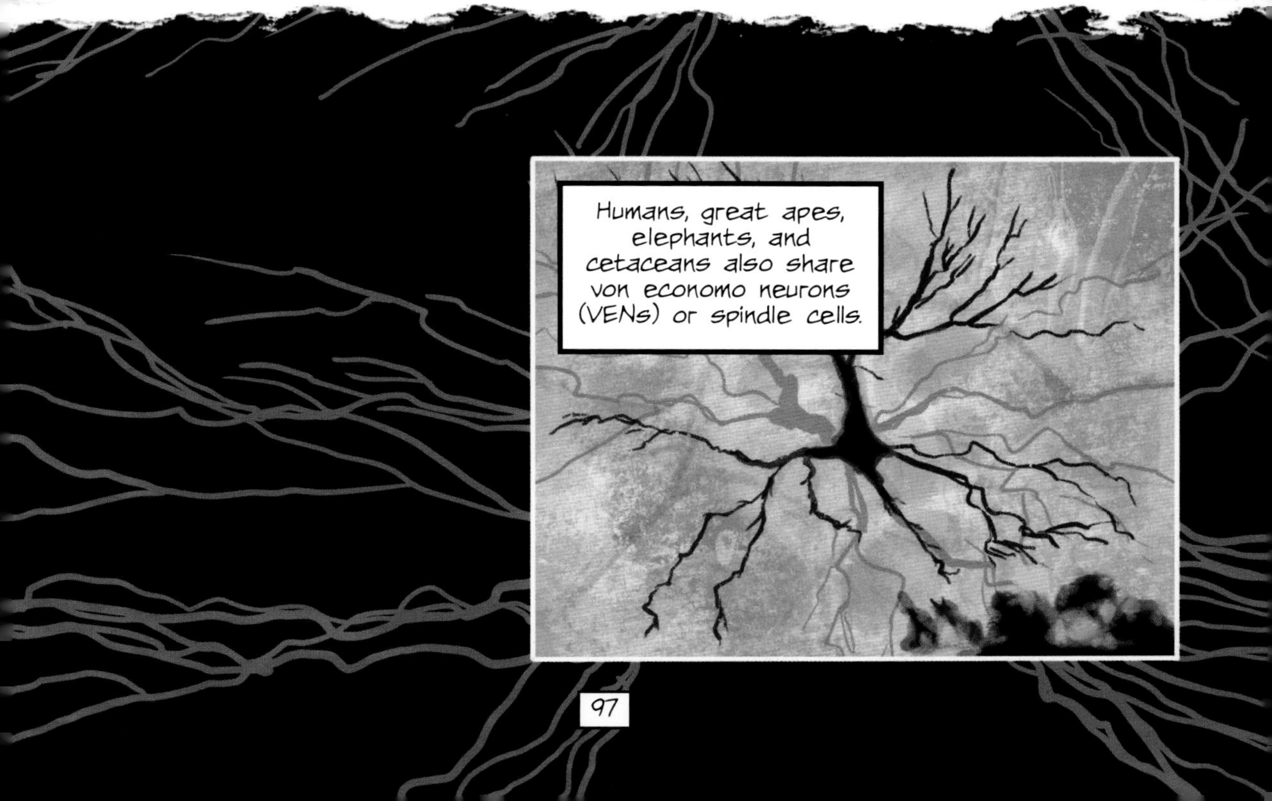
Humans, great apes, elephants, and cetaceans also share von economo neurons (VENs) or spindle cells.

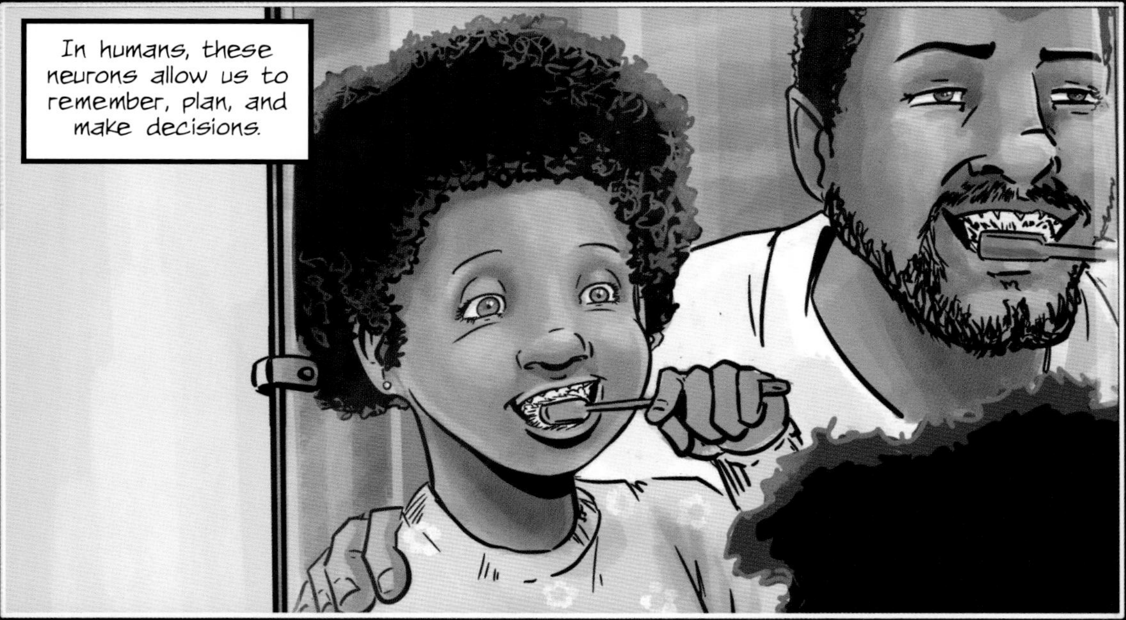
In humans, these neurons allow us to remember, plan, and make decisions.

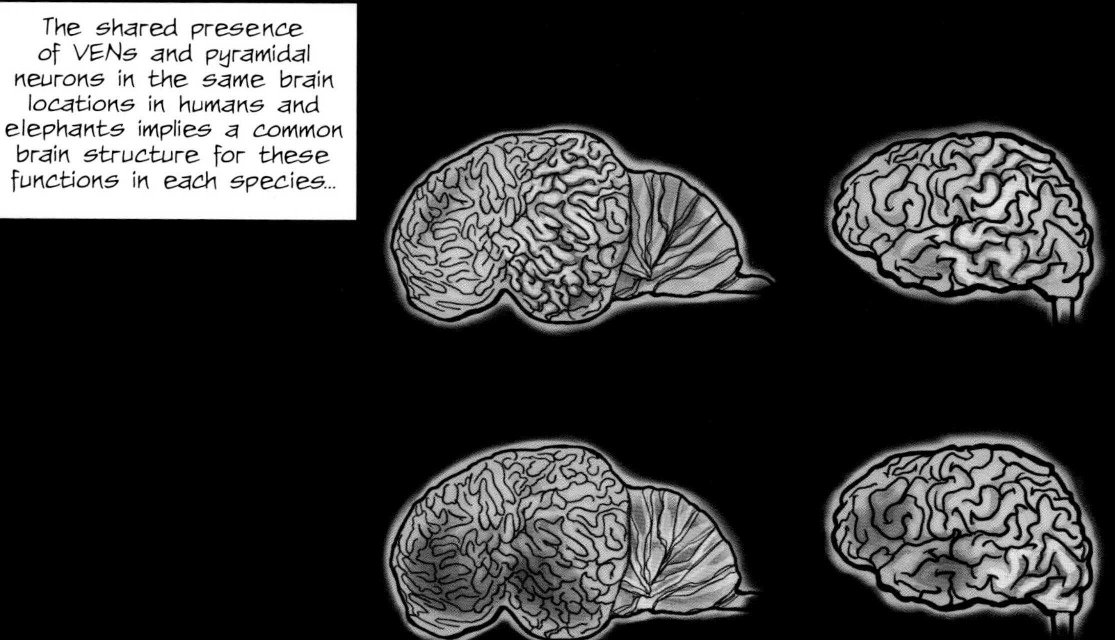
The shared presence of VENs and pyramidal neurons in the same brain locations in humans and elephants implies a common brain structure for these functions in each species...

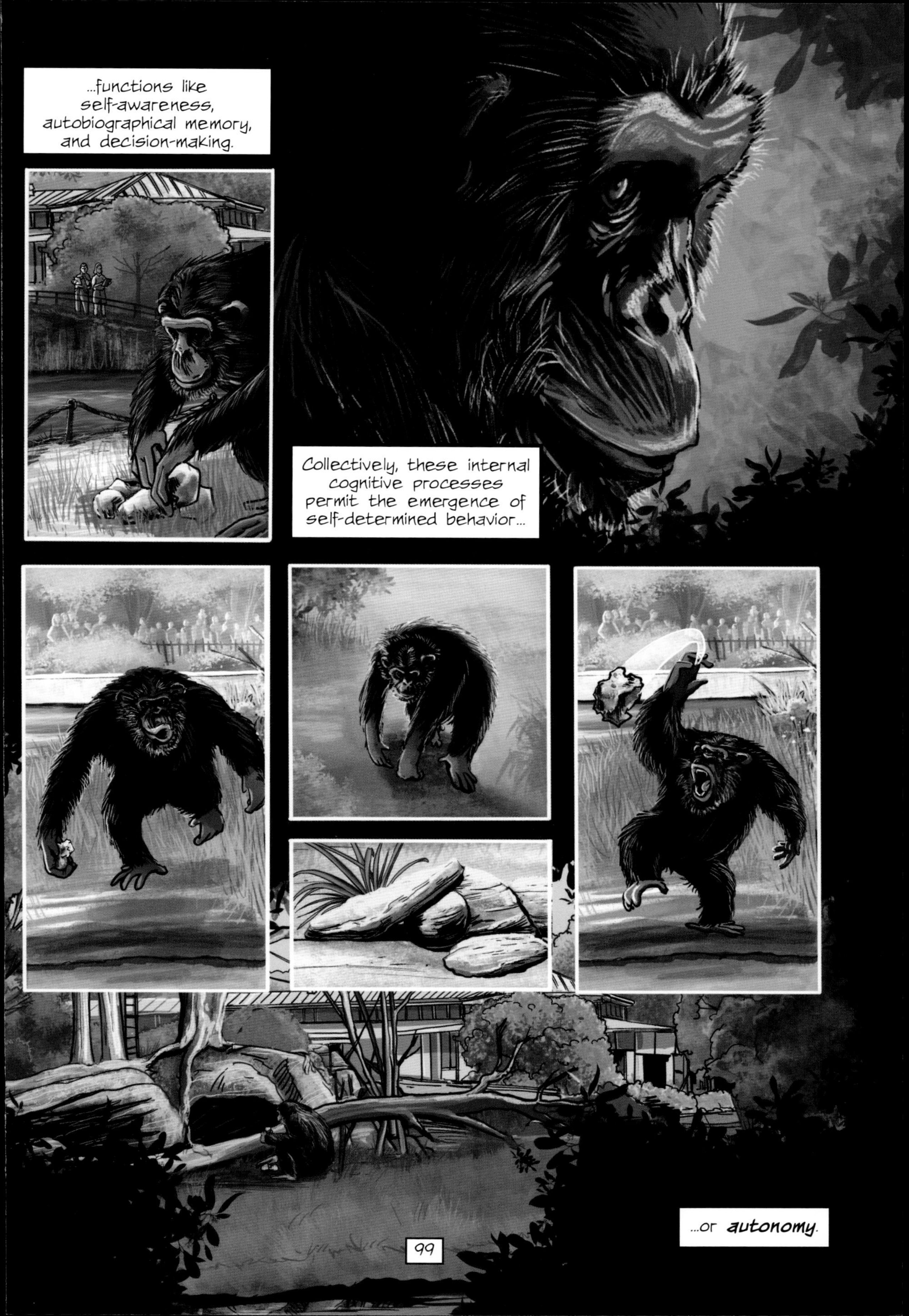

...functions like self-awareness, autobiographical memory, and decision-making.
Collectively, these internal cognitive processes permit the emergence of self-determined behavior...
...or **autonomy**.

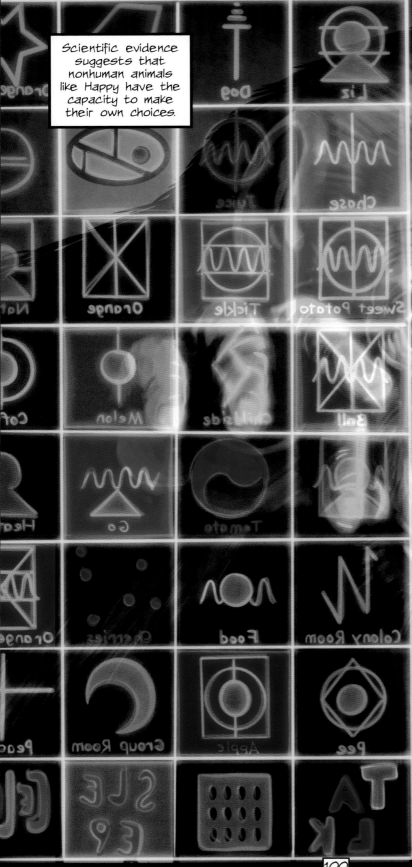

Scientific evidence suggests that nonhuman animals like Happy have the capacity to make their own choices.

Evidence also suggests that many of our complex nonhuman cousins share another feature of autonomy: *communication*.

CHAPTER 5: COMMUNICATION

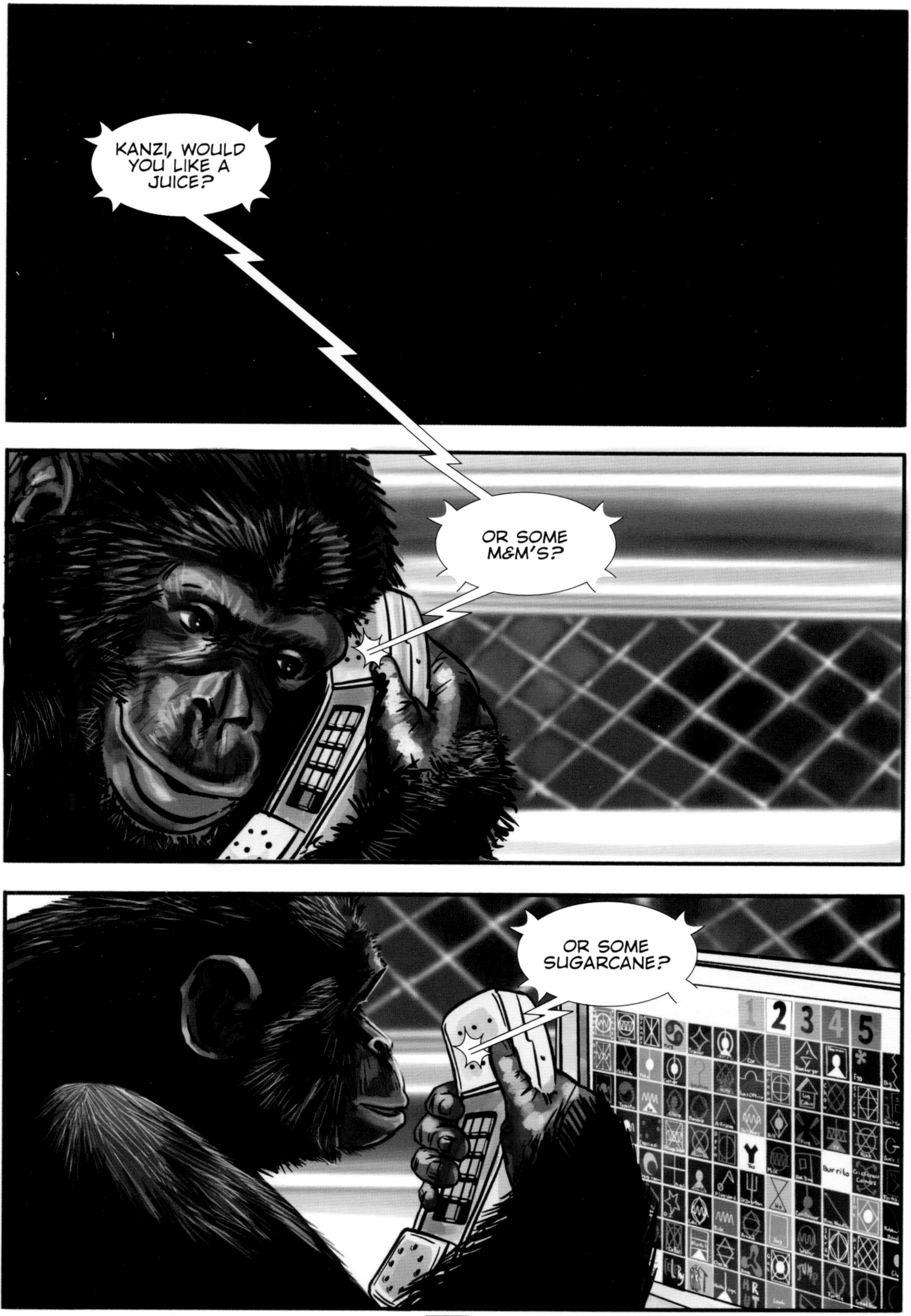
KANZI, WOULD YOU LIKE A JUICE?
OR SOME M&M'S?
OR SOME SUGARCANE?
1 2 3 4 5
Burrito

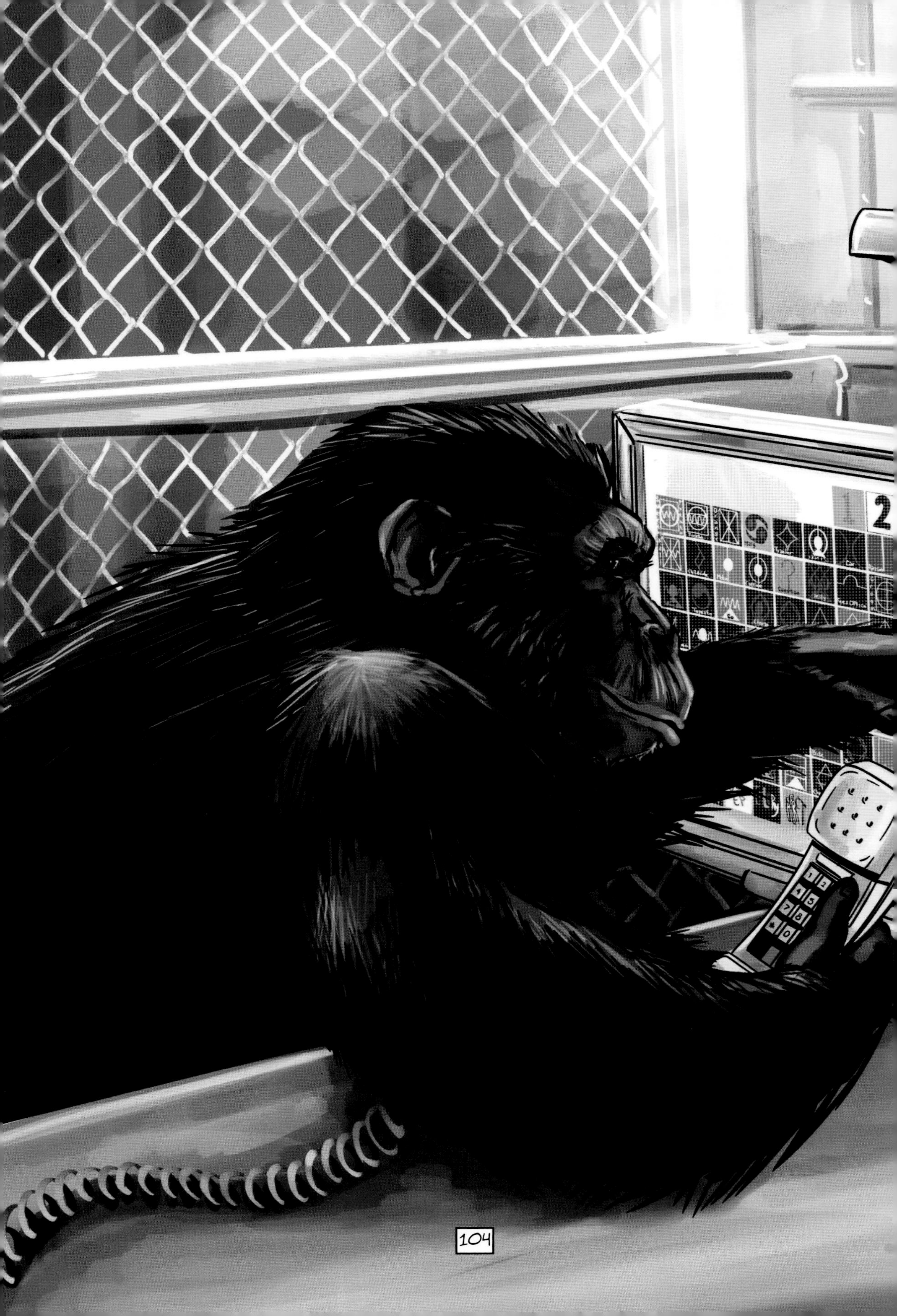

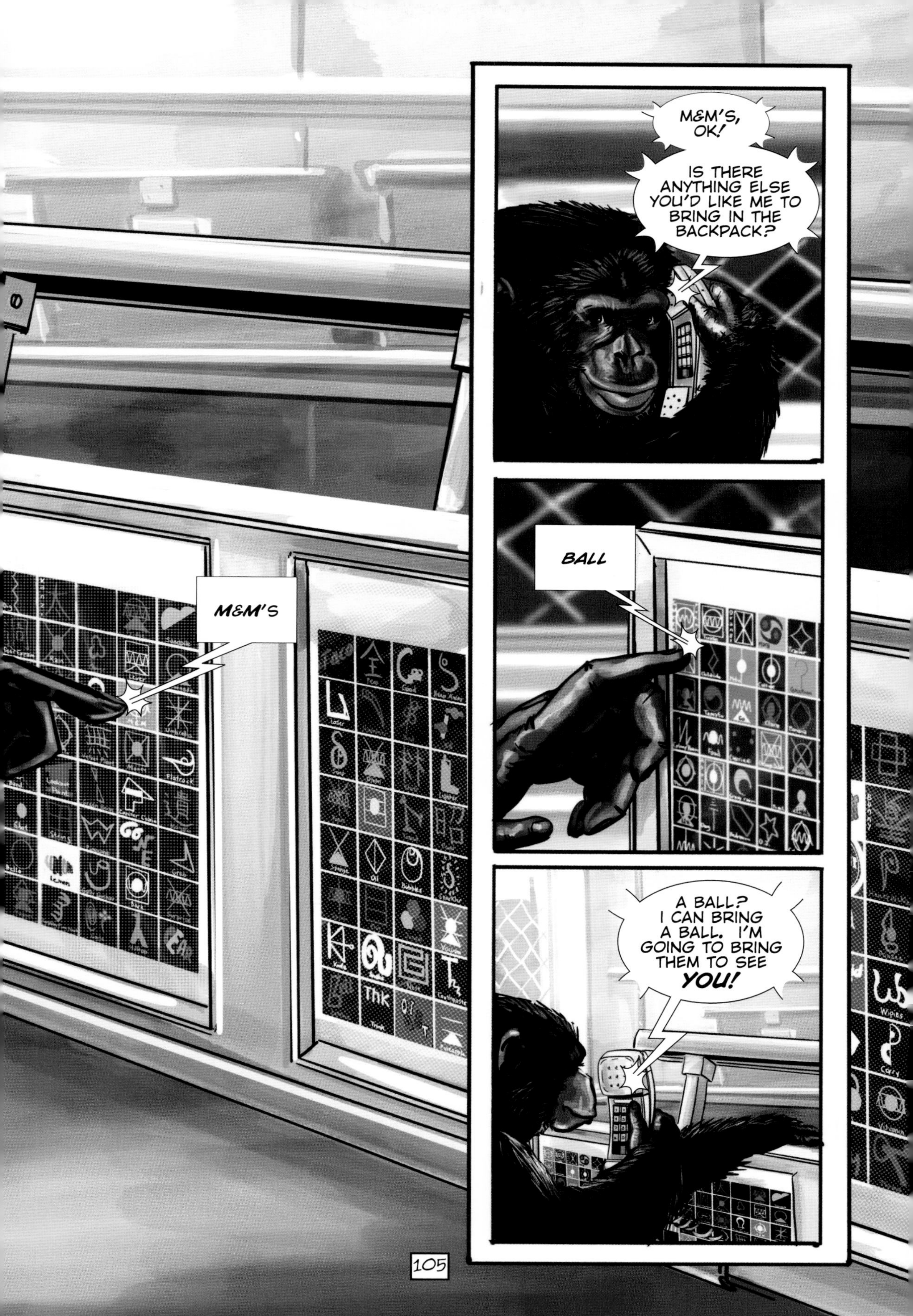
M&M'S
M&M'S, OK!
IS THERE ANYTHING ELSE YOU'D LIKE ME TO BRING IN THE BACKPACK?
BALL
A BALL? I CAN BRING A BALL. I'M GOING TO BRING THEM TO SEE YOU!

Kanzi, a bonobo from the Language Center at Georgia State University, had his priorities straight during his first phone call.

I THINK YOU HAVE TO UNTIE IT.

Footage and decades of peer-reviewed research have demonstrated that Kanzi understands an impressive amount of human language.

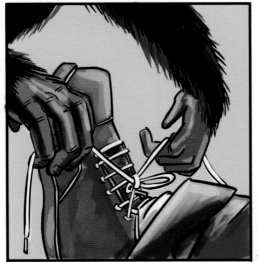

But his larynx would never let him simulate human speech.

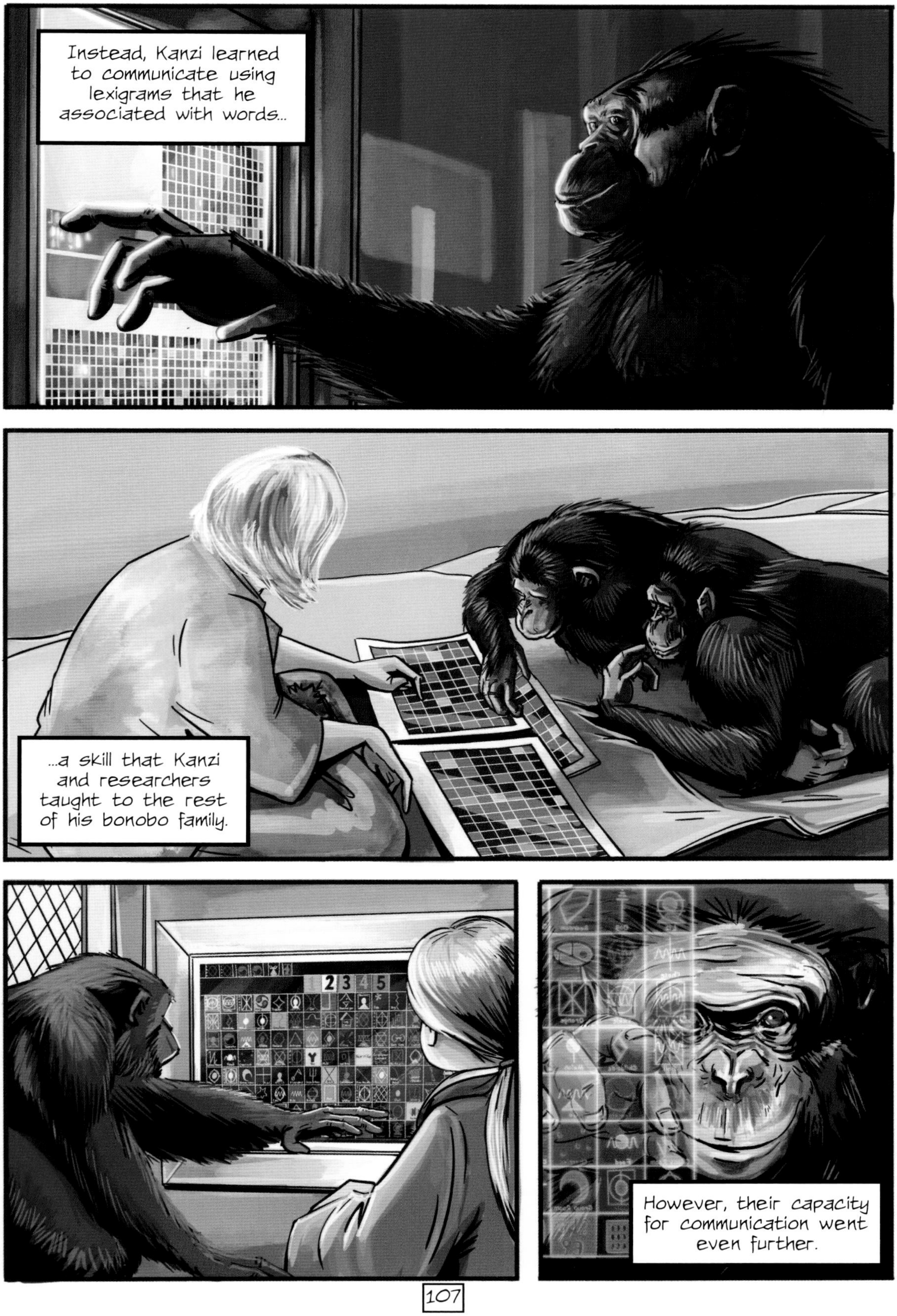
Instead, Kanzi learned to communicate using lexigrams that he associated with words...
...a skill that Kanzi and researchers taught to the rest of his bonobo family.
However, their capacity for communication went even further.

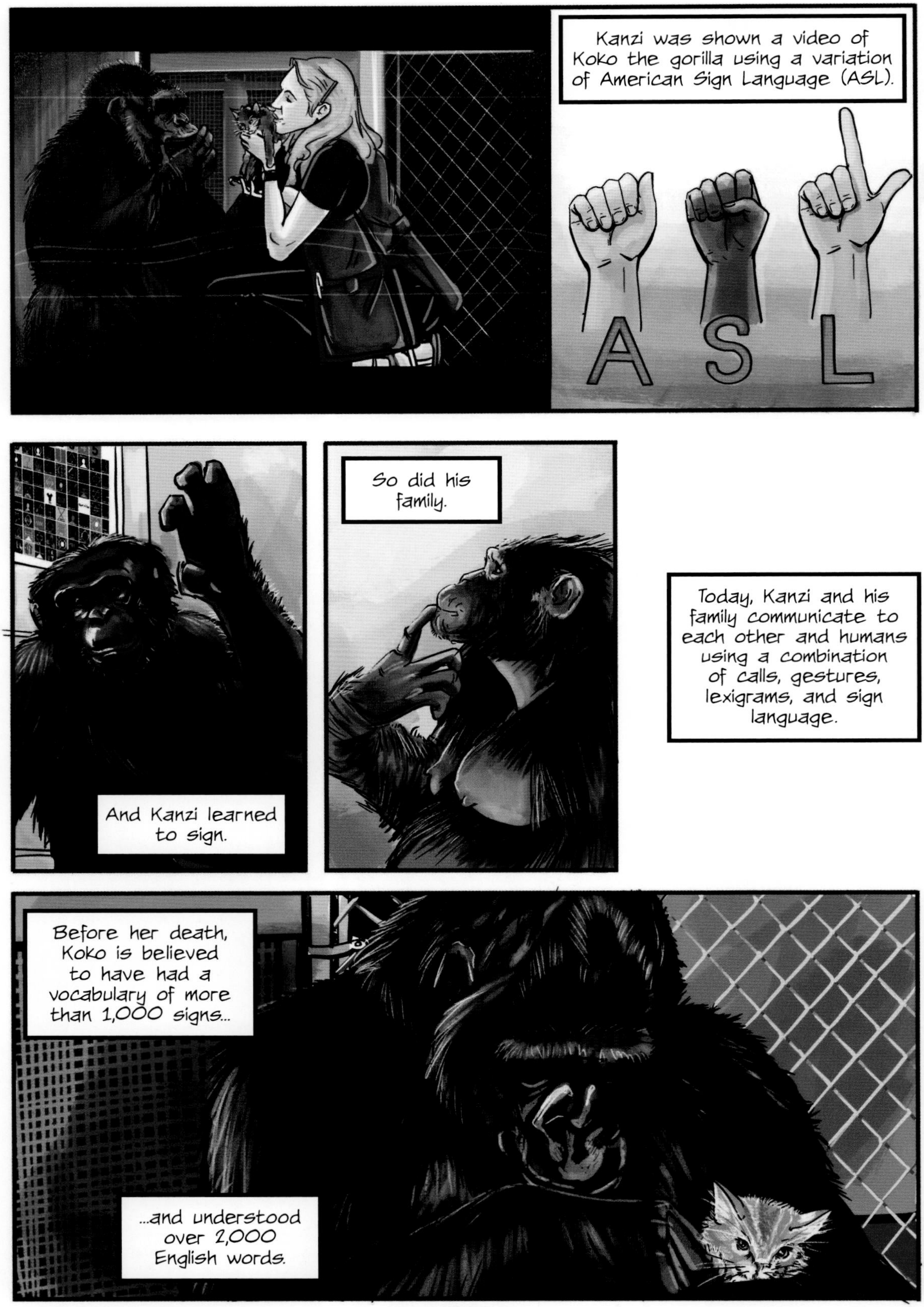
Kanzi was shown a video of Koko the gorilla using a variation of American Sign Language (ASL).
A S L
And Kanzi learned to sign.
So did his family.
Today, Kanzi and his family communicate to each other and humans using a combination of calls, gestures, lexigrams, and sign language.
Before her death, Koko is believed to have had a vocabulary of more than 1,000 signs...
...and understood over 2,000 English words.

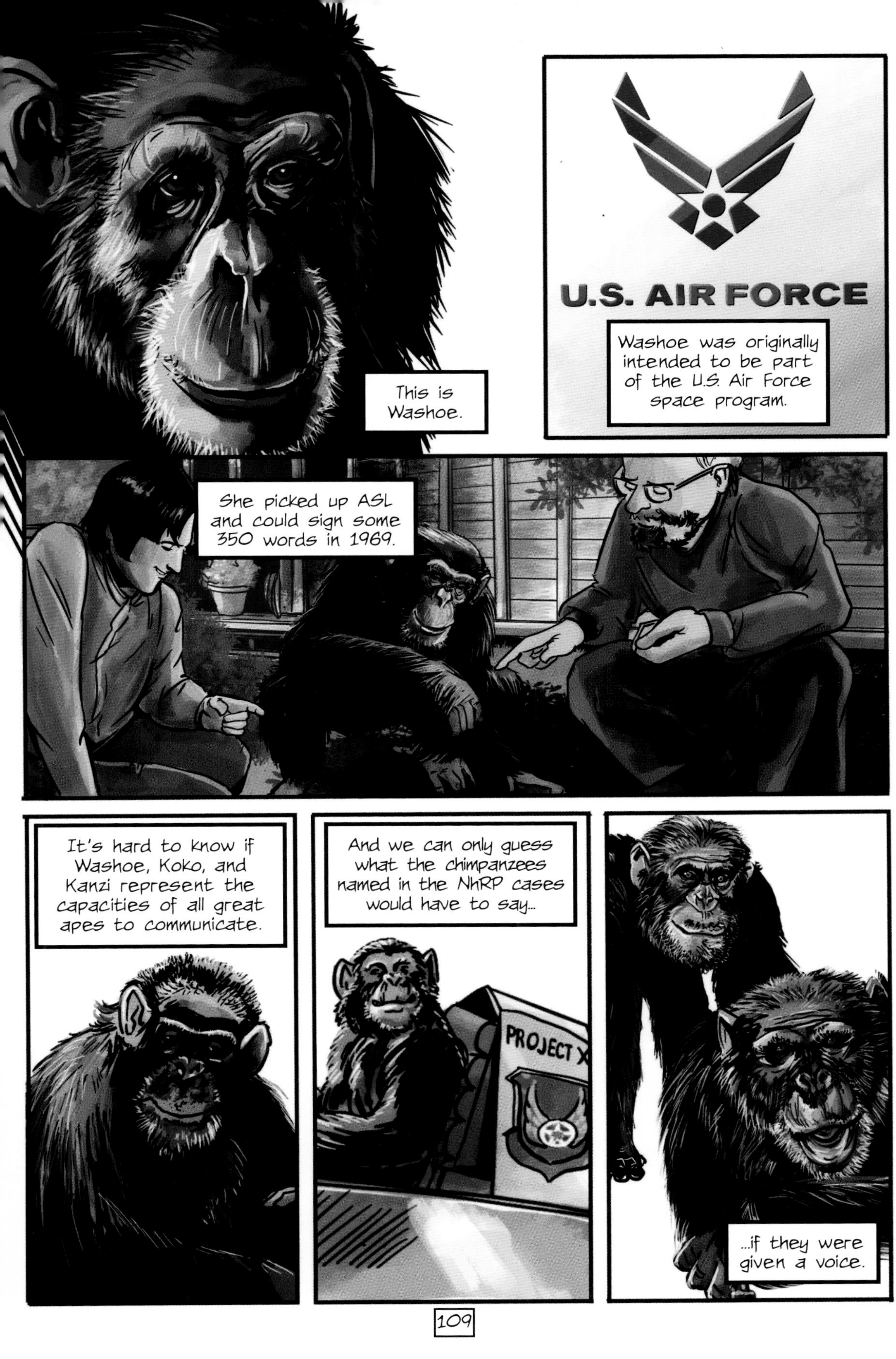

This is Washoe.
U.S. AIR FORCE
Washoe was originally intended to be part of the U.S. Air Force space program.
She picked up ASL and could sign some 350 words in 1969.
It's hard to know if Washoe, Koko, and Kanzi represent the capacities of all great apes to communicate.
And we can only guess what the chimpanzees named in the NhRP cases would have to say...
PROJECT X
...if they were given a voice.

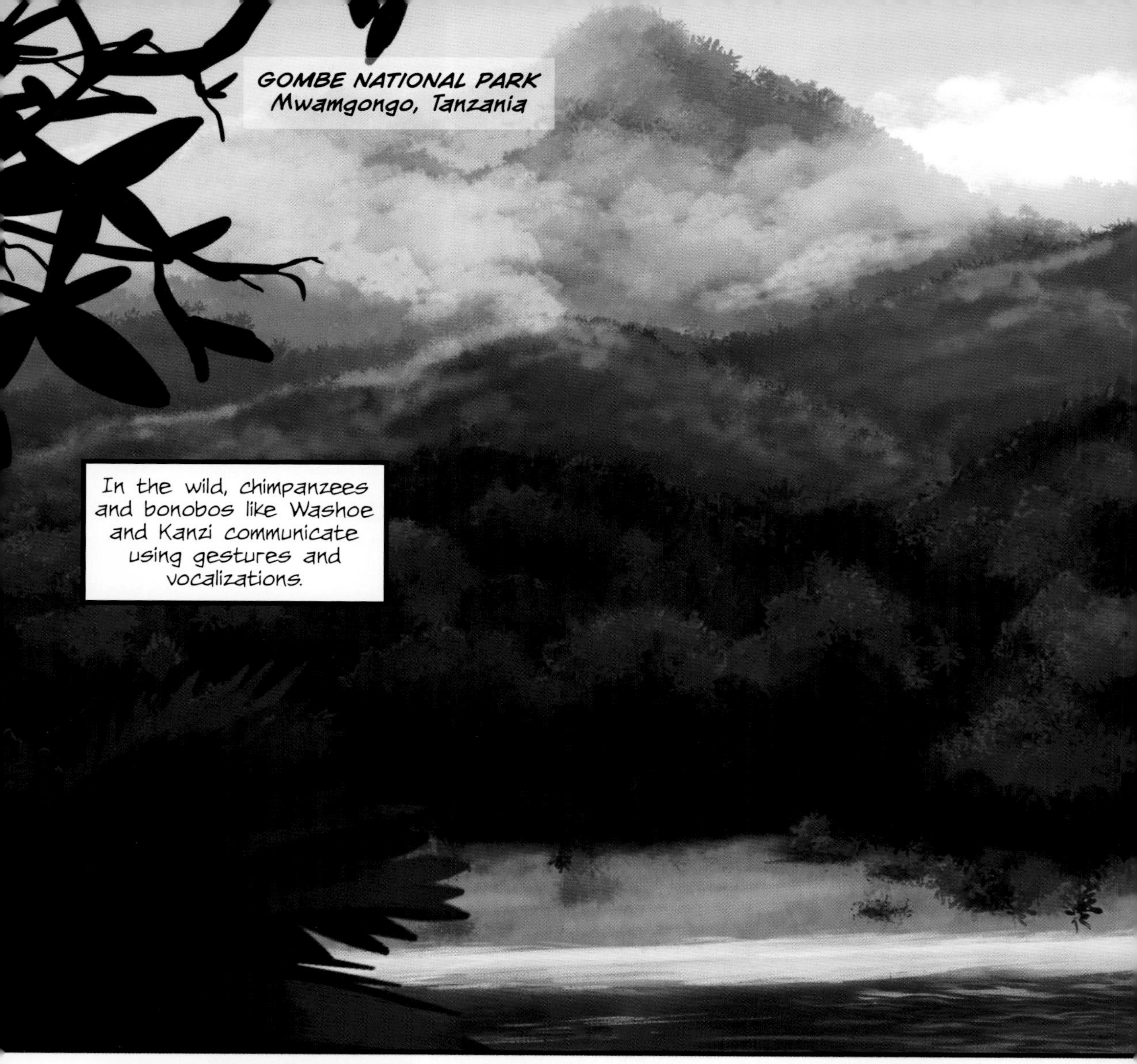
GOMBE NATIONAL PARK
Mwamgongo, Tanzania
In the wild, chimpanzees and bonobos like Washoe and Kanzi communicate using gestures and vocalizations.

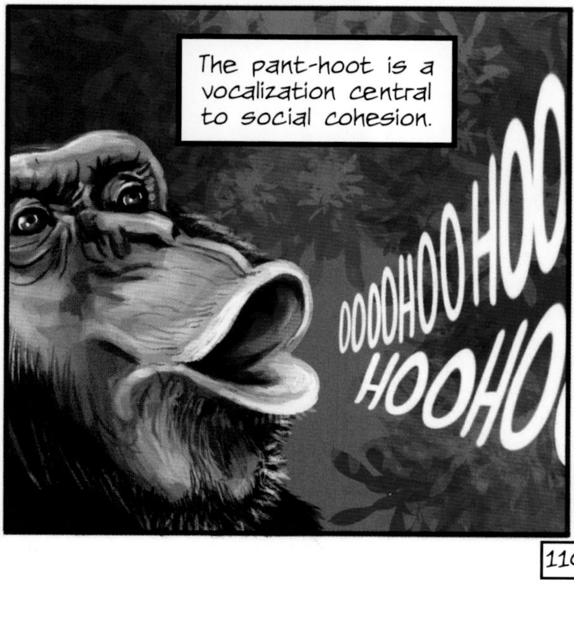
The pant-hoot is a vocalization central to social cohesion.
OOOOHOO HOO
HOOHO

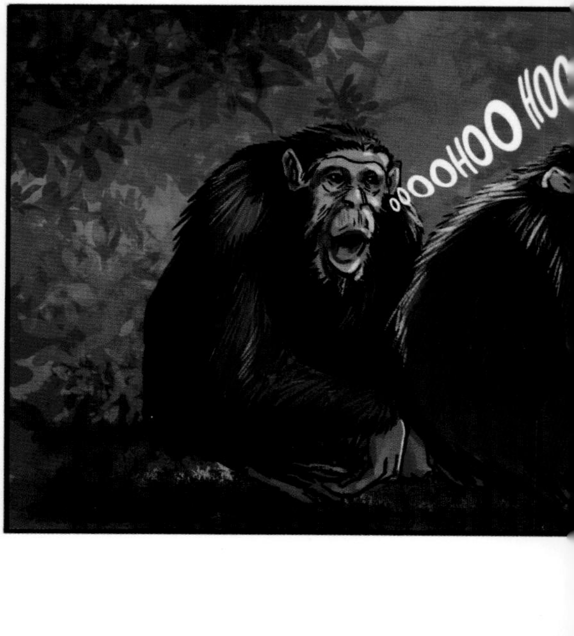
OOOOHOO HOO

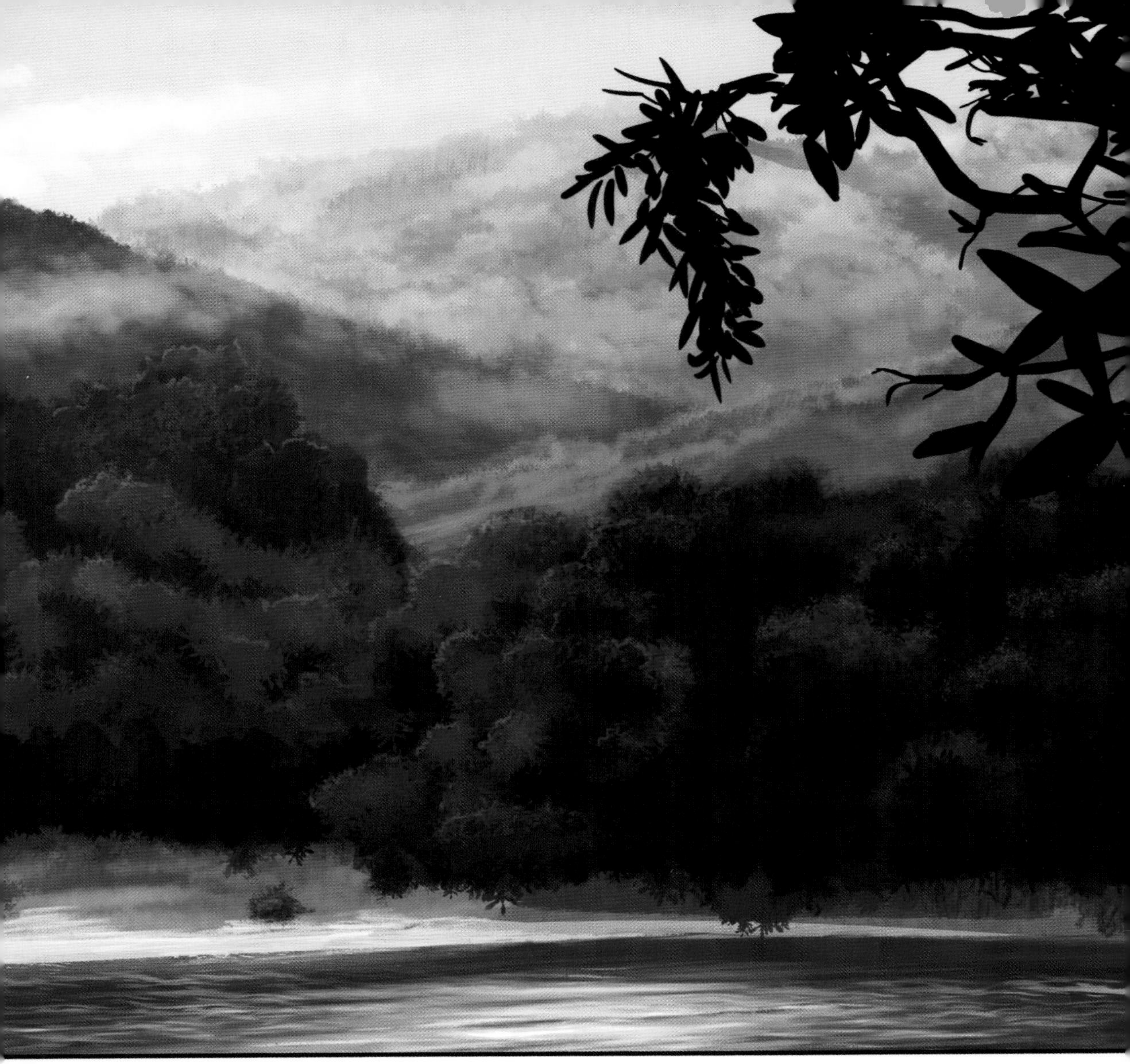

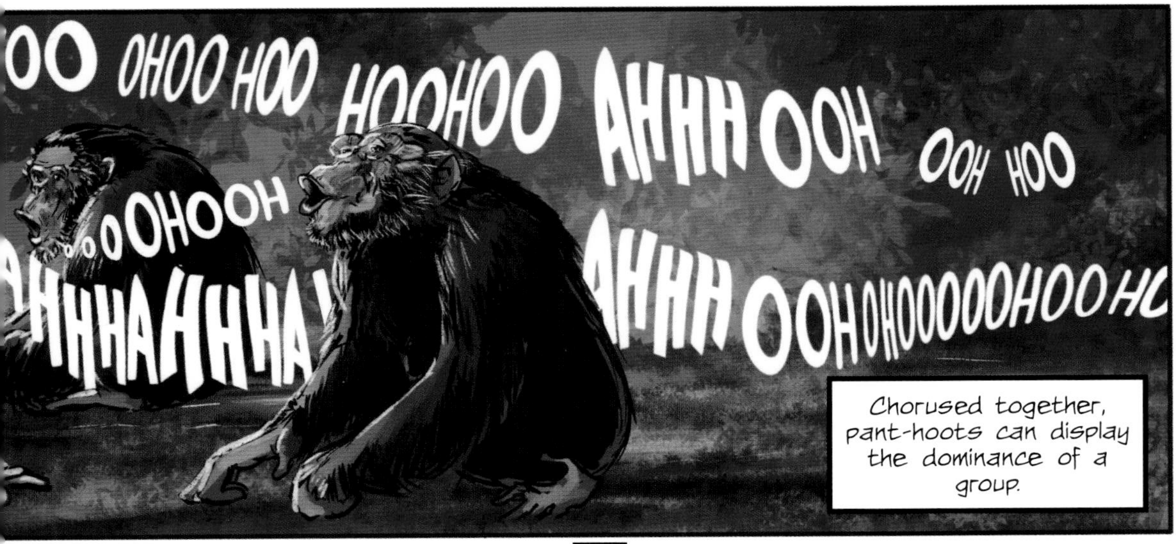
OO OHOO HOO HOOHOO AHHHOOH OOH HOO
OOOOHOOH
HHHAHHHA
AHHH OOHOHOOOOOHOOHO
Chorused together, pant-hoots can display the dominance of a group.

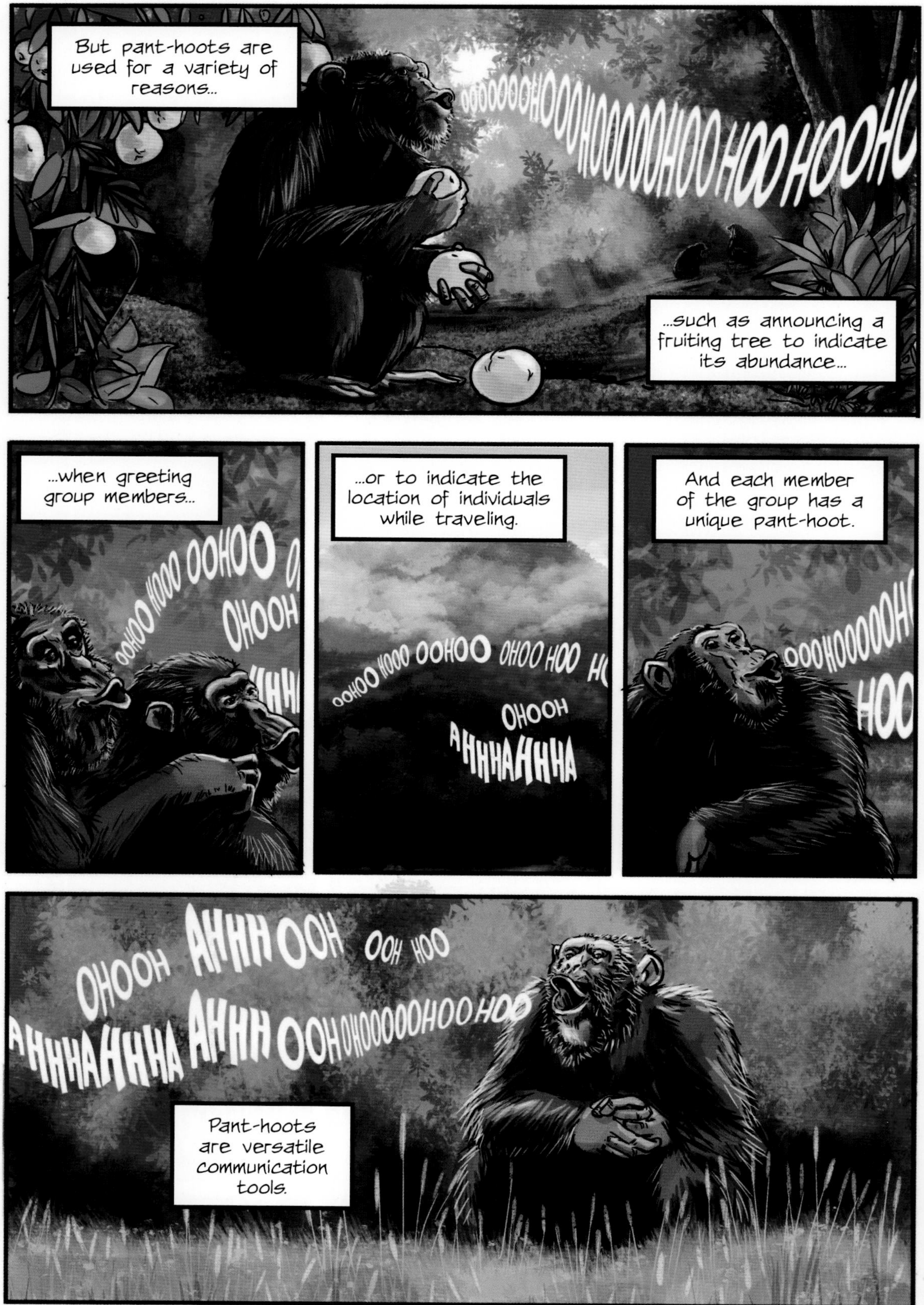

But pant-hoots are used for a variety of reasons...
OOOOOOOHOOOHOOOOOHOO HOO HOOH
...such as announcing a fruiting tree to indicate its abundance...
...when greeting group members...
OOHOO HOOO OOHOO
OHOOH
...or to indicate the location of individuals while traveling.
OOHOO HOOO OOHOO OHOO HOO
OHOOH
AHHHAHHHA
And each member of the group has a unique pant-hoot.
OOOHOOOOOOH
HOO
OHOOH AHHHOOH OOH HOO
AHHHAHHHA AHHH OOHOHOOOOOHOOHOO
Pant-hoots are versatile communication tools.

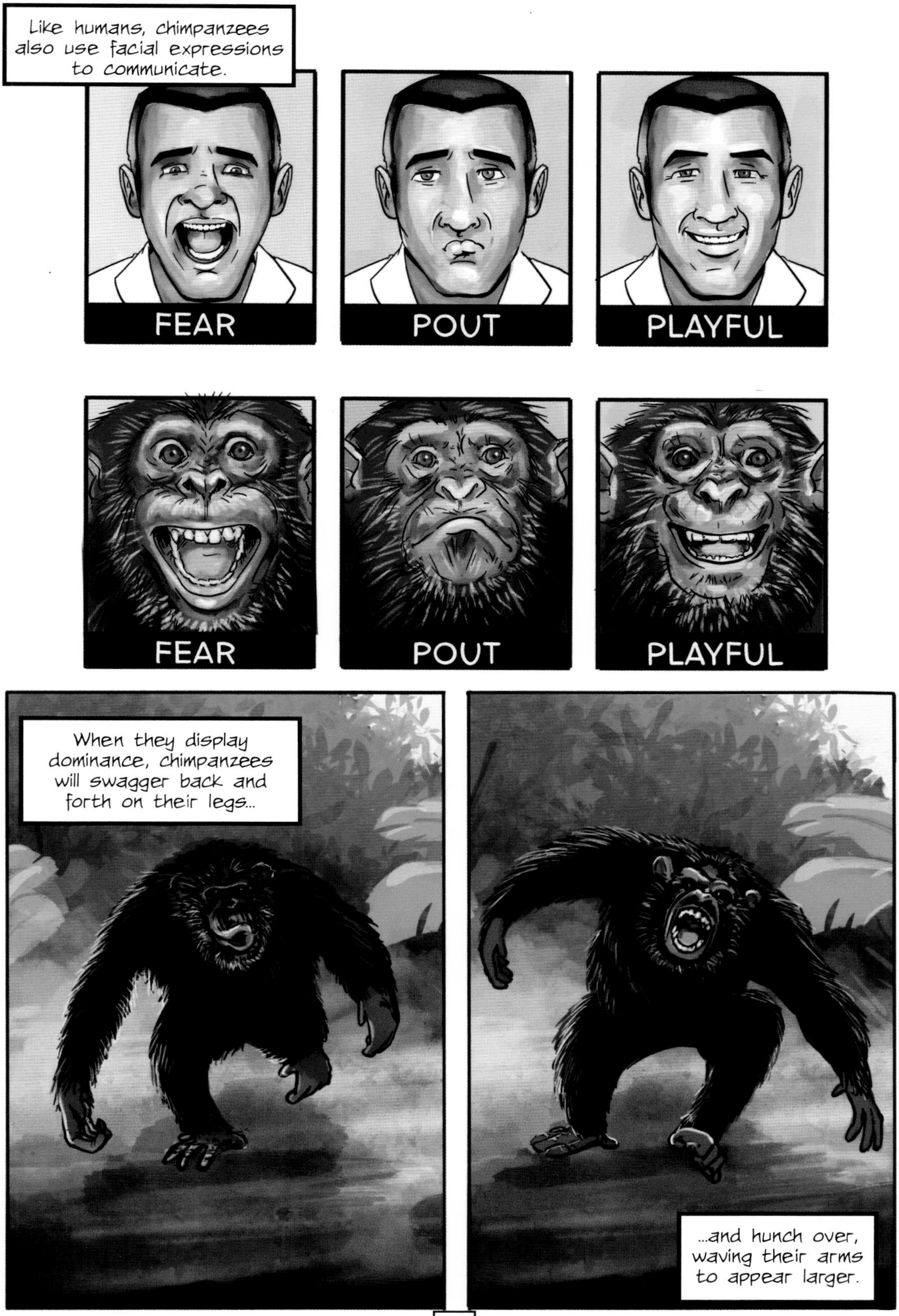
Like humans, chimpanzees also use facial expressions to communicate.
FEAR
POUT
PLAYFUL
FEAR
POUT
PLAYFUL
When they display dominance, chimpanzees will swagger back and forth on their legs...
...and hunch over, waving their arms to appear larger.

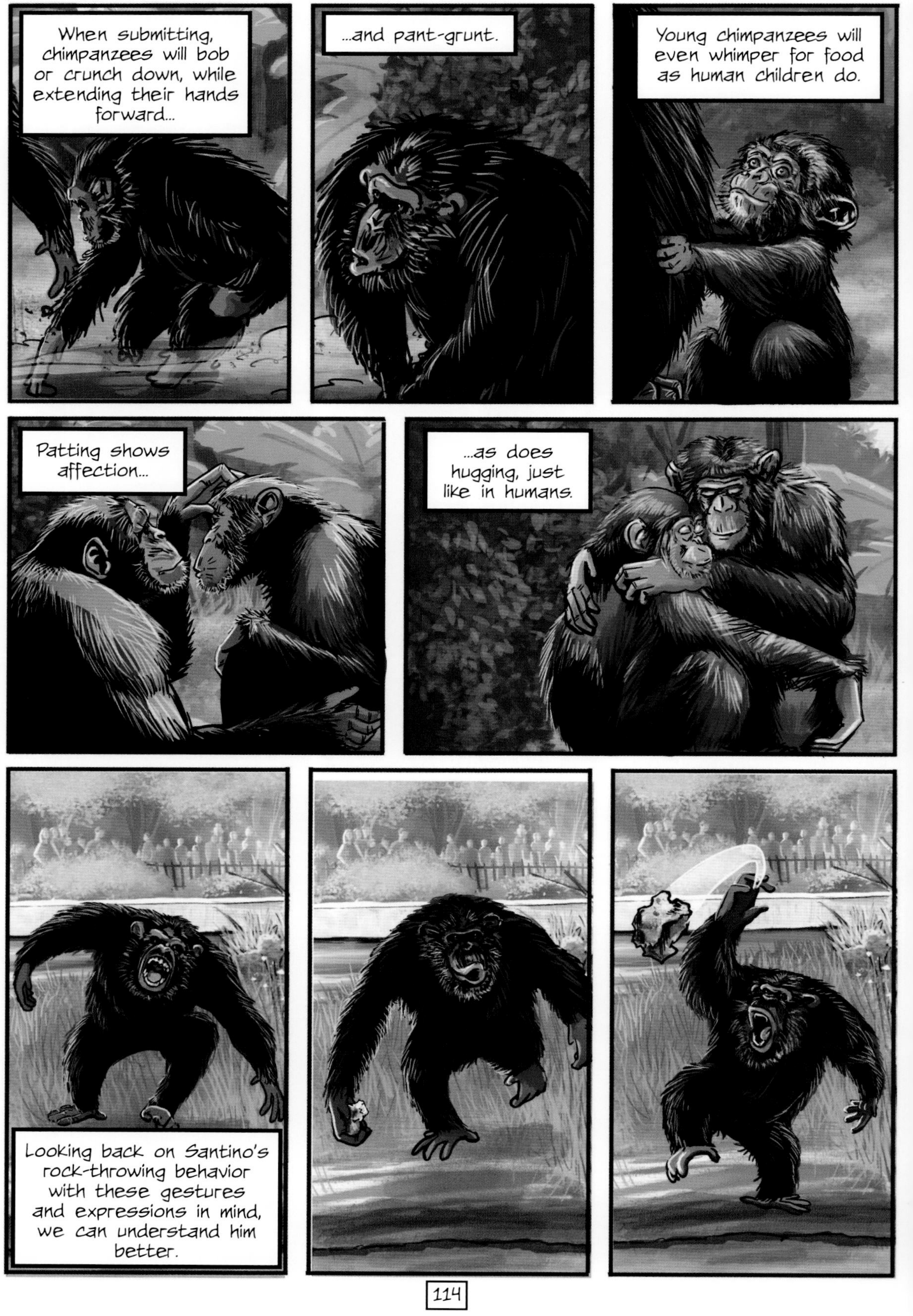
When submitting, chimpanzees will bob or crunch down, while extending their hands forward...
...and pant-grunt.
Young chimpanzees will even whimper for food as human children do.
Patting shows affection...
...as does hugging, just like in humans.
Looking back on Santino's rock-throwing behavior with these gestures and expressions in mind, we can understand him better.

DR. TETSURO MATSUZAWA
Primatologist

STATE OF NEW YORK
SUPREME COURT COUNTY OF FULTON

In the Matter of a Proceeding under Article 70 of
the CPLR for a Writ of Habeas Corpus
THE NONHUMAN RIGHTS PROJECT, INC.,
on behalf of TOMMY.

Petitioners,

v.

PATRICK C. LAVERY, individually a...
fficer of Circle L Trailer Sales, Inc, DL...
AVERY, and CIRCLE L TRAILER SA...

Respondents

AFFIDAVIT OF
TETSURO MATSUZAWA

COUNTRY OF INDIA
STATE OF MAHARASHTRA

Tetsuro Matsuzawa submitted an affidavit in the NhRP's chimpanzee cases.

He cited several of the shared anatomical features between chimpanzees and humans that are responsible for similarities in communication.

average mental abilities compared with other species of the same body size. **Both share similar**

circuits in the brain which are involved in language and communication (Gannon, Holloway,

Broadfield, and Braun, 1997; Taglialatele, Russell, Schaeffer, Hopkins, 2008; and see below).

Both have evolved large frontal lobes of the brain, which are intimately involved in the

capacities for insight and foreplanning (Semendeferi and Damasio, 2000). Both share a number

of highly specific cell types which are thought to be involved in higher-order thinking (see

below) and chimpanzee and human brains ...

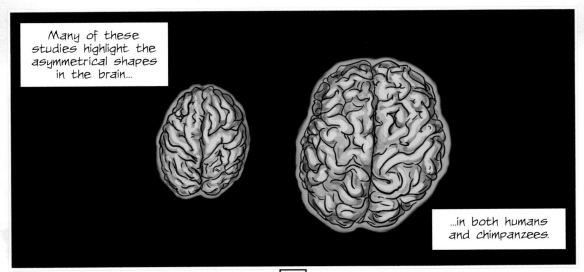

Many of these studies highlight the asymmetrical shapes in the brain...

...in both humans and chimpanzees.

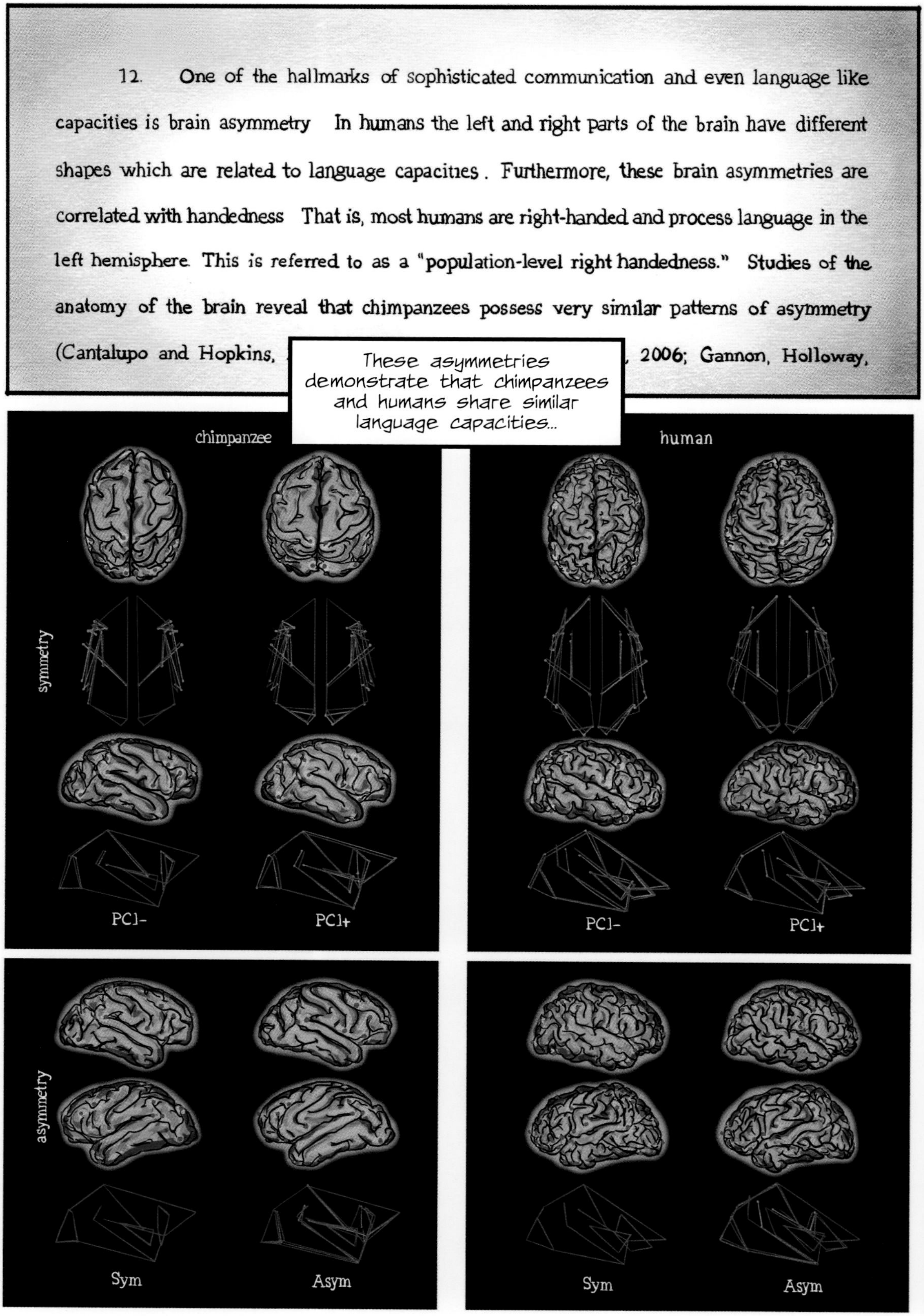
12. One of the hallmarks of sophisticated communication and even language like capacities is brain asymmetry In humans the left and right parts of the brain have different shapes which are related to language capacities . Furthermore, these brain asymmetries are correlated with handedness That is, most humans are right-handed and process language in the left hemisphere. This is referred to as a "population-level right handedness." Studies of the anatomy of the brain reveal that chimpanzees possess very similar patterns of asymmetry (Cantalupo and Hopkins,
, 2006; Gannon, Holloway,
These asymmetries demonstrate that chimpanzees and humans share similar language capacities...
chimpanzee
human
symmetry
PC1-
PC1+
PC1-
PC1+
asymmetry
Sym
Asym
Sym
Asym

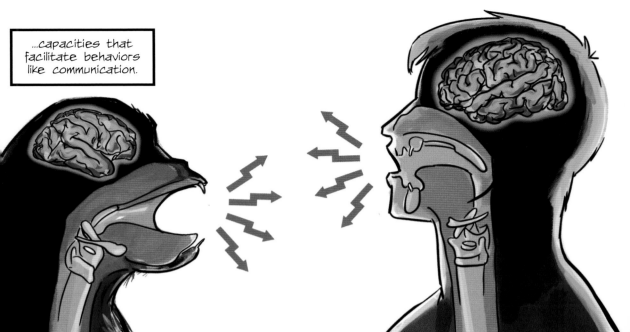

...capacities that facilitate behaviors like communication.

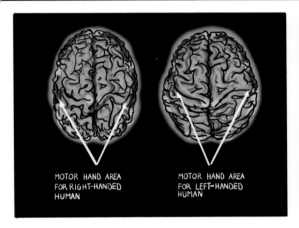

MOTOR HAND AREA FOR RIGHT-HANDED HUMAN

MOTOR HAND AREA FOR LEFT-HANDED HUMAN

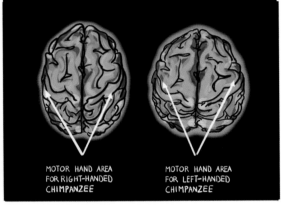

MOTOR HAND AREA FOR RIGHT-HANDED CHIMPANZEE

MOTOR HAND AREA FOR LEFT-HANDED CHIMPANZEE

(a)

AQ

rightwards

leftwards

5

0

-5

chimpanzees (c)
humans (h)

c h   c h   c h   c h   c h   c h
SF   IF   T   SP   IP   O

(b)

rightwards

leftwards

5

0

-5

c h   c h   c h   c h   c h   c h
FOS   PCS   CS   SyF   STS   LS

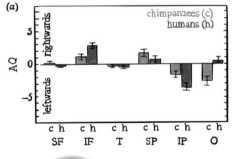

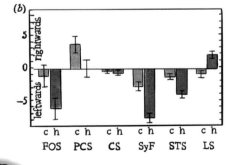

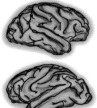

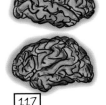

This probably relates to why Washoe, Koko, Kanzi, and his family can speak with humans using a shared language.

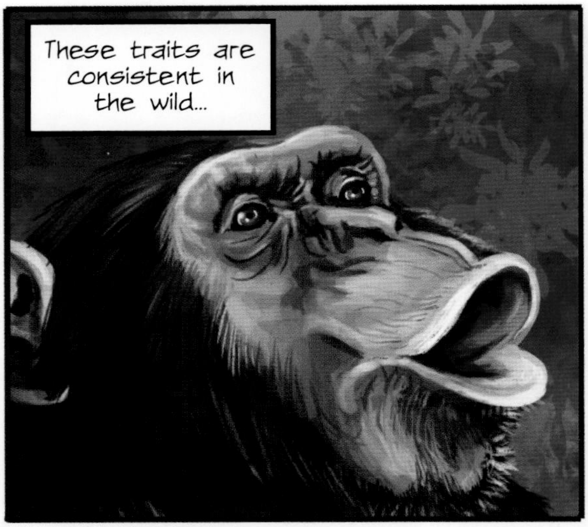
These traits are consistent in the wild...

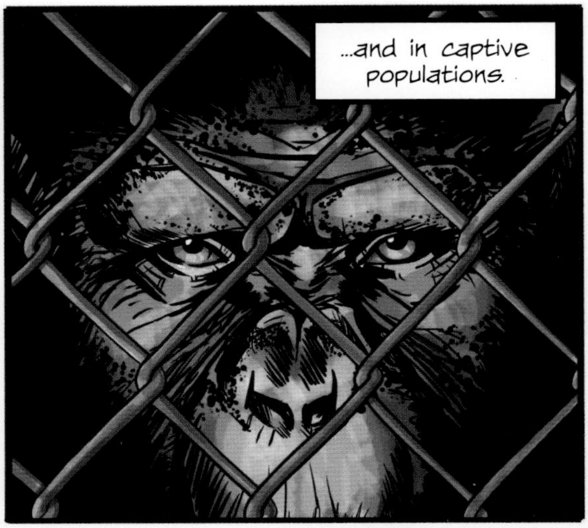
...and in captive populations.

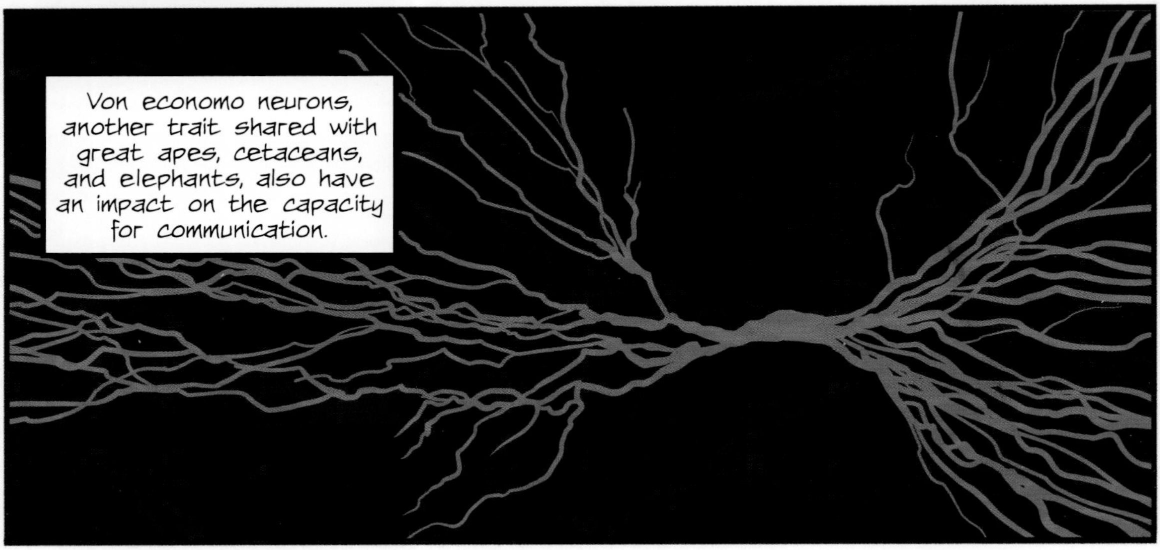
Von economo neurons, another trait shared with great apes, cetaceans, and elephants, also have an impact on the capacity for communication.

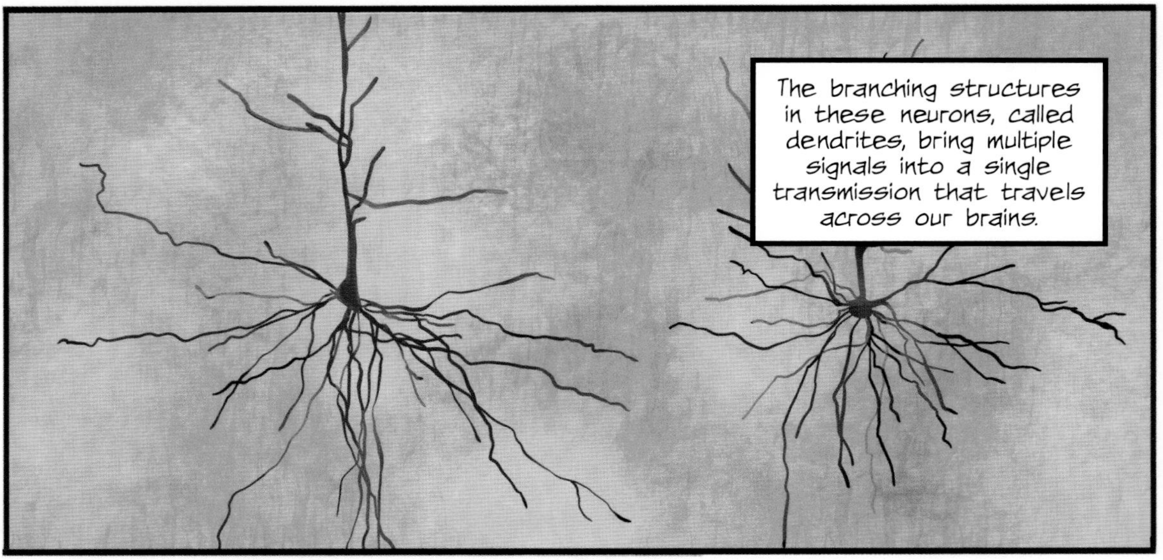
The branching structures in these neurons, called dendrites, bring multiple signals into a single transmission that travels across our brains.

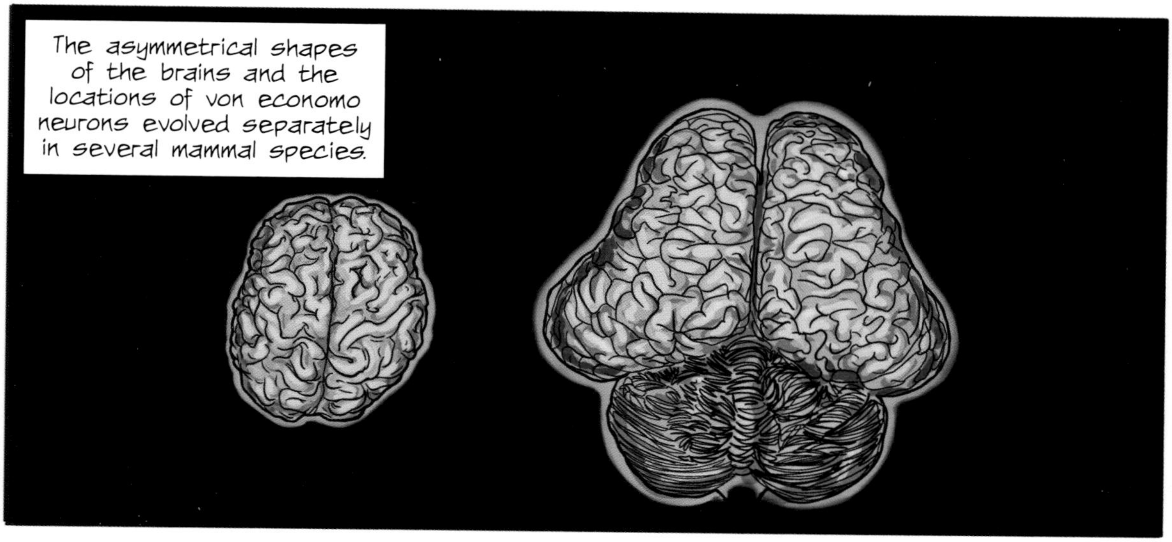
The asymmetrical shapes of the brains and the locations of von economo neurons evolved separately in several mammal species.

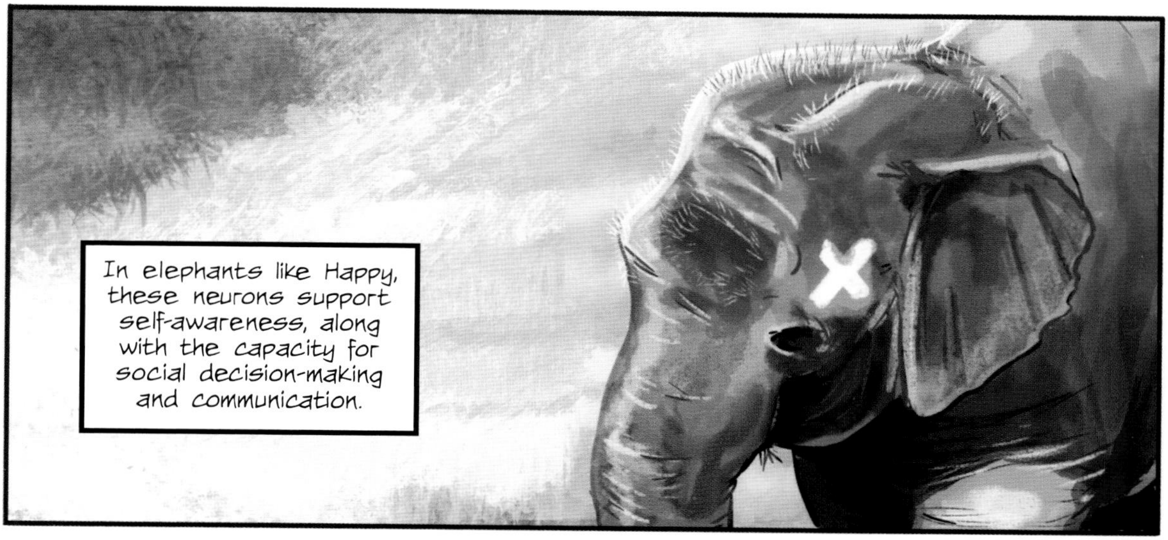
In elephants like Happy, these neurons support self-awareness, along with the capacity for social decision-making and communication.

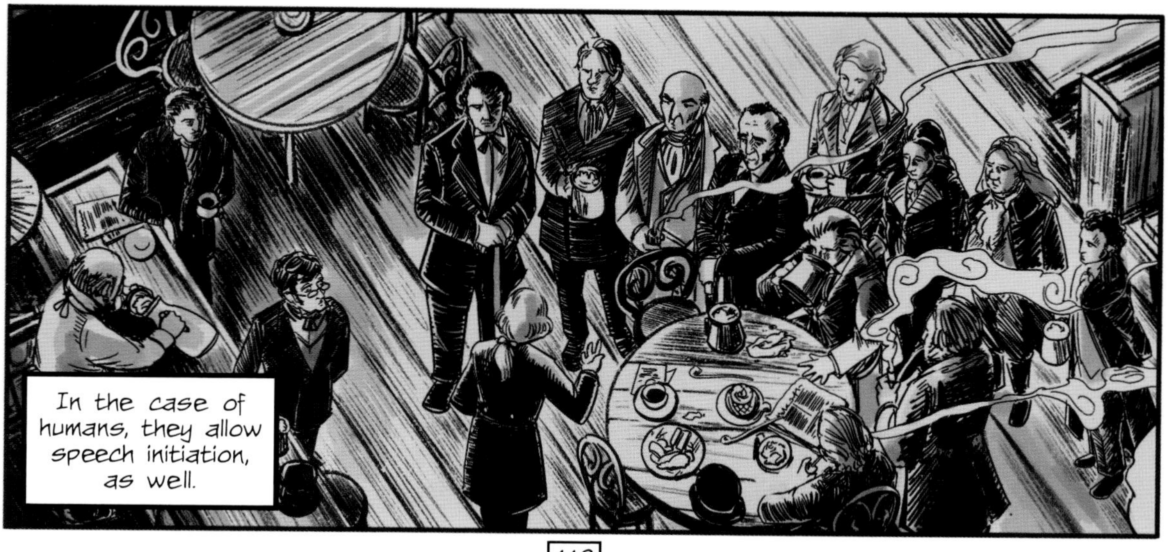
In the case of humans, they allow speech initiation, as well.

Together, these anatomical traits suggest that we share many of the same higher-order brain functions.
HUMAN
ELEPHANT
CHIMPANZEE
These shared traits don't factor into the Great Chain of Being.
Deus
Planta
But, they should inform us when re-examining how we classify nonhuman animals going forward.
INSTITVTIONES IVRIS CIVILIS

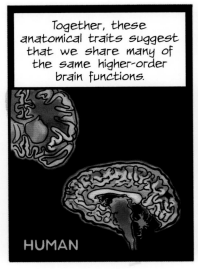

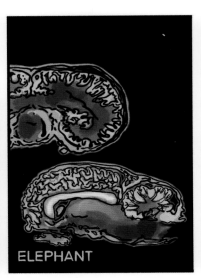

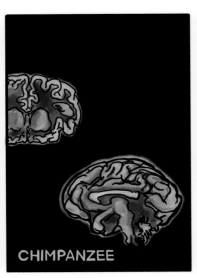

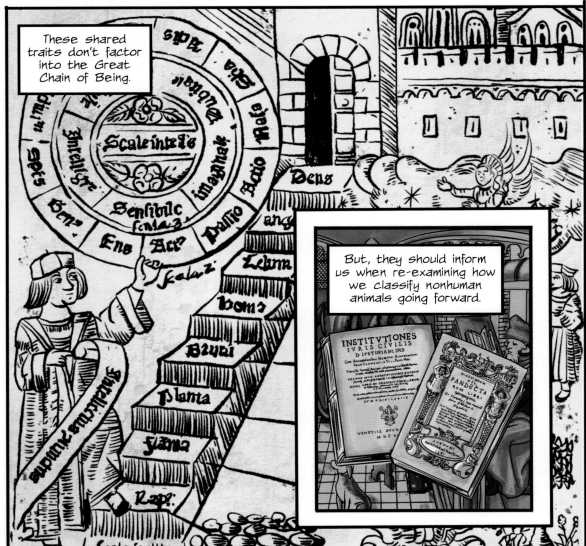

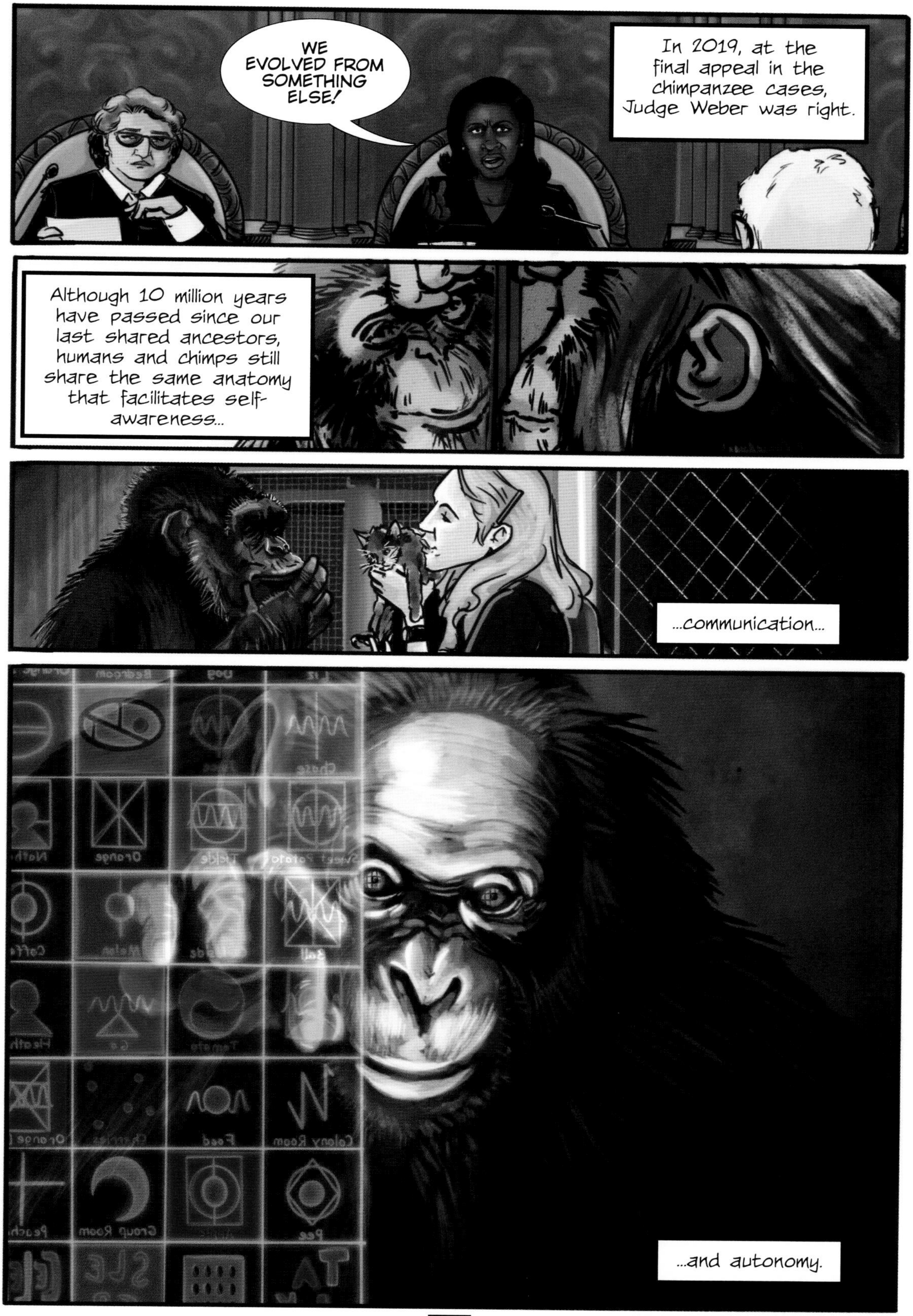
WE EVOLVED FROM SOMETHING ELSE!
In 2019, at the final appeal in the chimpanzee cases, Judge Weber was right.
Although 10 million years have passed since our last shared ancestors, humans and chimps still share the same anatomy that facilitates self-awareness...
...communication...
...and autonomy.

DR. JOYCE POOLE
Cofounder and Scientific Director of ElephantVoices

Joyce Poole, a behavioral scientist who has been studying elephants for 47 years, also submitted an affidavit on behalf of Happy.

Doctors Poole and Moss both confirmed that elephant communication is an expression of their autonomy...

have mental states (e.g., intentions).

Communication and social learning

42. Speech is a voluntary behaviour in humans, whereby a person can choose whether to utter words and thus communicate with another. Therefore speech and language reflect autonomous thinking and intentional behaviour. Elephants also intentionally use their vocalisations to share knowledge and information with others (Poole 2011). Females and dependents call to emphasise and reinforce their social units, and to coordinate movement. Male elephants primarily commu...

...just like it is with primates and humans.

● elephantvoices.org

WHAT YOU CAN DO   SIGN THE ELEPHANT CHARTER   DONATE   Search ...

ELEPHANTVOICES

About | Studies & | The Elephant Ethogram | Elephant Communication | Elephant Sense & Sociality | Elephants & Ethics | Threats to Elephants | Elephants in Captivity | Multimedia Resources | Support NOW!

Through her research, Dr. Poole has amassed a database of elephant communication.

The Elephant Ethogram

A Library of African Elephant Behavior. The Elephant Ethogram is a fully searchable database on www.elephantvoices.org with annotated video, audio and images. Copyright: ElephantVoices

▷ Start watching

Visit The Elephant Ethogram - with over 3000 video clips, sounds and photos

122

Communication and social learning

42. Speech is a voluntary behaviour in humans, whereby a person can choose whether to utter words and thus communicate with another. Therefore speech and language reflect autonomous thinking and intentional behaviour. Elephants also intentionally use their vocalisations to share knowledge and information with others (Poole 2011). Females and dependents call to emphasise and reinforce their social units and to coordinate movement. Male elephants primarily communicate about their sexual status, rank and identity, though like females they also use calls to coordinate

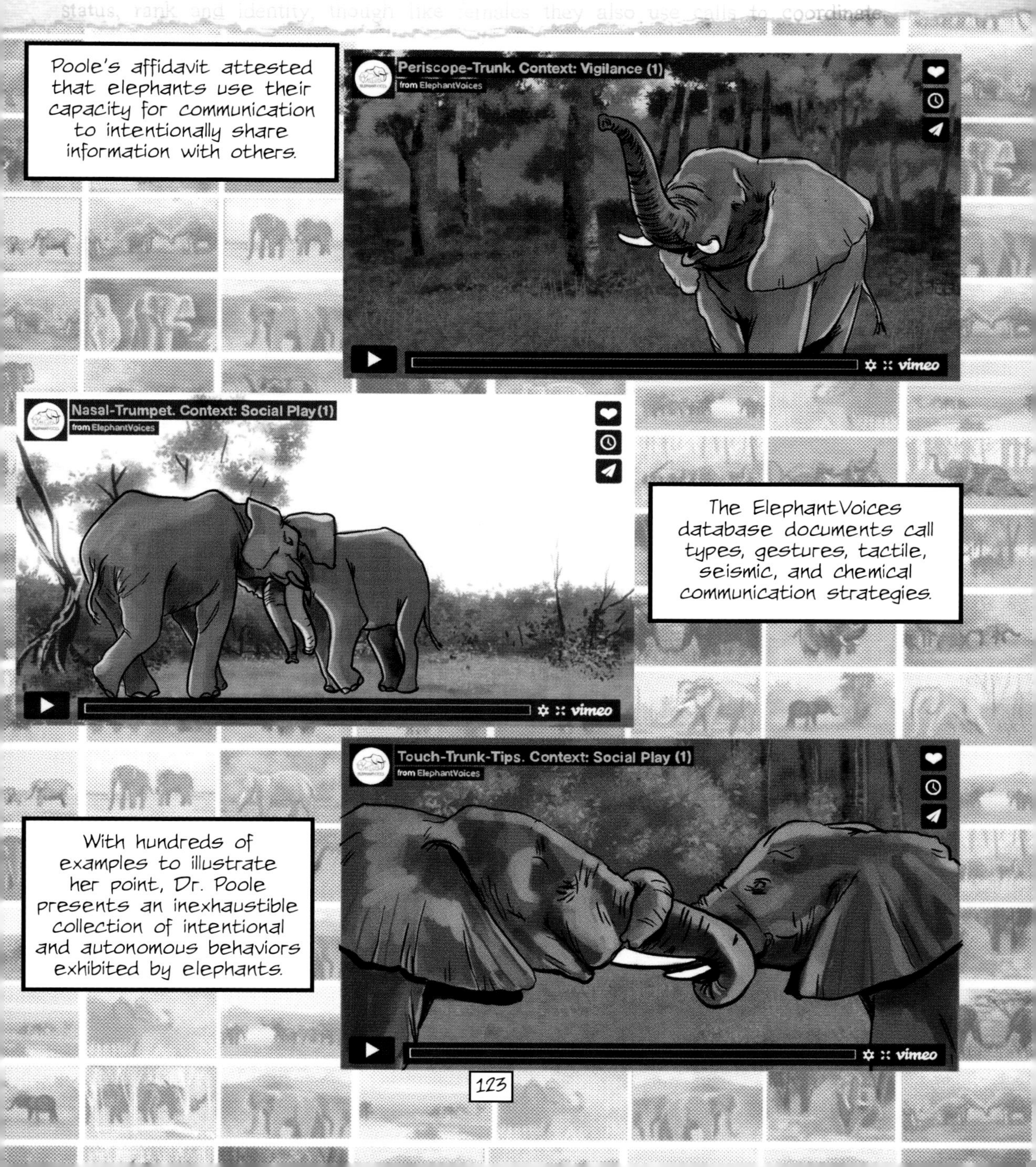

Poole's affidavit attested that elephants use their capacity for communication to intentionally share information with others.
Periscope-Trunk. Context: Vigilance (1)
from ElephantVoices
vimeo
Nasal-Trumpet. Context: Social Play (1)
from ElephantVoices
vimeo
The ElephantVoices database documents call types, gestures, tactile, seismic, and chemical communication strategies.
Touch-Trunk-Tips. Context: Social Play (1)
from ElephantVoices
vimeo
With hundreds of examples to illustrate her point, Dr. Poole presents an inexhaustible collection of intentional and autonomous behaviors exhibited by elephants.

Dr. Poole's affidavit demonstrated several examples:

"I have observed that a member of a family will use the axis of her body to point in the direction she wishes to go..."

"...and then vocalize, every couple of minutes, with a specific call known as a 'lets go' rumble..."

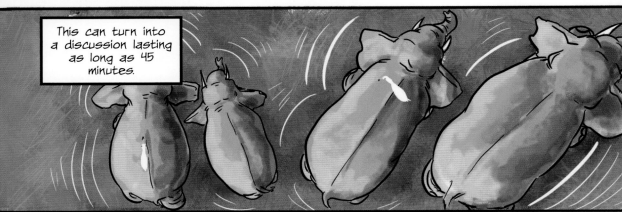

This can turn into a discussion lasting as long as 45 minutes.

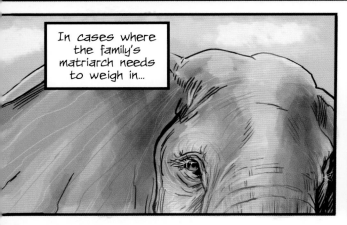

In cases where the family's matriarch needs to weigh in...

...members will adopt a waiting posture.

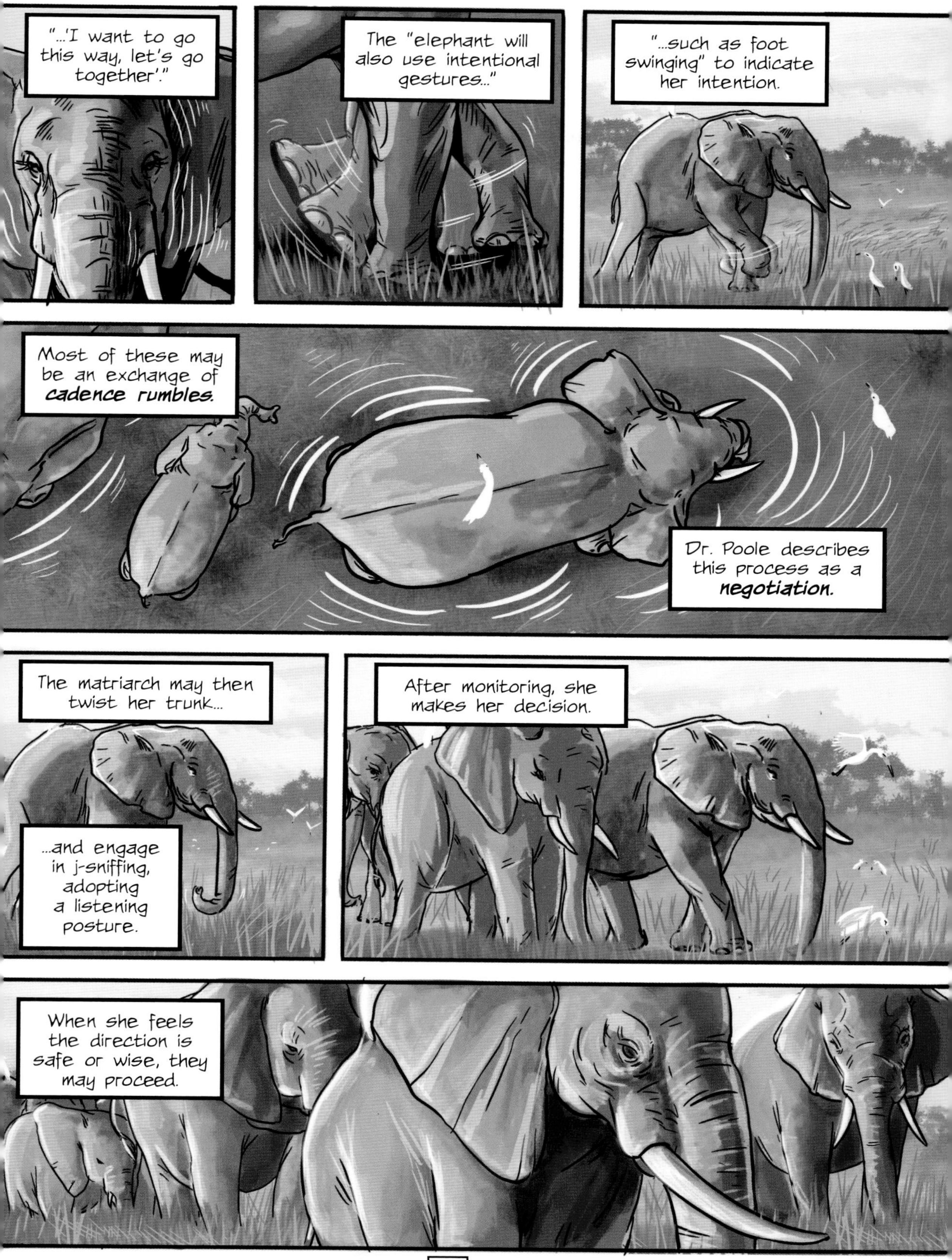
"...'I want to go this way, let's go together'."
The "elephant will also use intentional gestures..."
"...such as foot swinging" to indicate her intention.
Most of these may be an exchange of **cadence rumbles.**
Dr. Poole describes this process as a **negotiation.**
The matriarch may then twist her trunk...
...and engage in j-sniffing, adopting a listening posture.
After monitoring, she makes her decision.
When she feels the direction is safe or wise, they may proceed.

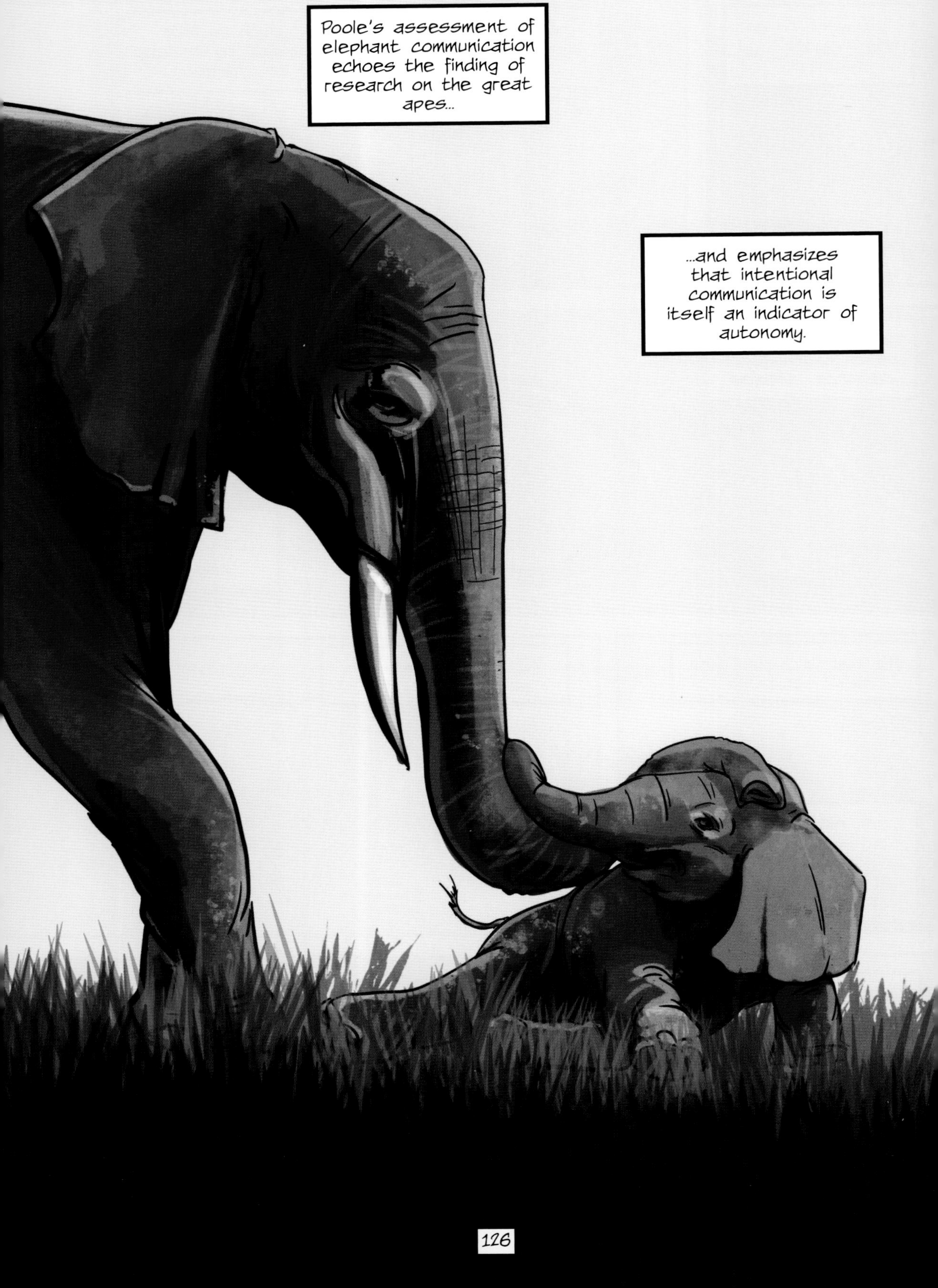
Poole's assessment of elephant communication echoes the finding of research on the great apes...
...and emphasizes that intentional communication is itself an indicator of autonomy.

THING

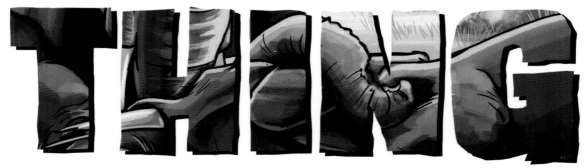

CHAPTER 6: EMPATHY

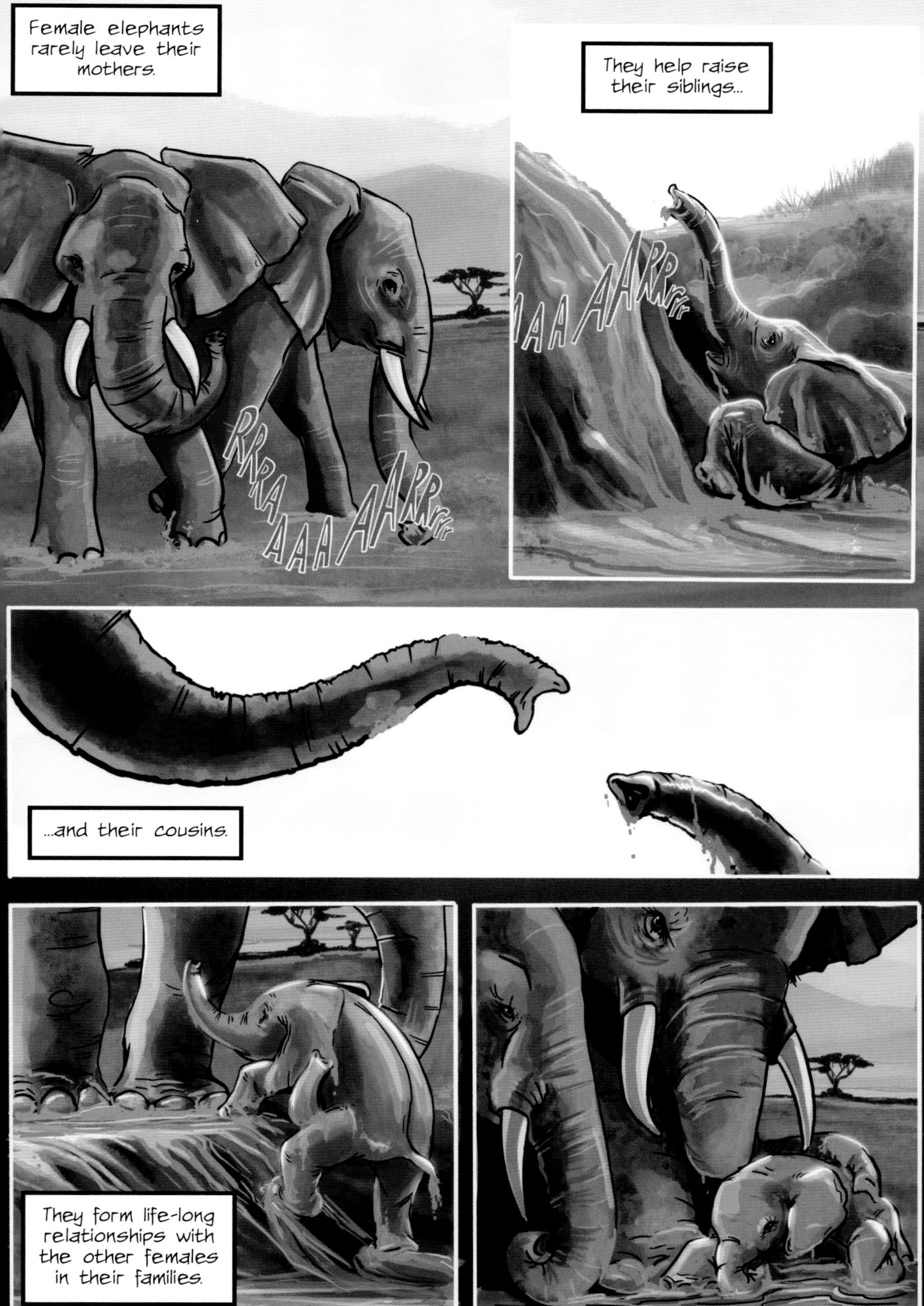
Female elephants rarely leave their mothers.
RRRRAAAAAARRrrr
They help raise their siblings...
AAAAARRrrr
...and their cousins.
They form life-long relationships with the other females in their families.

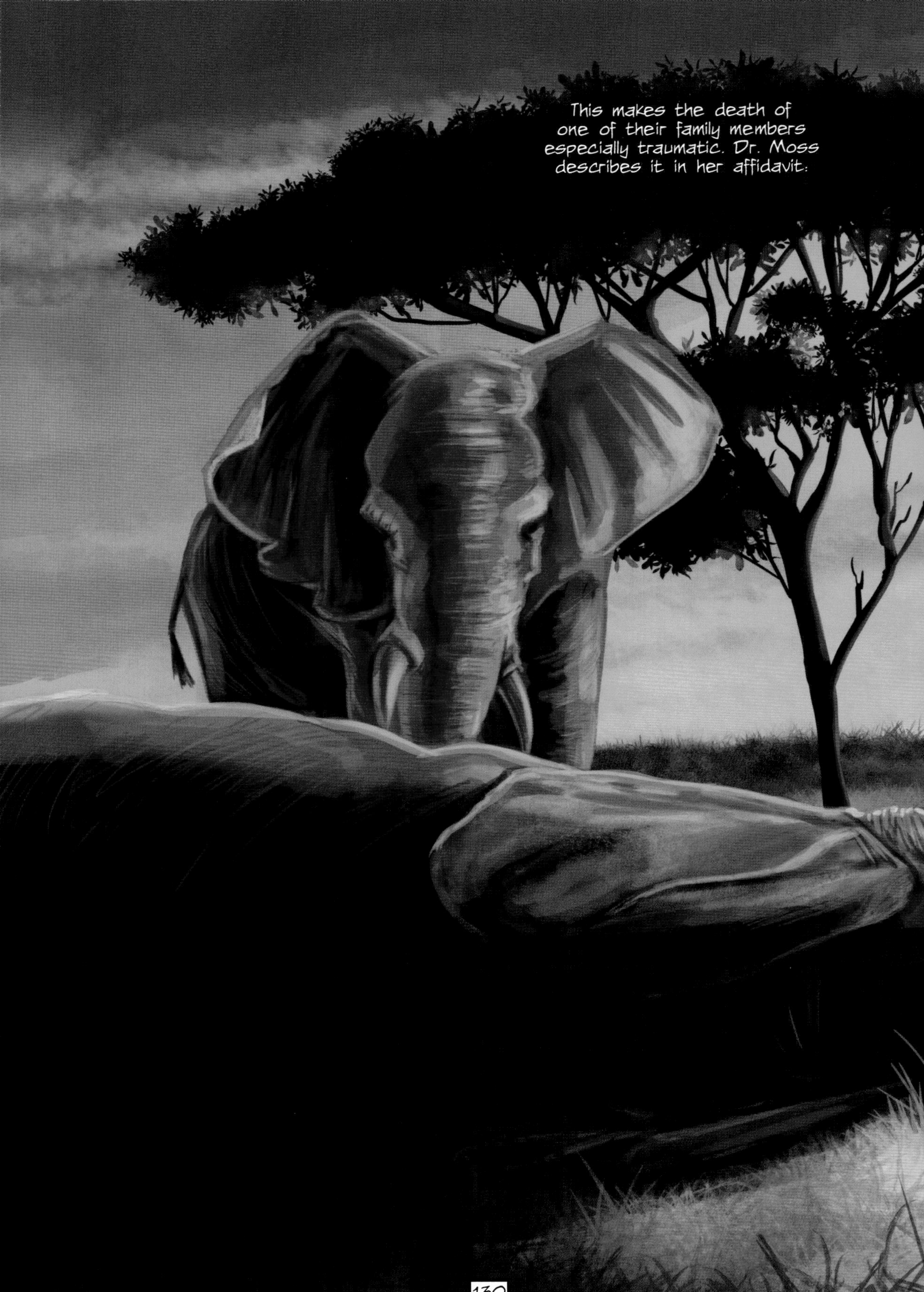
This makes the death of one of their family members especially traumatic. Dr. Moss describes it in her affidavit:

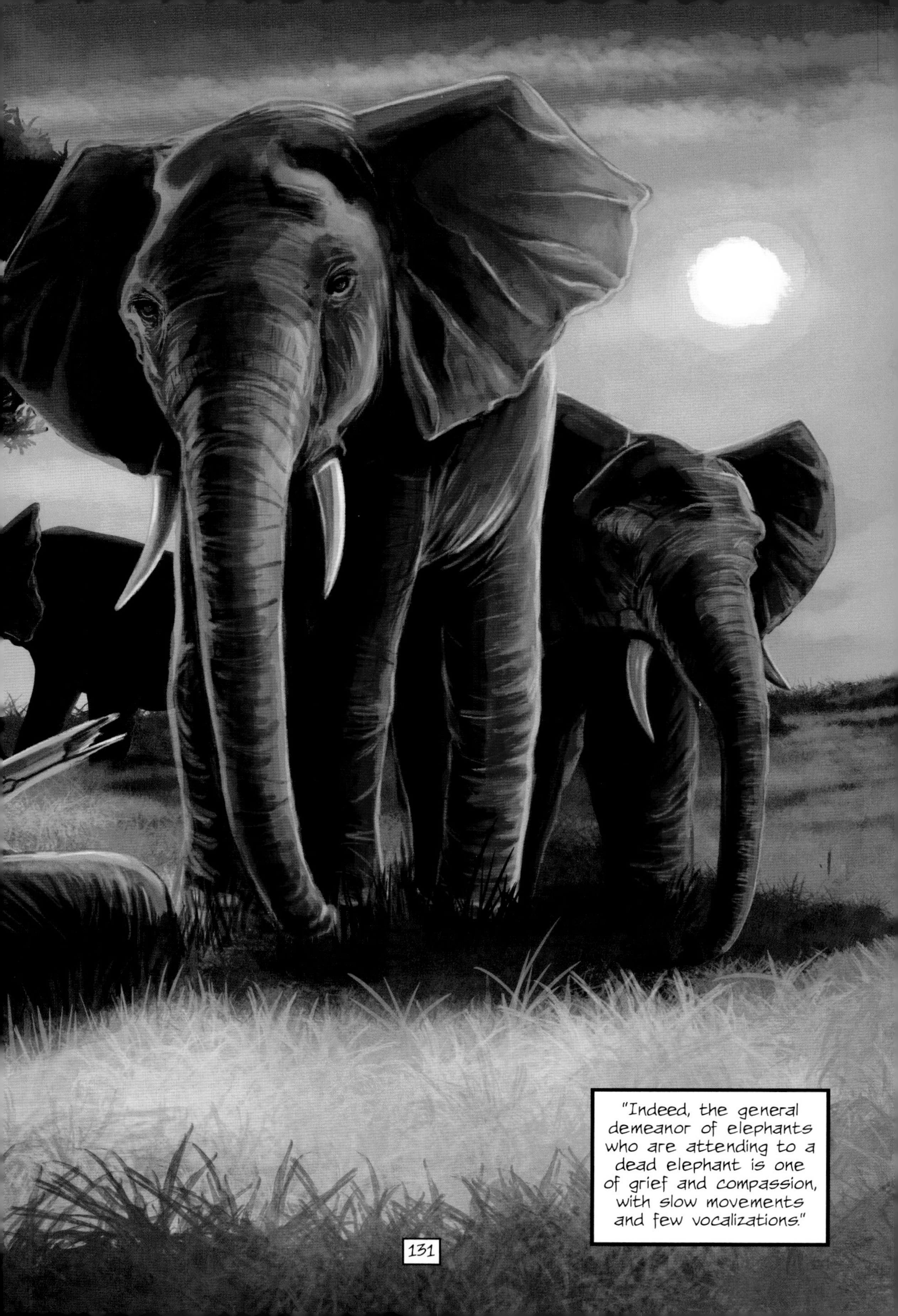
"Indeed, the general demeanor of elephants who are attending to a dead elephant is one of grief and compassion, with slow movements and few vocalizations."

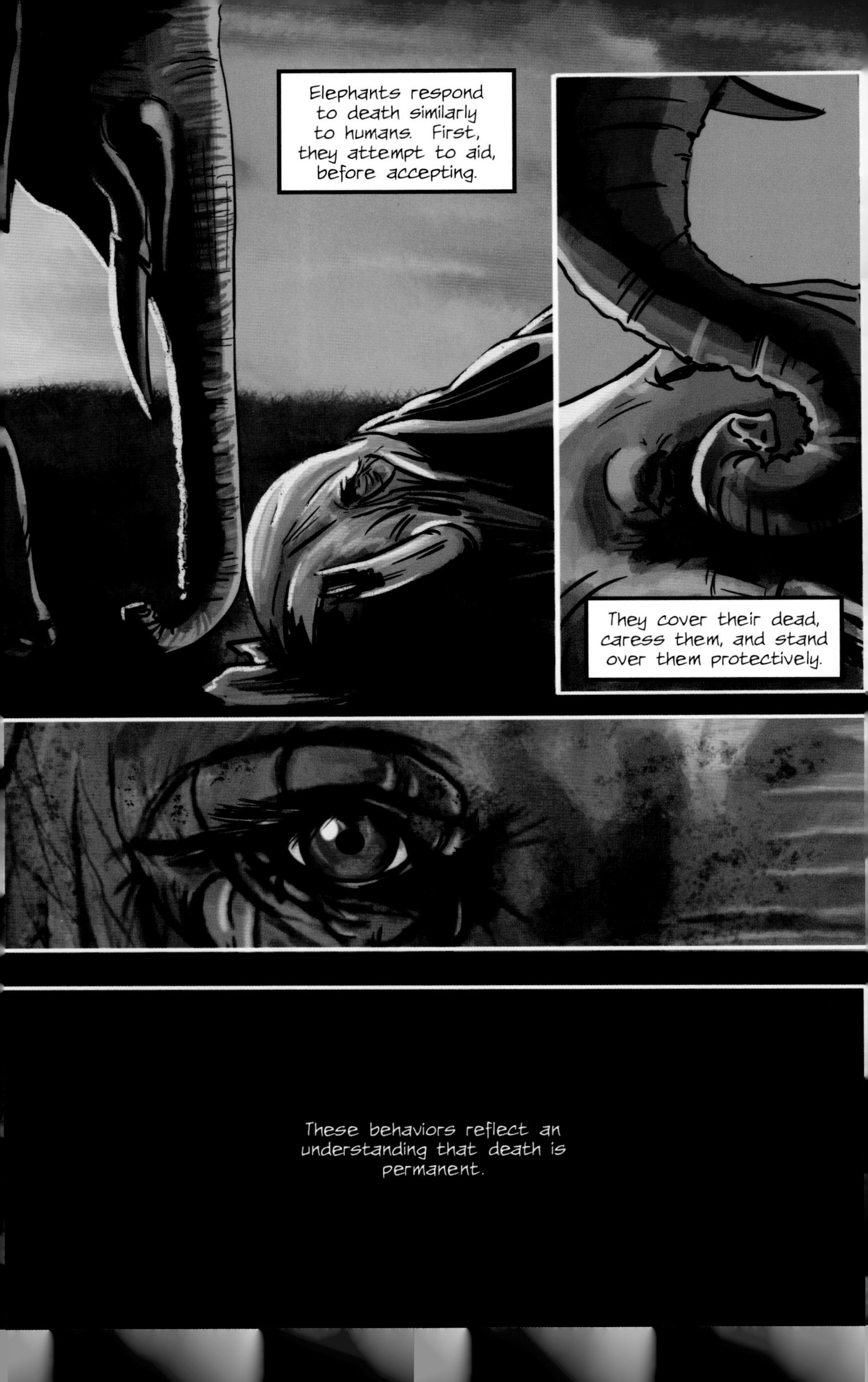
Elephants respond to death similarly to humans. First, they attempt to aid, before accepting.
They cover their dead, caress them, and stand over them protectively.
These behaviors reflect an understanding that death is permanent.

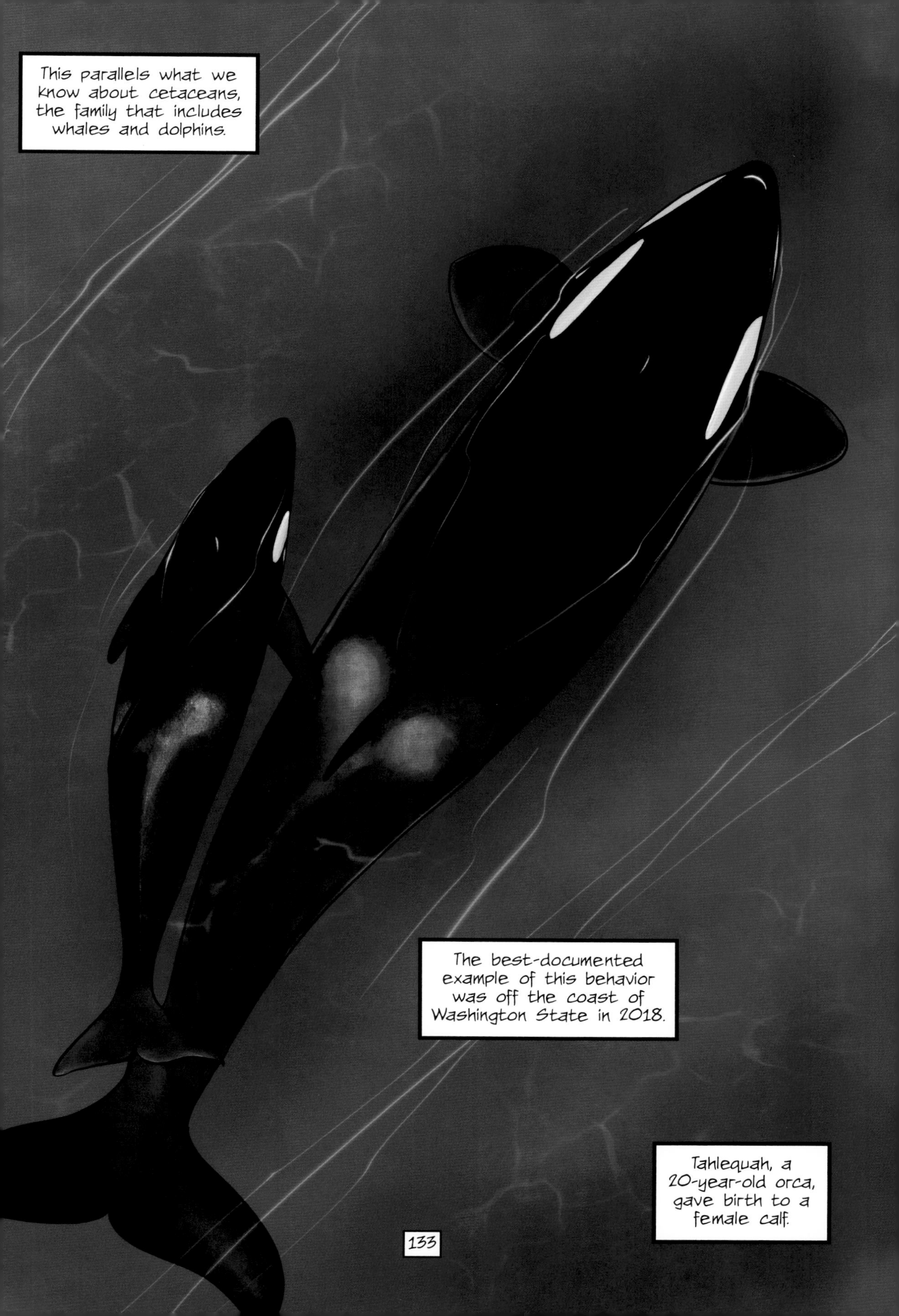
This parallels what we know about cetaceans, the family that includes whales and dolphins.
The best-documented example of this behavior was off the coast of Washington State in 2018.
Tahlequah, a 20-year-old orca, gave birth to a female calf.

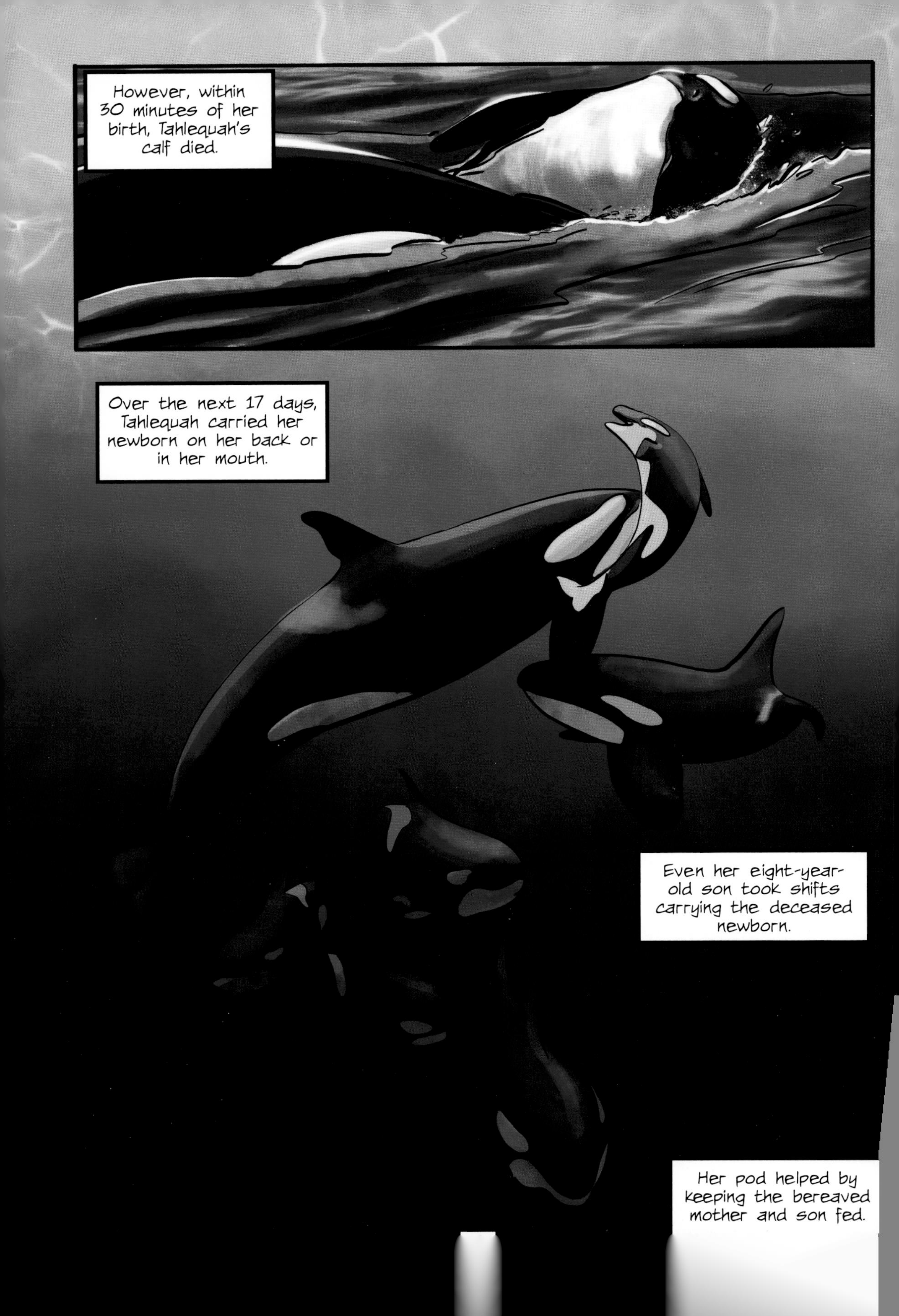
However, within 30 minutes of her birth, Tahlequah's calf died.
Over the next 17 days, Tahlequah carried her newborn on her back or in her mouth.
Even her eight-year-old son took shifts carrying the deceased newborn.
Her pod helped by keeping the bereaved mother and son fed.

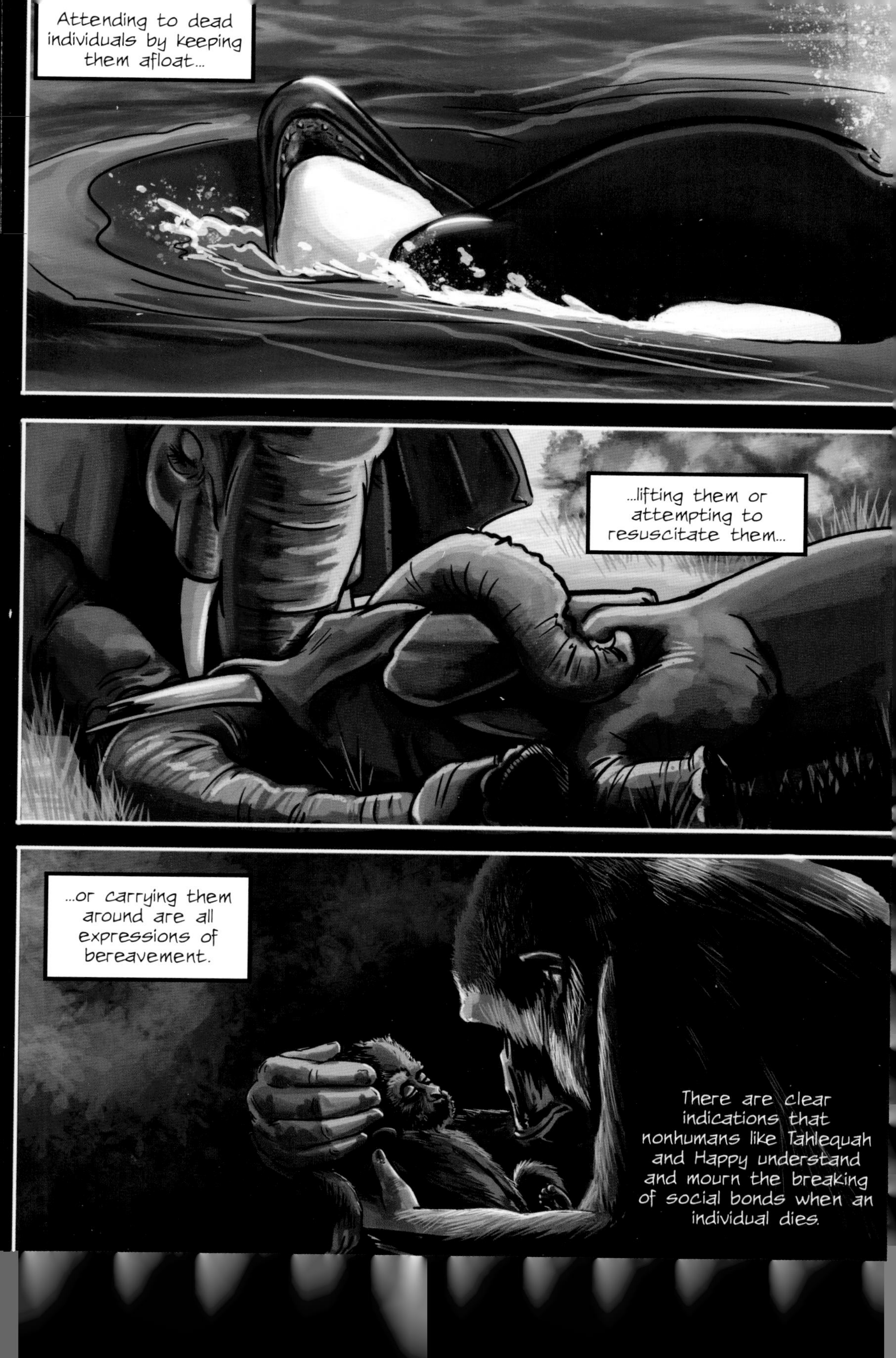
Attending to dead individuals by keeping them afloat...
...lifting them or attempting to resuscitate them...
...or carrying them around are all expressions of bereavement.
There are clear indications that nonhumans like Tahlequah and Happy understand and mourn the breaking of social bonds when an individual dies.

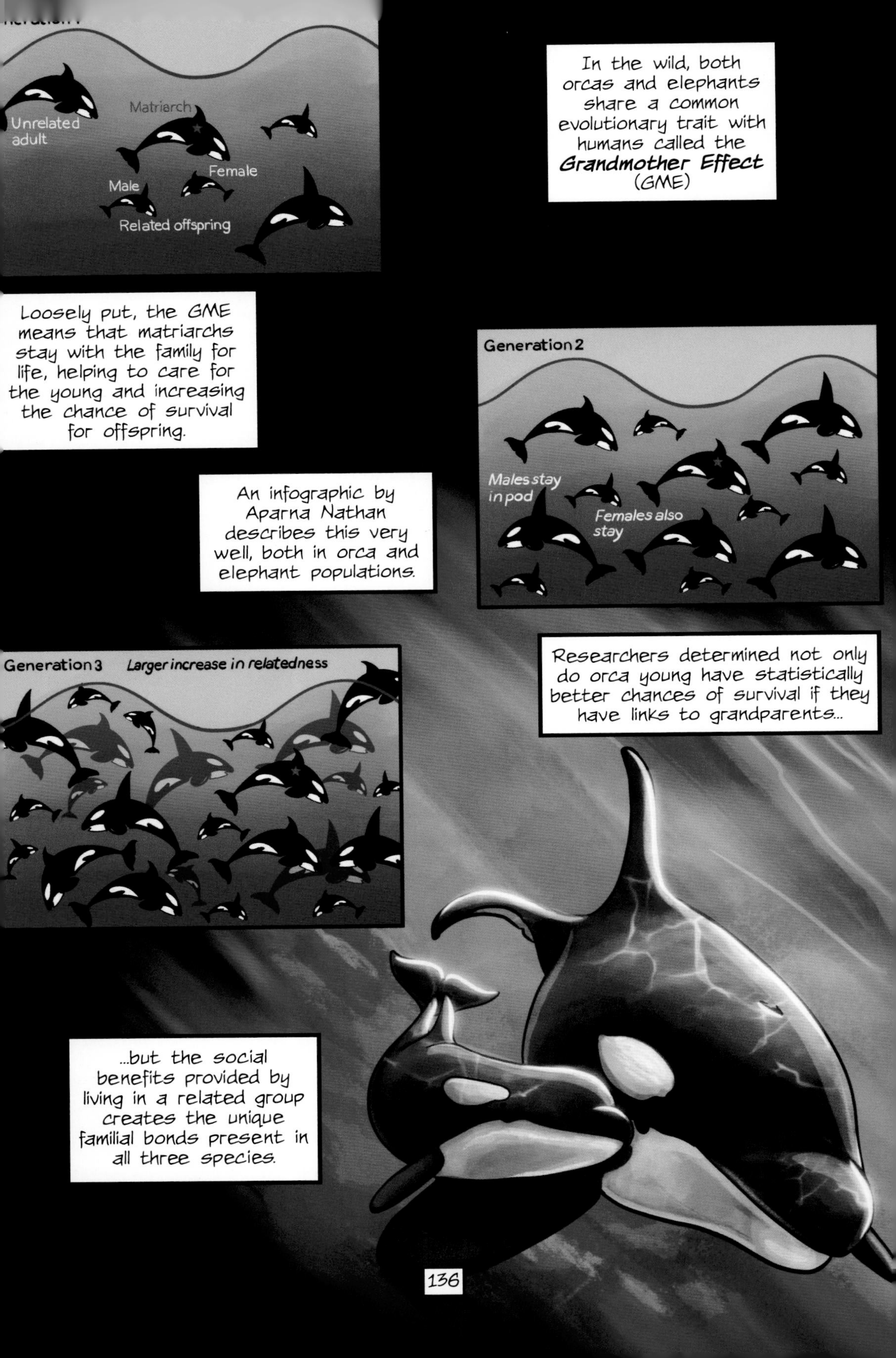

Unrelated adult
Matriarch
Female
Male
Related offspring
In the wild, both orcas and elephants share a common evolutionary trait with humans called the **Grandmother Effect** (GME)
Loosely put, the GME means that matriarchs stay with the family for life, helping to care for the young and increasing the chance of survival for offspring.
Generation 2
Males stay in pod
Females also stay
An infographic by Aparna Nathan describes this very well, both in orca and elephant populations.
Generation 3
Larger increase in relatedness
Researchers determined not only do orca young have statistically better chances of survival if they have links to grandparents...
...but the social benefits provided by living in a related group creates the unique familial bonds present in all three species.

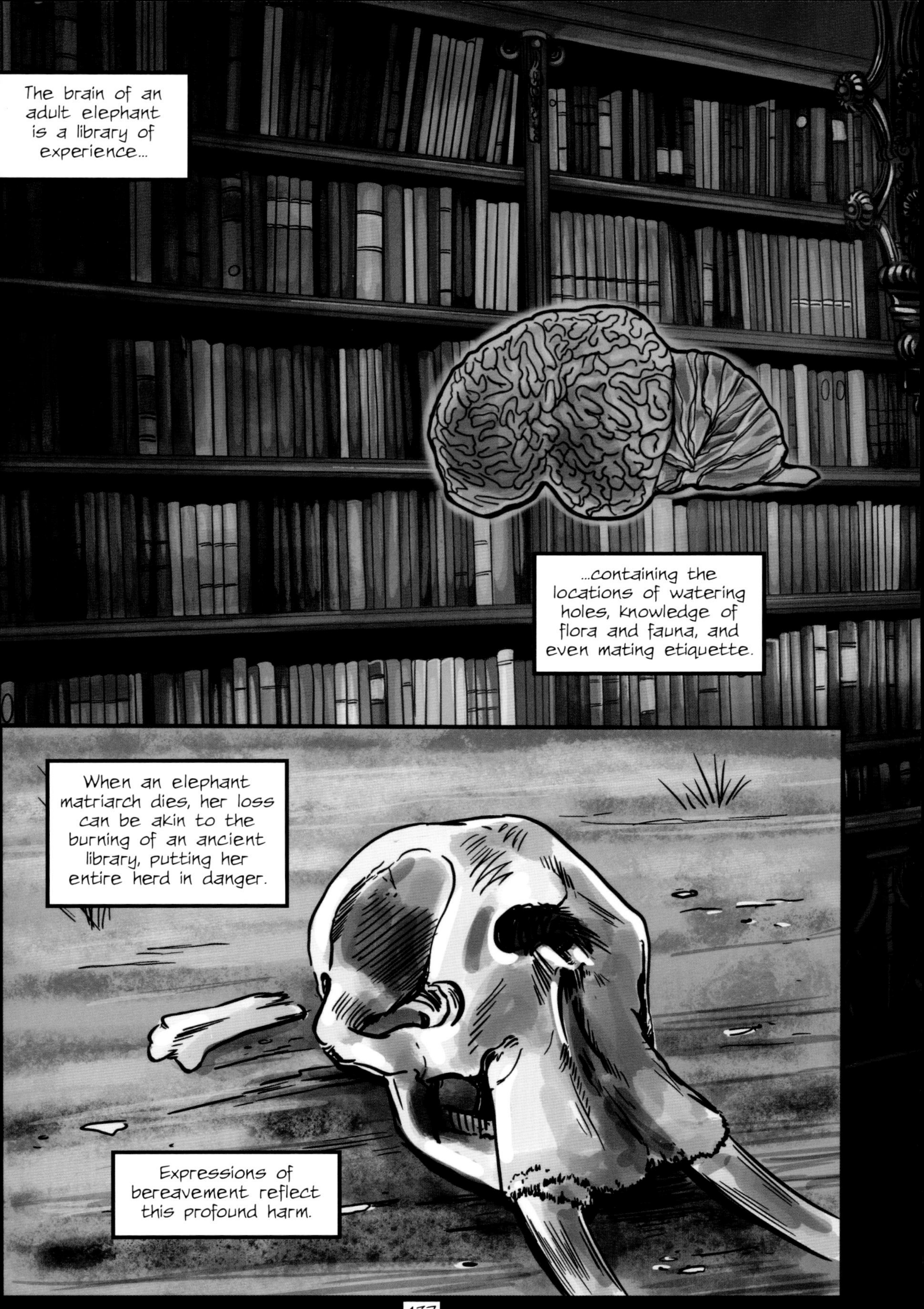
The brain of an adult elephant is a library of experience...
...containing the locations of watering holes, knowledge of flora and fauna, and even mating etiquette.
When an elephant matriarch dies, her loss can be akin to the burning of an ancient library, putting her entire herd in danger.
Expressions of bereavement reflect this profound harm.

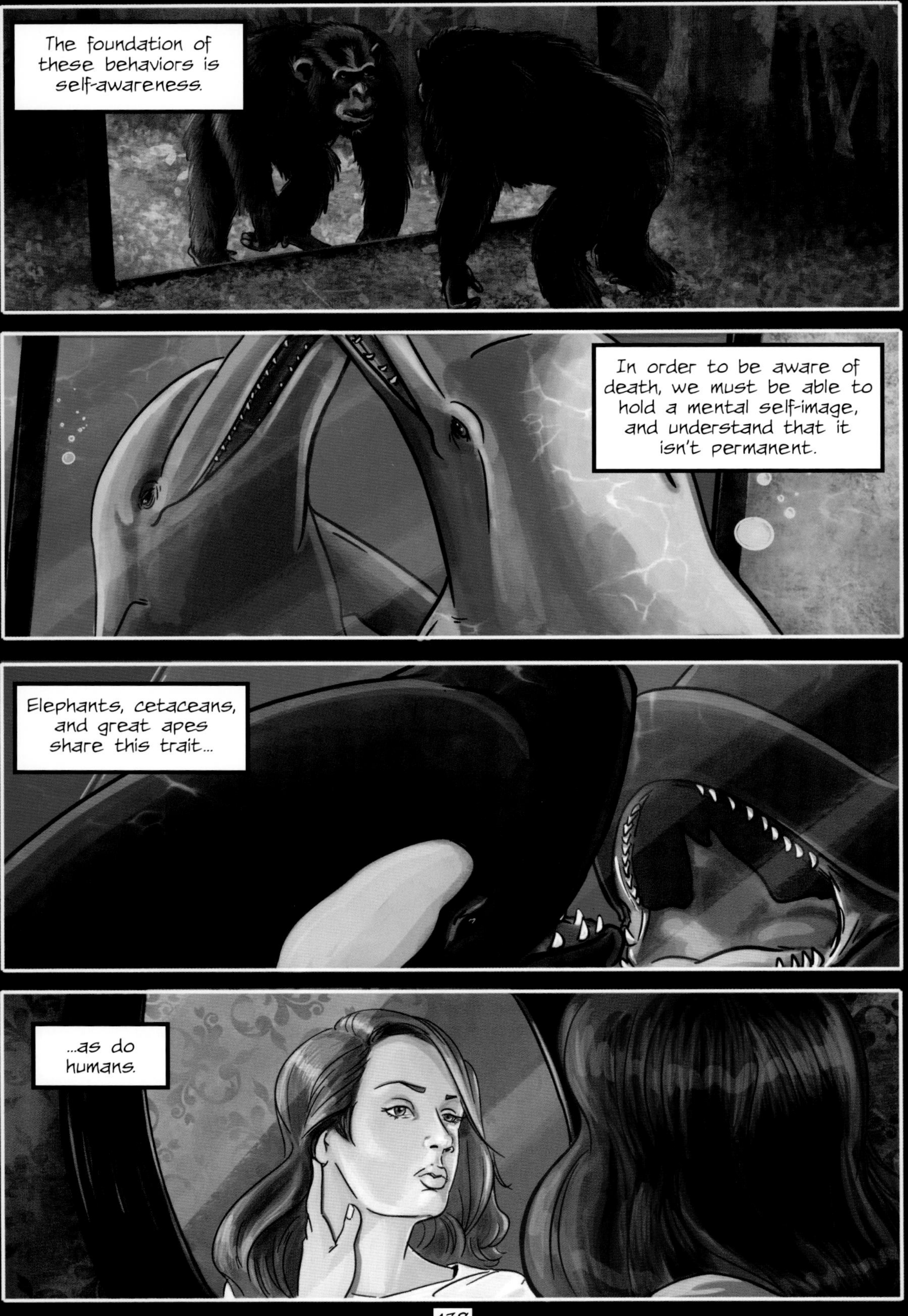

The foundation of these behaviors is self-awareness.
In order to be aware of death, we must be able to hold a mental self-image, and understand that it isn't permanent.
Elephants, cetaceans, and great apes share this trait...
...as do humans.

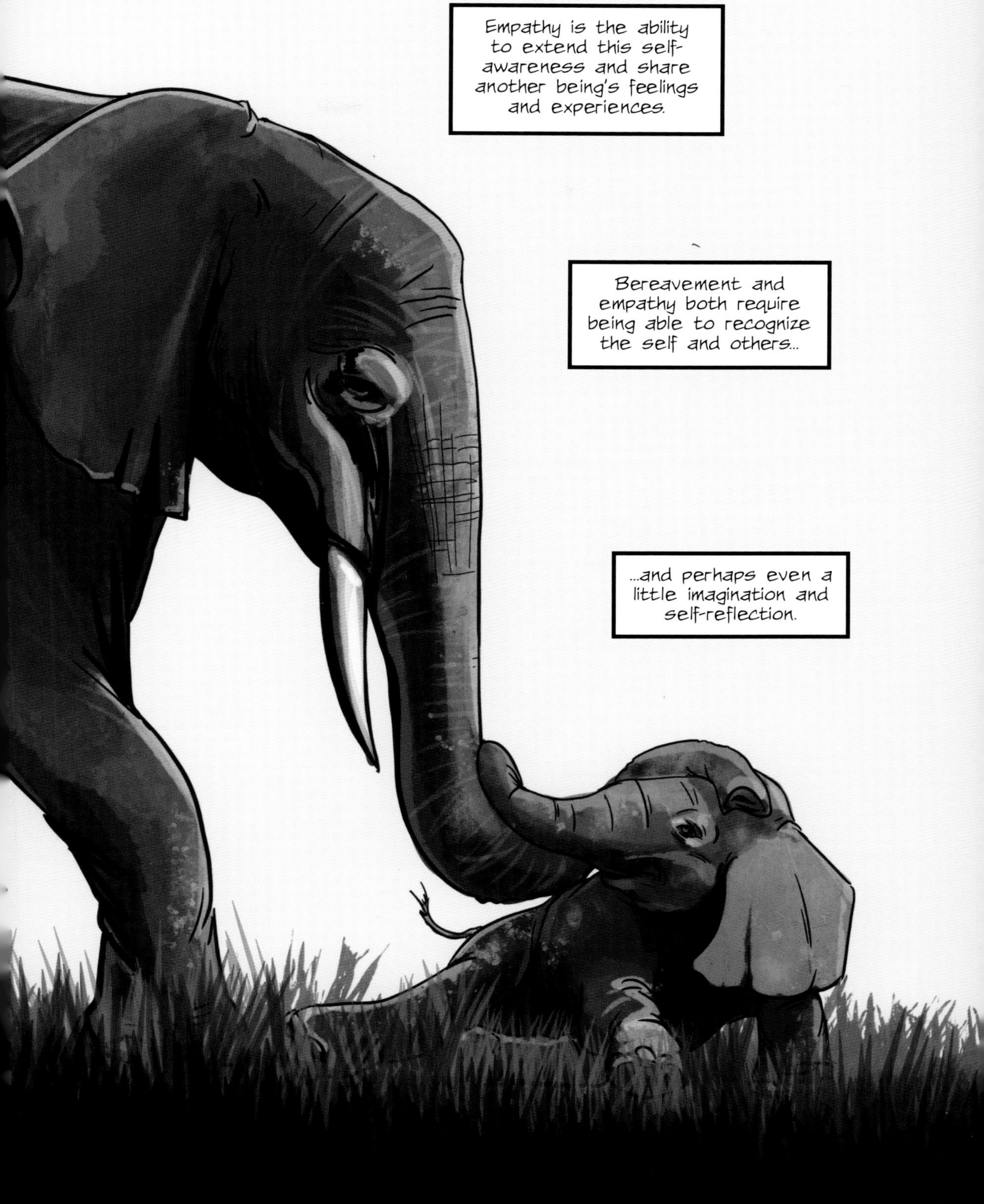
Empathy is the ability to extend this self-awareness and share another being's feelings and experiences.
Bereavement and empathy both require being able to recognize the self and others...
...and perhaps even a little imagination and self-reflection.

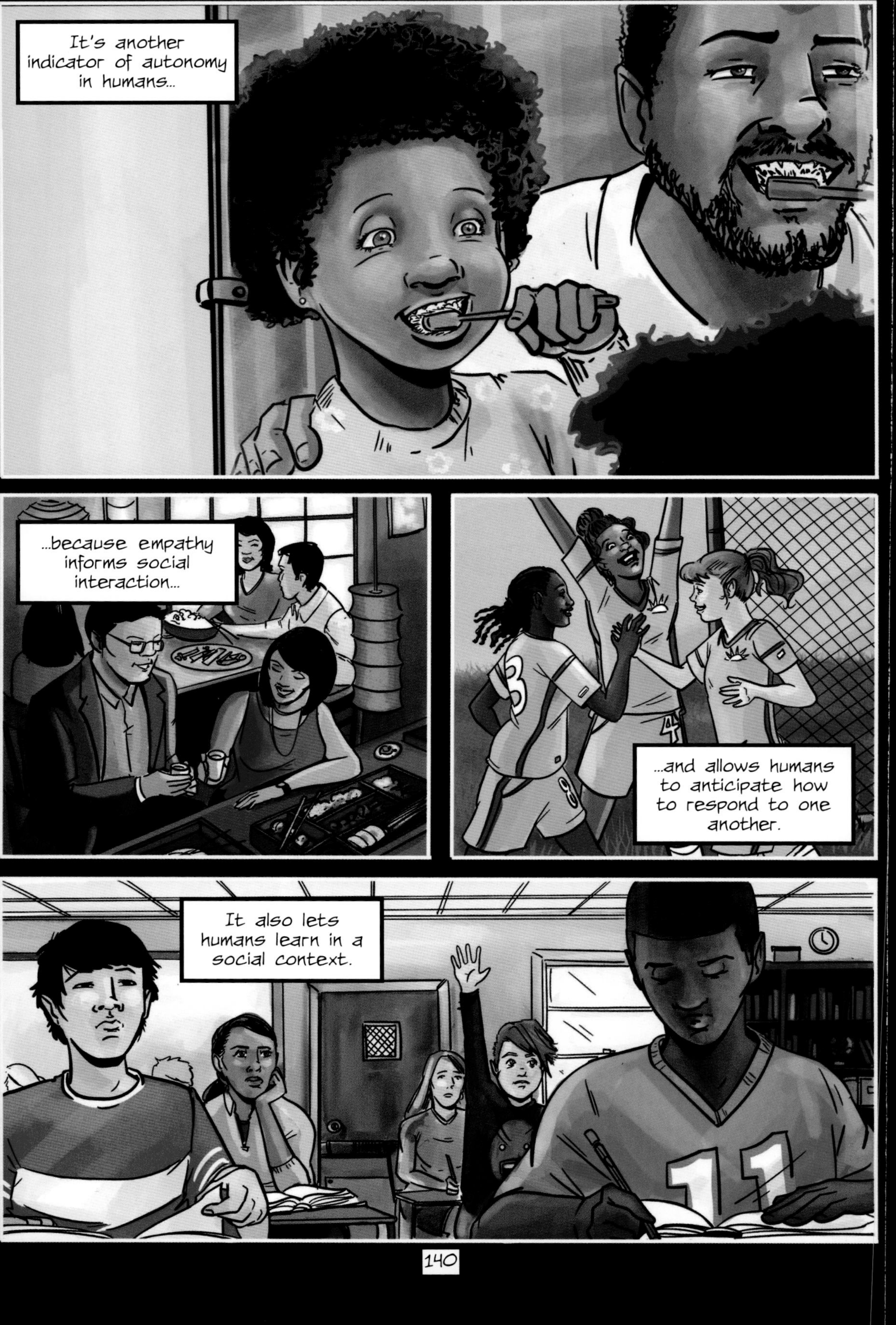
It's another indicator of autonomy in humans...
...because empathy informs social interaction...
...and allows humans to anticipate how to respond to one another.
It also lets humans learn in a social context.

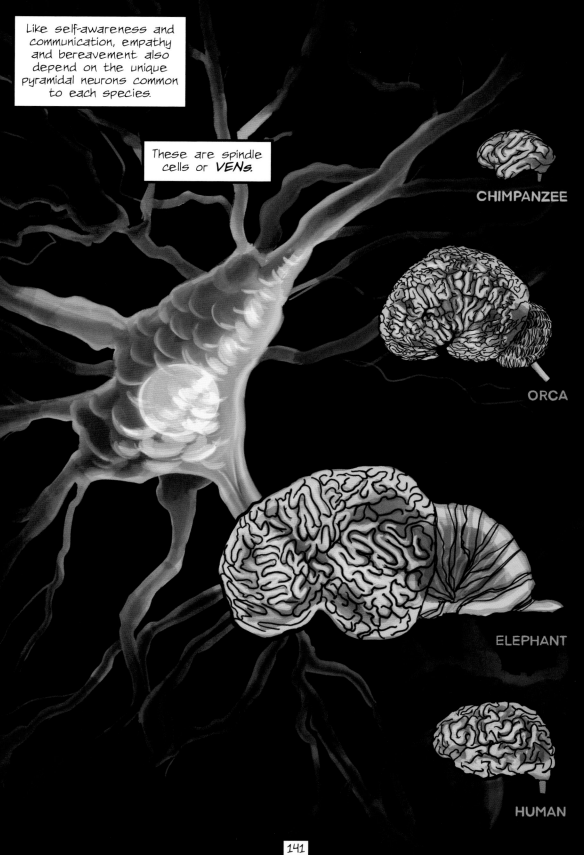

Like self-awareness and communication, empathy and bereavement also depend on the unique pyramidal neurons common to each species.

These are spindle cells or *VENs*.

CHIMPANZEE

ORCA

ELEPHANT

HUMAN

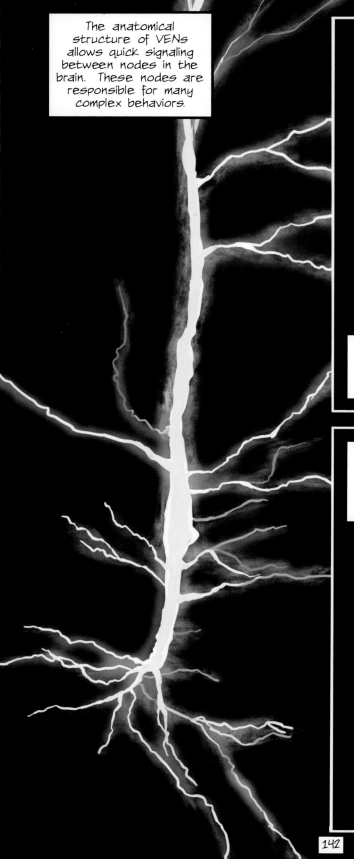

The anatomical structure of VENs allows quick signaling between nodes in the brain. These nodes are responsible for many complex behaviors.

Together, these nodes, neurons, and their features create the *salience network.*

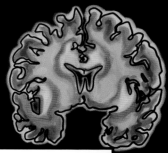

The salience network integrates sensory, emotional, and cognitive information...

...and adaptively guides complex functions like self-awareness, communication, social behavior, and emotions.

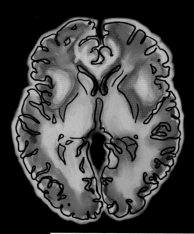

In humans, we associate these behaviors with autonomy.

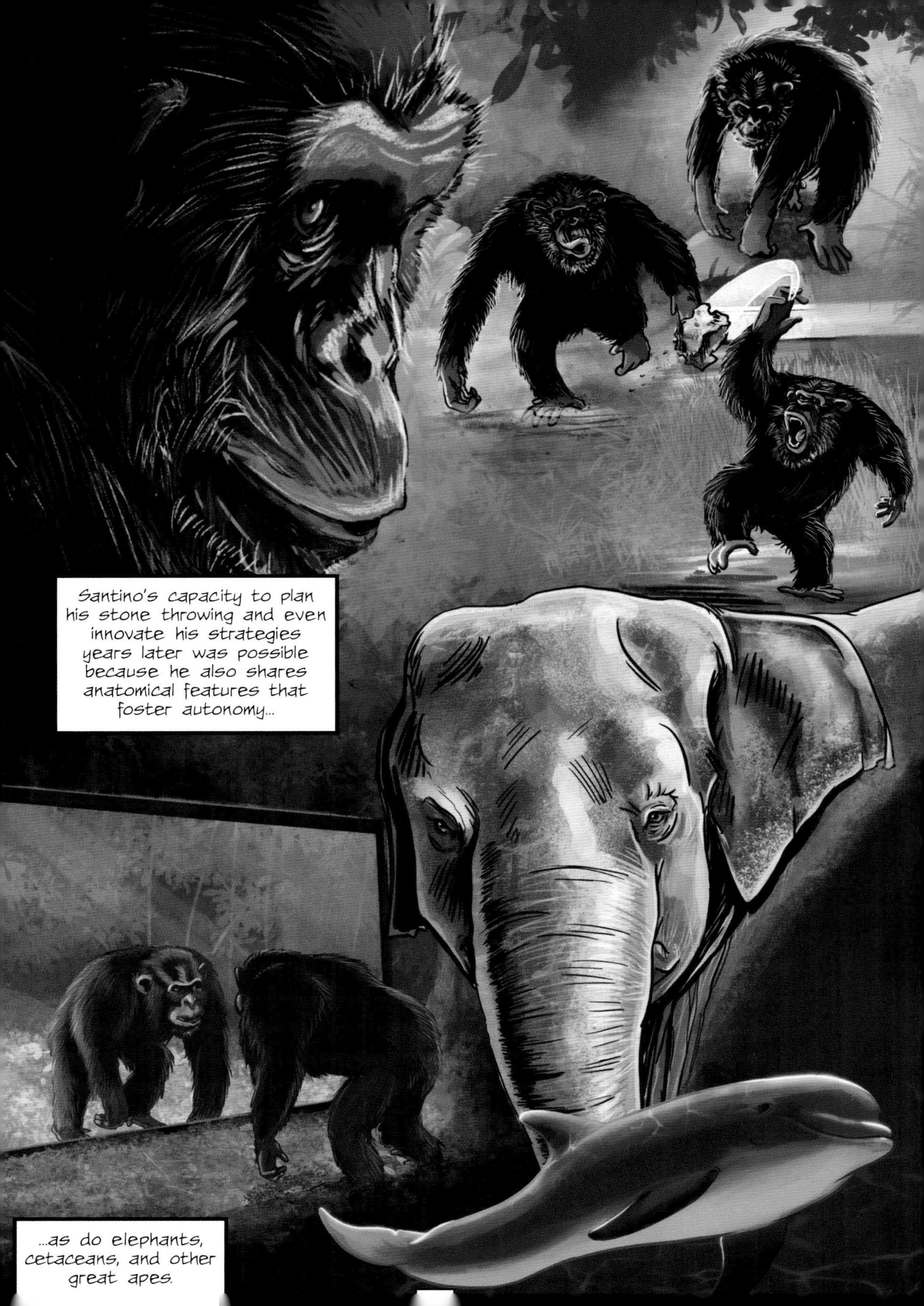
Santino's capacity to plan his stone throwing and even innovate his strategies years later was possible because he also shares anatomical features that foster autonomy...
...as do elephants, cetaceans, and other great apes.

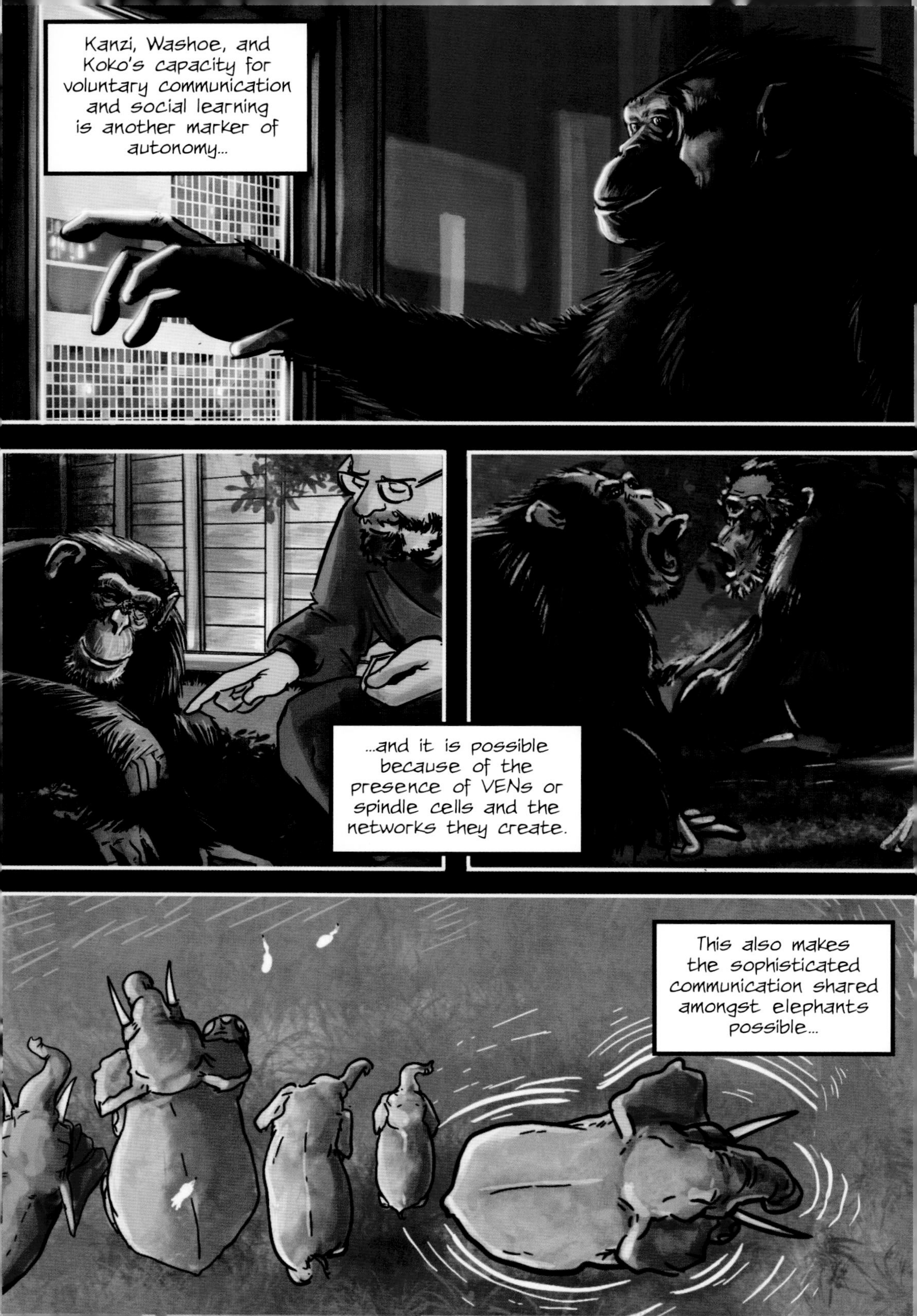
Kanzi, Washoe, and Koko's capacity for voluntary communication and social learning is another marker of autonomy...
...and it is possible because of the presence of VENs or spindle cells and the networks they create.
This also makes the sophisticated communication shared amongst elephants possible...

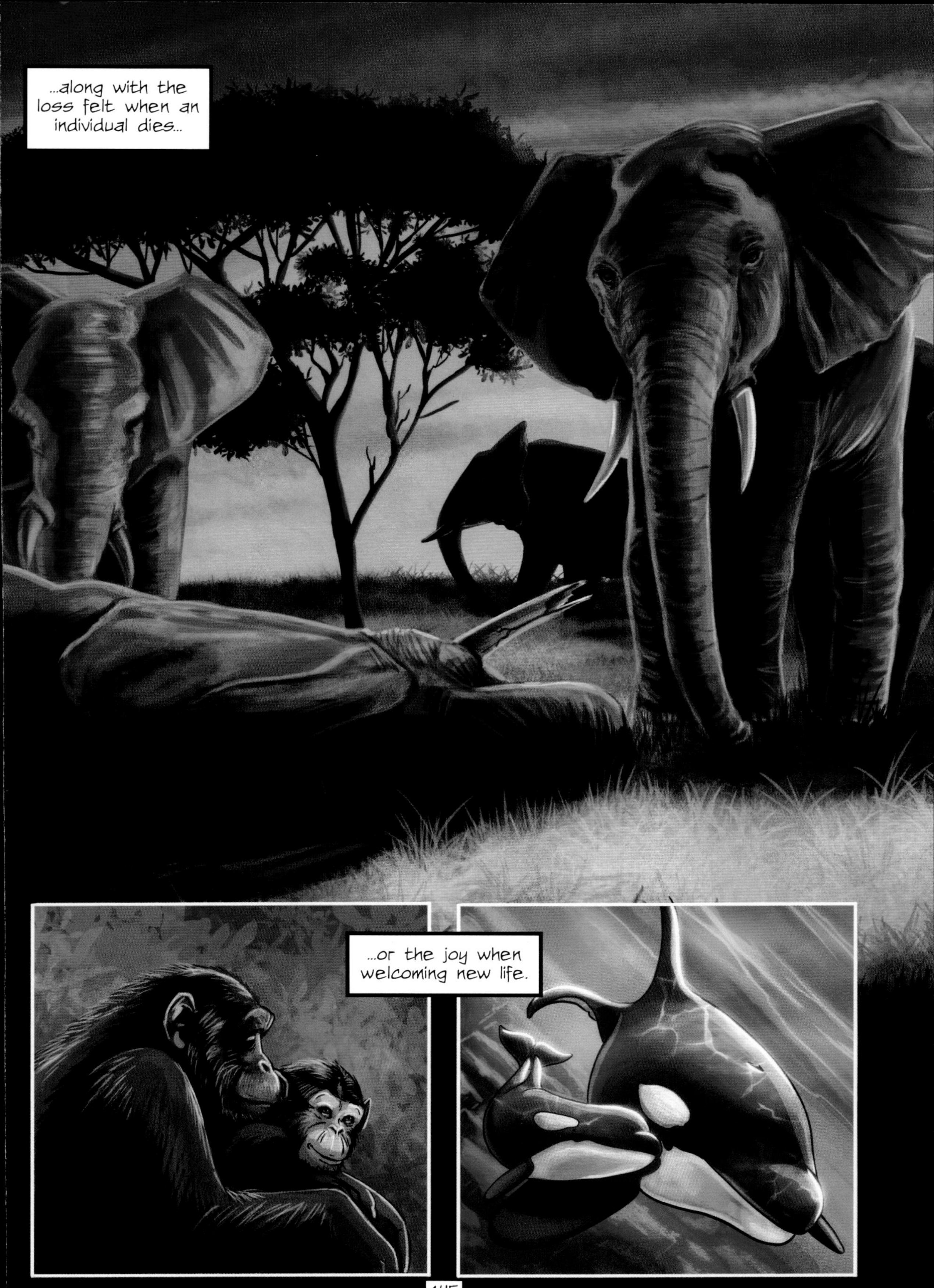
...along with the loss felt when an individual dies...
...or the joy when welcoming new life.

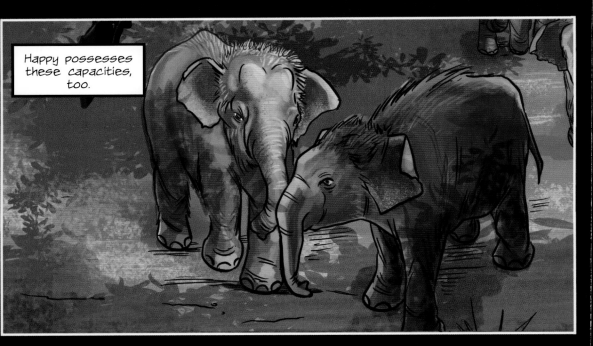

Happy possesses these capacities, too.

She may recall her capture or training...

...and her life-long relationship with Grumpy.

And she likely mourned Grumpy's death.

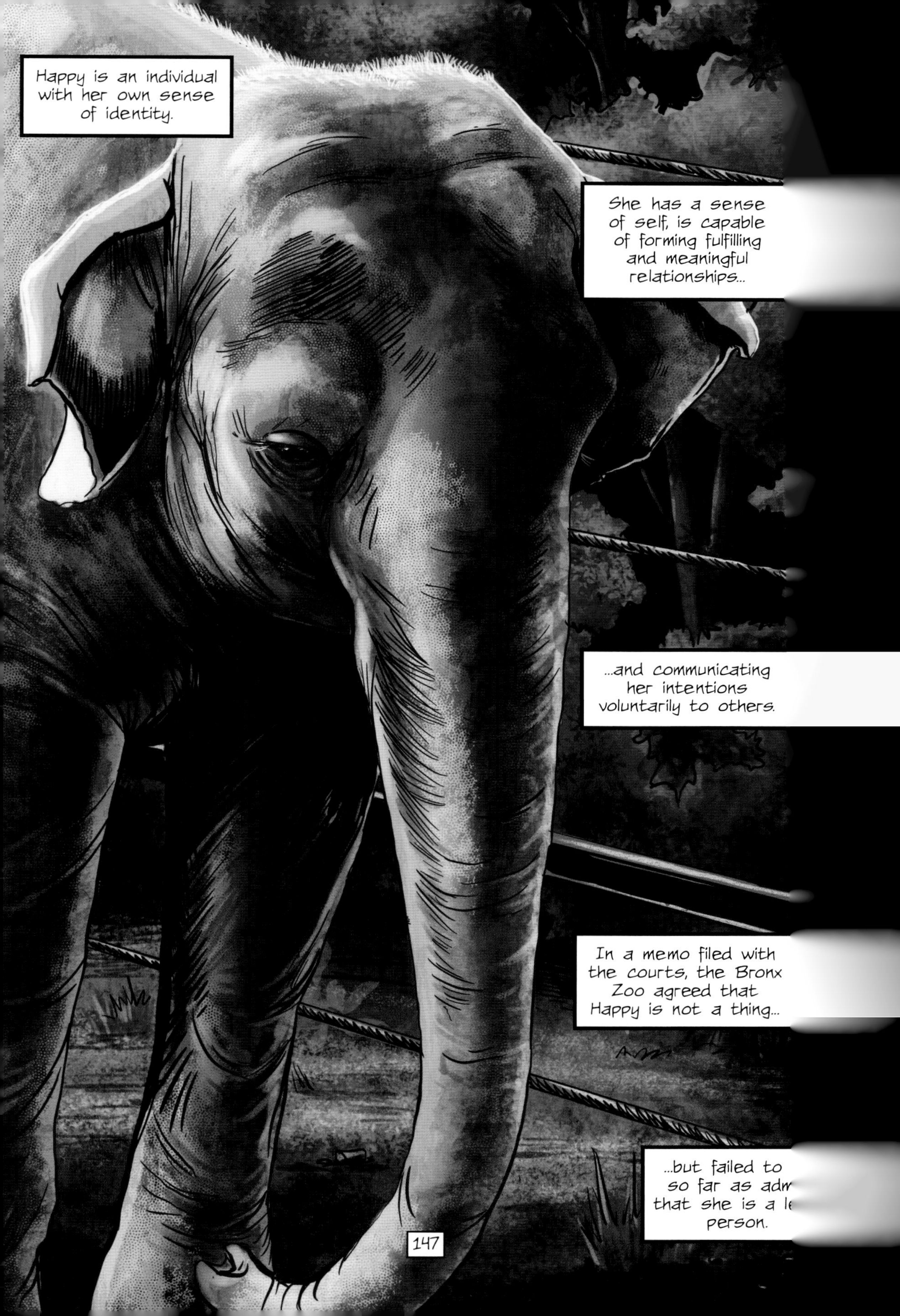
Happy is an individual with her own sense of identity.
She has a sense of self, is capable of forming fulfilling and meaningful relationships...
...and communicating her intentions voluntarily to others.
In a memo filed with the courts, the Bronx Zoo agreed that Happy is not a thing...
...but failed to so far as adm that she is a le person.

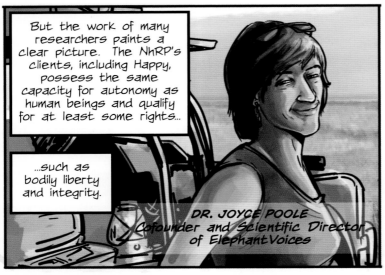

But the work of many researchers paints a clear picture. The NhRP's clients, including Happy, possess the same capacity for autonomy as human beings and qualify for at least some rights...

...such as bodily liberty and integrity.

DR. JOYCE POOLE
Cofounder and Scientific Director of ElephantVoices

DR. CYNTHIA MOSS
Director of the Amboseli Elephant Research Project

DR. MARY LEE JENSVOLD
Experimental Psychology Central Washington University

DR. WILLIAM C. MCGREW
Emeritus Professor of Evolutionary Primatology University of Cambridge

DR. JAMES R. ANDERSON
Professor of Psychology Kyoto University

DR. CHRISTOPHER BOESCH
Director of Department of Primatology Max Planck Institute for Evolutionary Anthropology

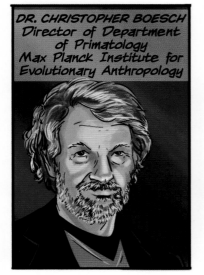

DR. MATHIAS OSVATH
Associate Professor of Cognitive Zoology at Lund University

DR. SUE SAVAGE RUMBAUGH
Scientific Advisor at The Great Ape Project

148

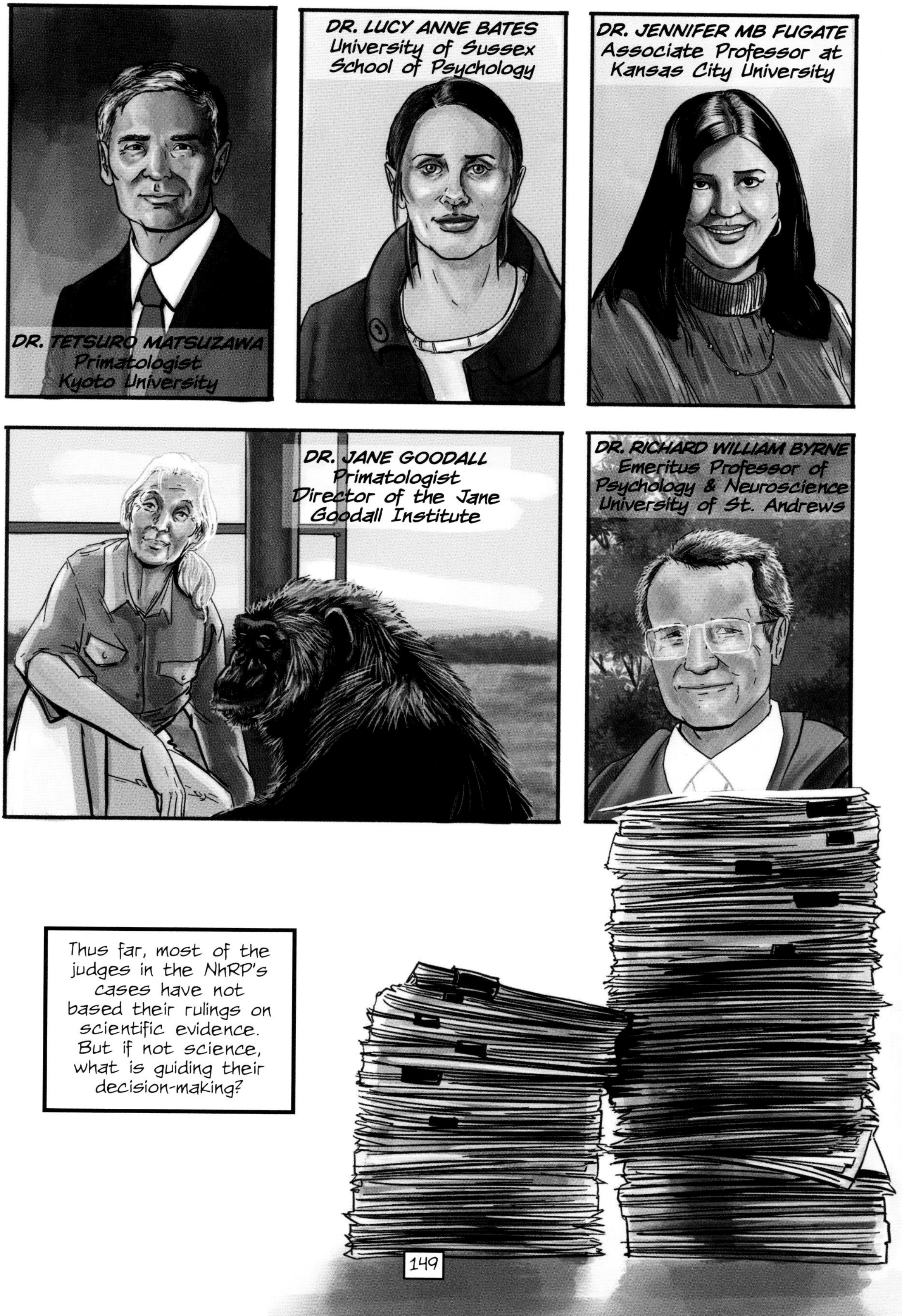
DR. TETSURO MATSUZAWA
Primatologist
Kyoto University
DR. LUCY ANNE BATES
University of Sussex
School of Psychology
DR. JENNIFER MB FUGATE
Associate Professor at
Kansas City University
DR. JANE GOODALL
Primatologist
Director of the Jane
Goodall Institute
DR. RICHARD WILLIAM BYRNE
Emeritus Professor of
Psychology & Neuroscience
University of St. Andrews
Thus far, most of the judges in the NhRP's cases have not based their rulings on scientific evidence. But if not science, what is guiding their decision-making?

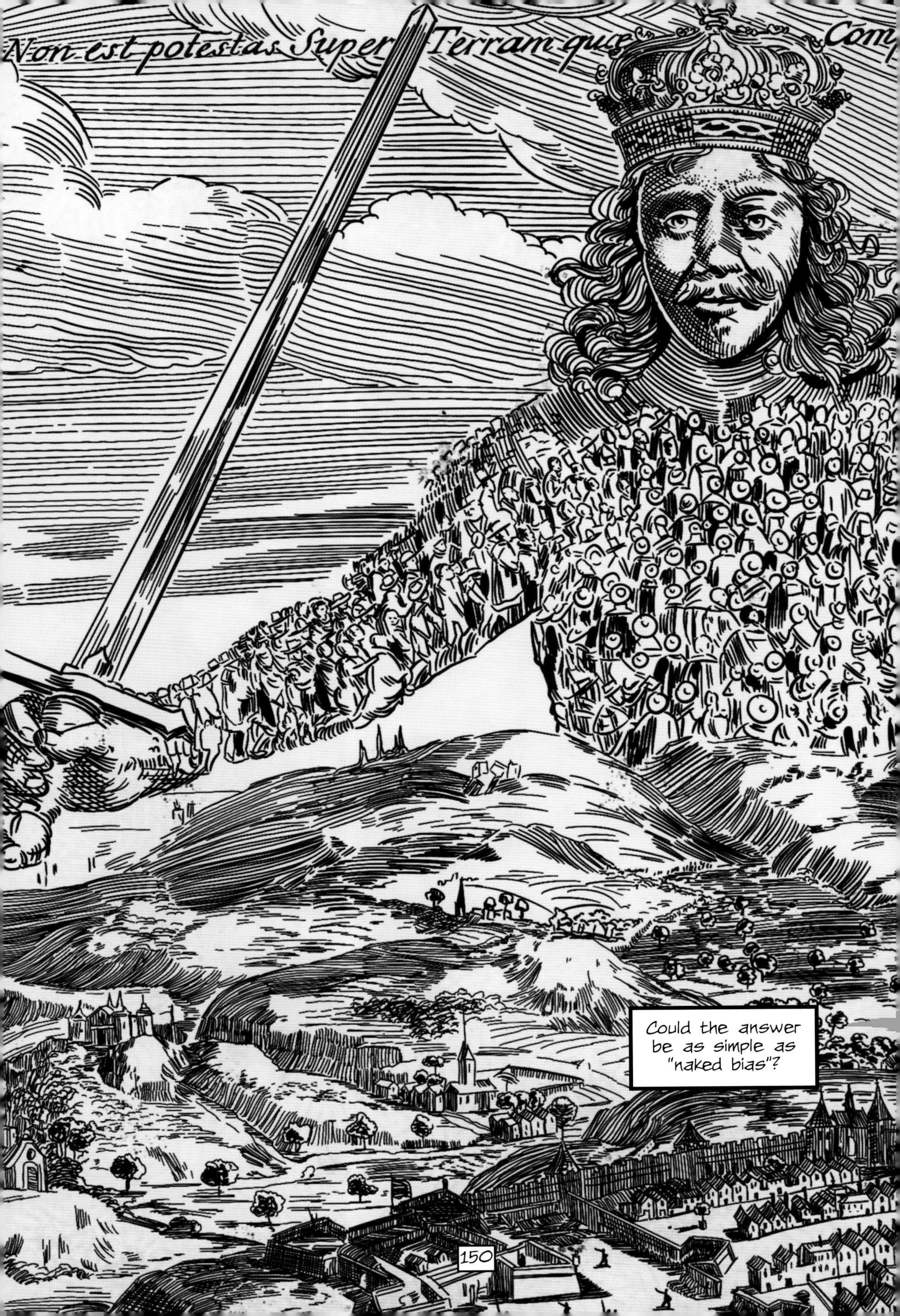
Non est potestas Super Terram quæ Com
Could the answer be as simple as "naked bias"?

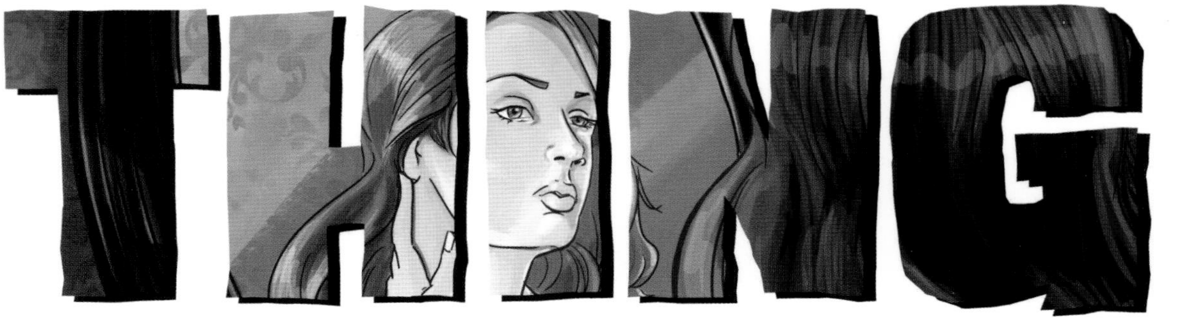
THING

CHAPTER 7: NAKED BIAS

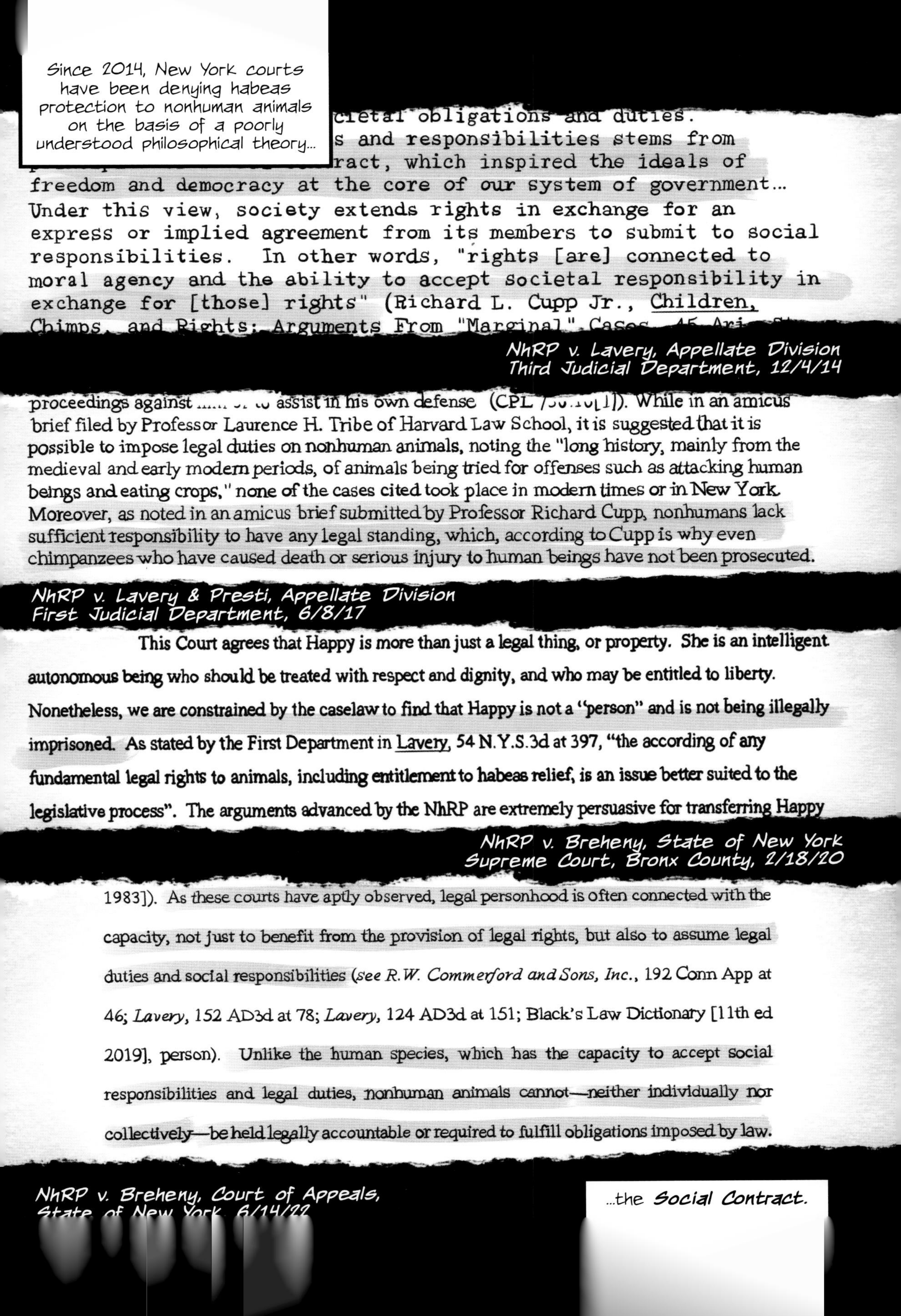
Since 2014, New York courts have been denying habeas protection to nonhuman animals on the basis of a poorly understood philosophical theory...
cietal obligations and duties.
s and responsibilities stems from
ract, which inspired the ideals of
freedom and democracy at the core of our system of government... Under this view, society extends rights in exchange for an express or implied agreement from its members to submit to social responsibilities. In other words, "rights [are] connected to moral agency and the ability to accept societal responsibility in exchange for [those] rights" (Richard L. Cupp Jr., Children, Chimps, and Rights: Arguments From "Marginal" Cases
NhRP v. Lavery, Appellate Division
Third Judicial Department, 12/4/14
proceedings against … assist in his own defense (CPL …). While in an amicus brief filed by Professor Laurence H. Tribe of Harvard Law School, it is suggested that it is possible to impose legal duties on nonhuman animals, noting the "long history, mainly from the medieval and early modern periods, of animals being tried for offenses such as attacking human beings and eating crops," none of the cases cited took place in modern times or in New York. Moreover, as noted in an amicus brief submitted by Professor Richard Cupp, nonhumans lack sufficient responsibility to have any legal standing, which, according to Cupp is why even chimpanzees who have caused death or serious injury to human beings have not been prosecuted.
NhRP v. Lavery & Presti, Appellate Division
First Judicial Department, 6/8/17
This Court agrees that Happy is more than just a legal thing, or property. She is an intelligent autonomous being who should be treated with respect and dignity, and who may be entitled to liberty. Nonetheless, we are constrained by the caselaw to find that Happy is not a "person" and is not being illegally imprisoned. As stated by the First Department in Lavery, 54 N.Y.S.3d at 397, "the according of any fundamental legal rights to animals, including entitlement to habeas relief, is an issue better suited to the legislative process". The arguments advanced by the NhRP are extremely persuasive for transferring Happy
NhRP v. Breheny, State of New York
Supreme Court, Bronx County, 2/18/20
1983]). As these courts have aptly observed, legal personhood is often connected with the capacity, not just to benefit from the provision of legal rights, but also to assume legal duties and social responsibilities (see R.W. Commerford and Sons, Inc., 192 Conn App at 46; Lavery, 152 AD3d at 78; Lavery, 124 AD3d at 151; Black's Law Dictionary [11th ed 2019], person). Unlike the human species, which has the capacity to accept social responsibilities and legal duties, nonhuman animals cannot—neither individually nor collectively—be held legally accountable or required to fulfill obligations imposed by law.
NhRP v. Breheny, Court of Appeals,
State of New York, 6/14/22
...the Social Contract.

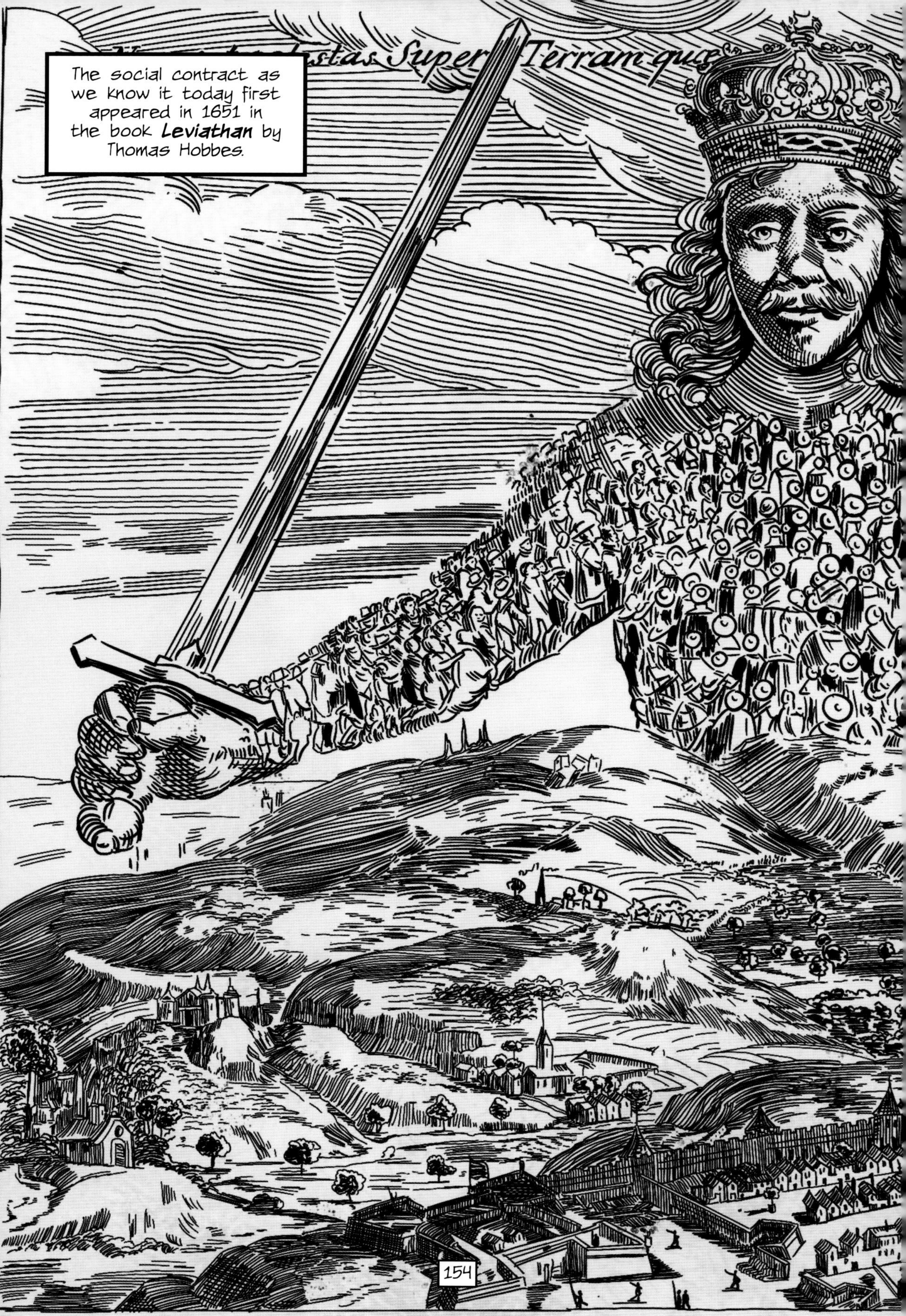
The social contract as we know it today first appeared in 1651 in the book **Leviathan** by Thomas Hobbes.

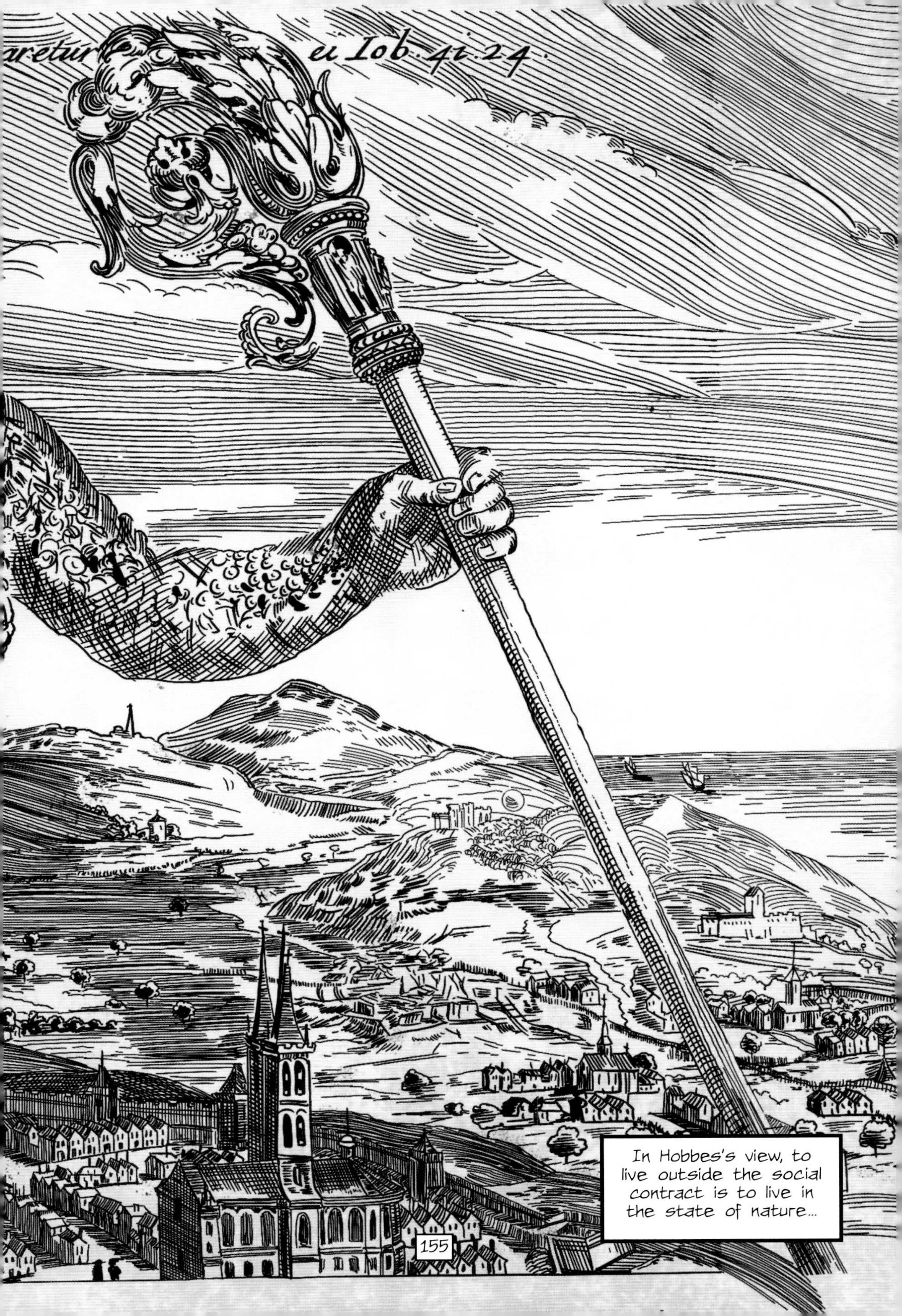
aretur ... ei Iob. 41. 24.
In Hobbes's view, to live outside the social contract is to live in the state of nature...

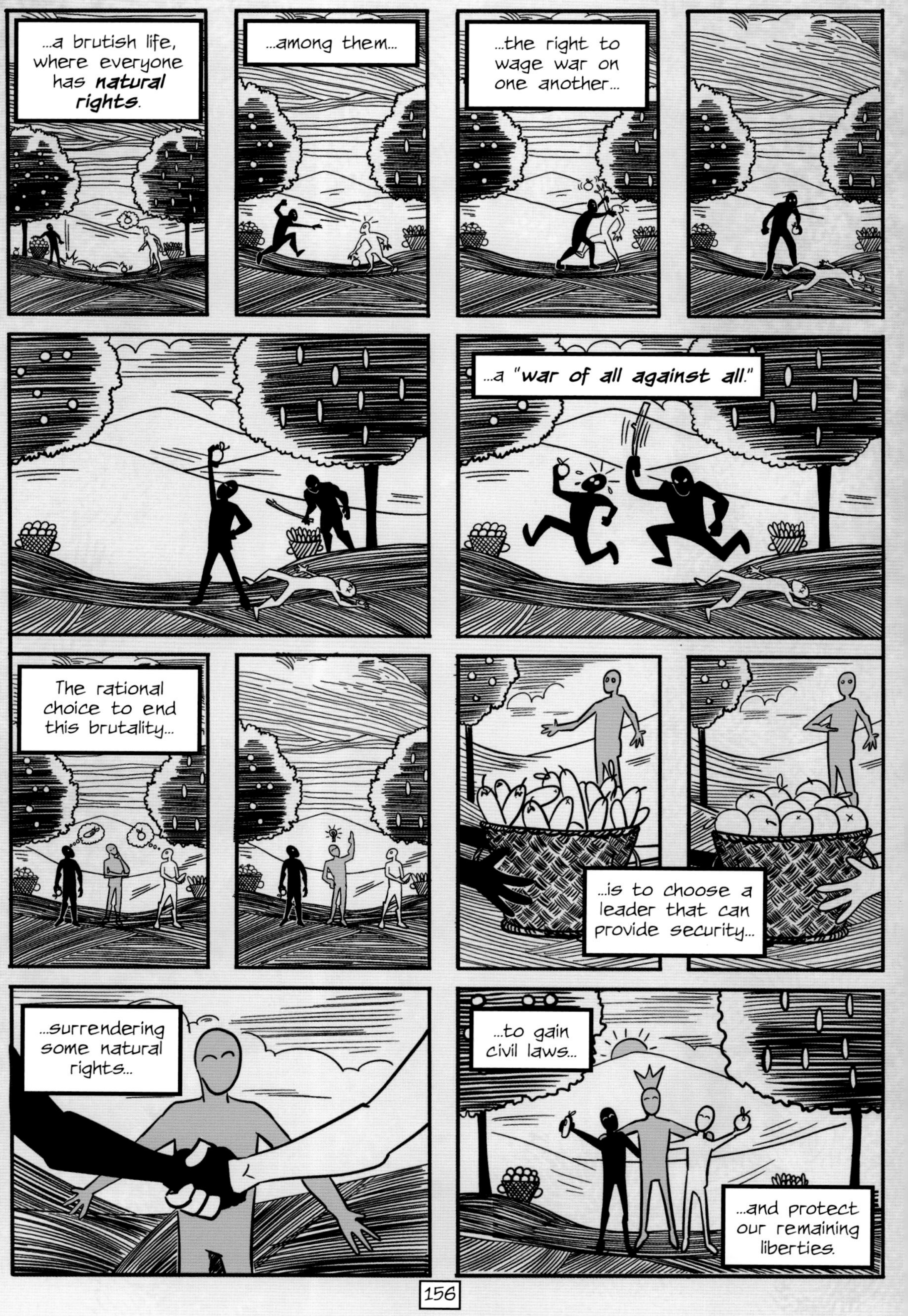
...a brutish life, where everyone has **natural rights**.
...among them...
...the right to wage war on one another...
...a "**war of all against all**."
The rational choice to end this brutality...
...is to choose a leader that can provide security...
...surrendering some natural rights...
...to gain civil laws...
...and protect our remaining liberties.

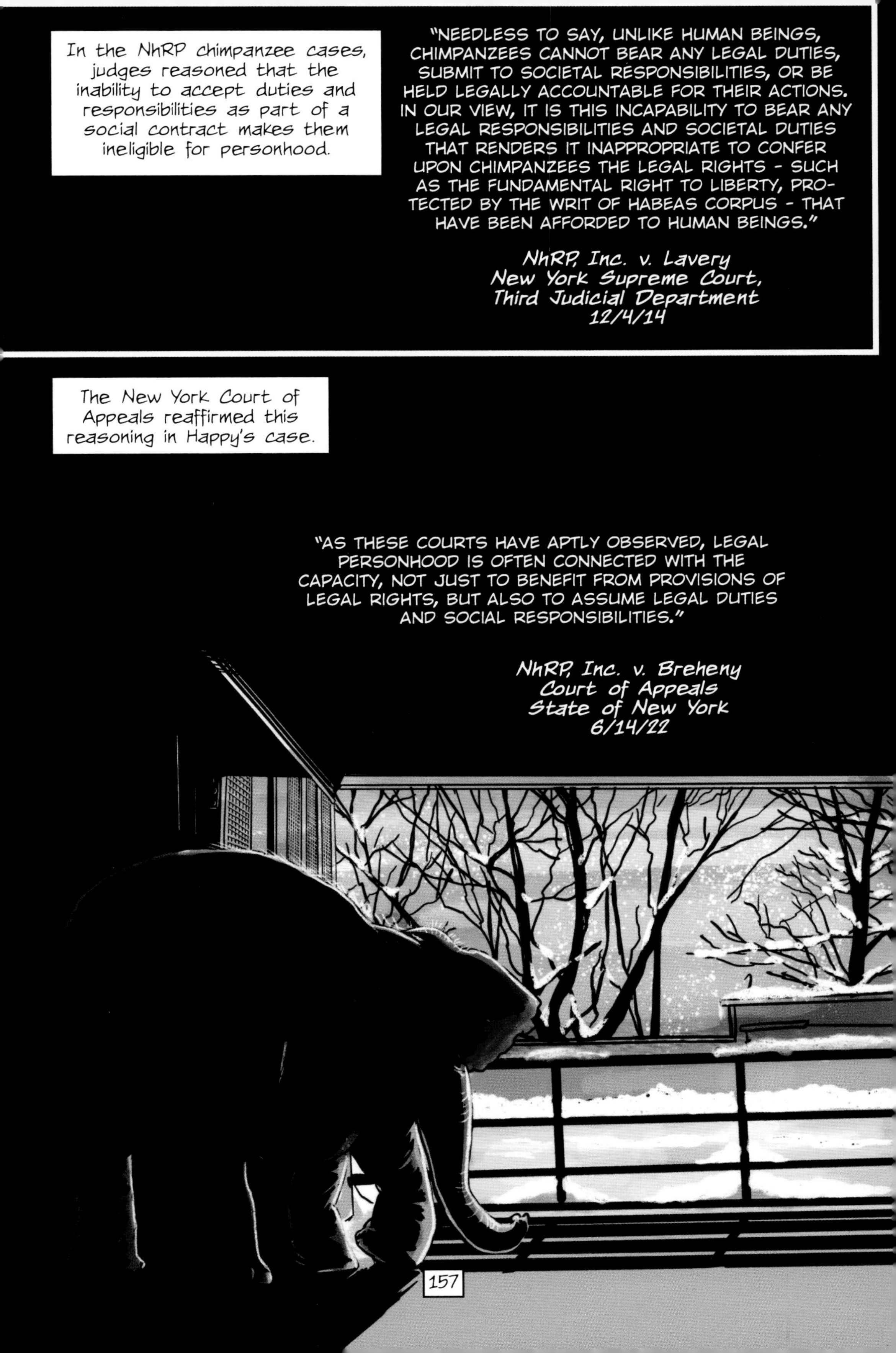
In the NhRP chimpanzee cases, judges reasoned that the inability to accept duties and responsibilities as part of a social contract makes them ineligible for personhood.
"NEEDLESS TO SAY, UNLIKE HUMAN BEINGS, CHIMPANZEES CANNOT BEAR ANY LEGAL DUTIES, SUBMIT TO SOCIETAL RESPONSIBILITIES, OR BE HELD LEGALLY ACCOUNTABLE FOR THEIR ACTIONS. IN OUR VIEW, IT IS THIS INCAPABILITY TO BEAR ANY LEGAL RESPONSIBILITIES AND SOCIETAL DUTIES THAT RENDERS IT INAPPROPRIATE TO CONFER UPON CHIMPANZEES THE LEGAL RIGHTS - SUCH AS THE FUNDAMENTAL RIGHT TO LIBERTY, PROTECTED BY THE WRIT OF HABEAS CORPUS - THAT HAVE BEEN AFFORDED TO HUMAN BEINGS."
NhRP, Inc. v. Lavery
New York Supreme Court,
Third Judicial Department
12/4/14
The New York Court of Appeals reaffirmed this reasoning in Happy's case.
"AS THESE COURTS HAVE APTLY OBSERVED, LEGAL PERSONHOOD IS OFTEN CONNECTED WITH THE CAPACITY, NOT JUST TO BENEFIT FROM PROVISIONS OF LEGAL RIGHTS, BUT ALSO TO ASSUME LEGAL DUTIES AND SOCIAL RESPONSIBILITIES."
NhRP, Inc. v. Breheny
Court of Appeals
State of New York
6/14/22

A group of philosophers submitted amicus briefs to refute this reasoning in each case.

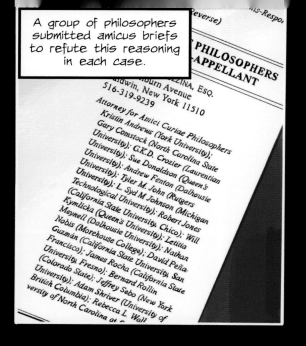

(Reverse)
...ns-Respo...

PHILOSOPHERS
-APPELLANT

...ZINA, ESQ.
...dwin, ...ourn Avenue
516-319-9239
Baldwin, New York 11510
Attorney for Amici Curiae Philosophers
Kristin Andrews (York University);
Gary Comstock (North Carolina State
University); G.K.D. Crozier (Laurentian
University); Sue Donaldson (Queen's
University); Andrew Fenton (Dalhousie
University); Tyler M. John (Rutgers
Technological University); Robert Jones
(California State University, Chico); Will
Kymlicka (Queen's University); Letitia
Meynell (Dalhousie University); Nathan
Nobis (Morehouse College); Letitia
Guzmán (California State University, San
Francisco); James Rocha (California State
University, Fresno); Bernard Rollin
(Colorado State); Jeffrey Sebo (New York
University); Adam Shriver (University of
British Columbia); Rebecca L. Wall
versity of North Carolina at ...

As the philosophers explain it...

...the courts have it "backwards."

What we acquire with a social contract, according to Hobbes, are law and morality, not rights. In fact, in the act of creating a social contract, we give up nearly all of our rights, save one: the right to life. And what we receive in exchange for giving up all these rights are not new rights, but rather security in the form of the protection of the sovereign.

When we enter the social contract, we give up most of our natural rights.

Jean-Jacques Rousseau, a political philosopher and author of **The Social Contract,** also saw things differently than the New York courts.

Rousseau explicitly rejected the idea that the social contract gives rights to persons, proclaiming, "Man is born free, and everywhere he is in chains" (Rousseau 1762, Book 1, Chapter 1). These chains, for Rousseau, are self-imposed, forged by ourselves, when we give up our natural rights and freedoms and place ourselves under the authority of another. The social contract 'chains' us. We find a similar argument in Hobbes.

Hobbes and Rousseau's views on natural rights reflect the legal tradition going all the way back to Roman law.

But the law of nations is common to the whole human race; for nations have settled certain things for themselves as occasion and the necessities of human life required. for instance, wars arose, and then followed captivity and slavery, which are contrary to the law of nature; for by the law of nature all men from the beginning were born free. the law of nations again is the source of almost all contracts; for instance, sale, hire, partnership, deposit, loan for consumption, and very many others.
— Institutes of Justinian

The social contract does not make persons, because an individual must be a person to enter into a contract.

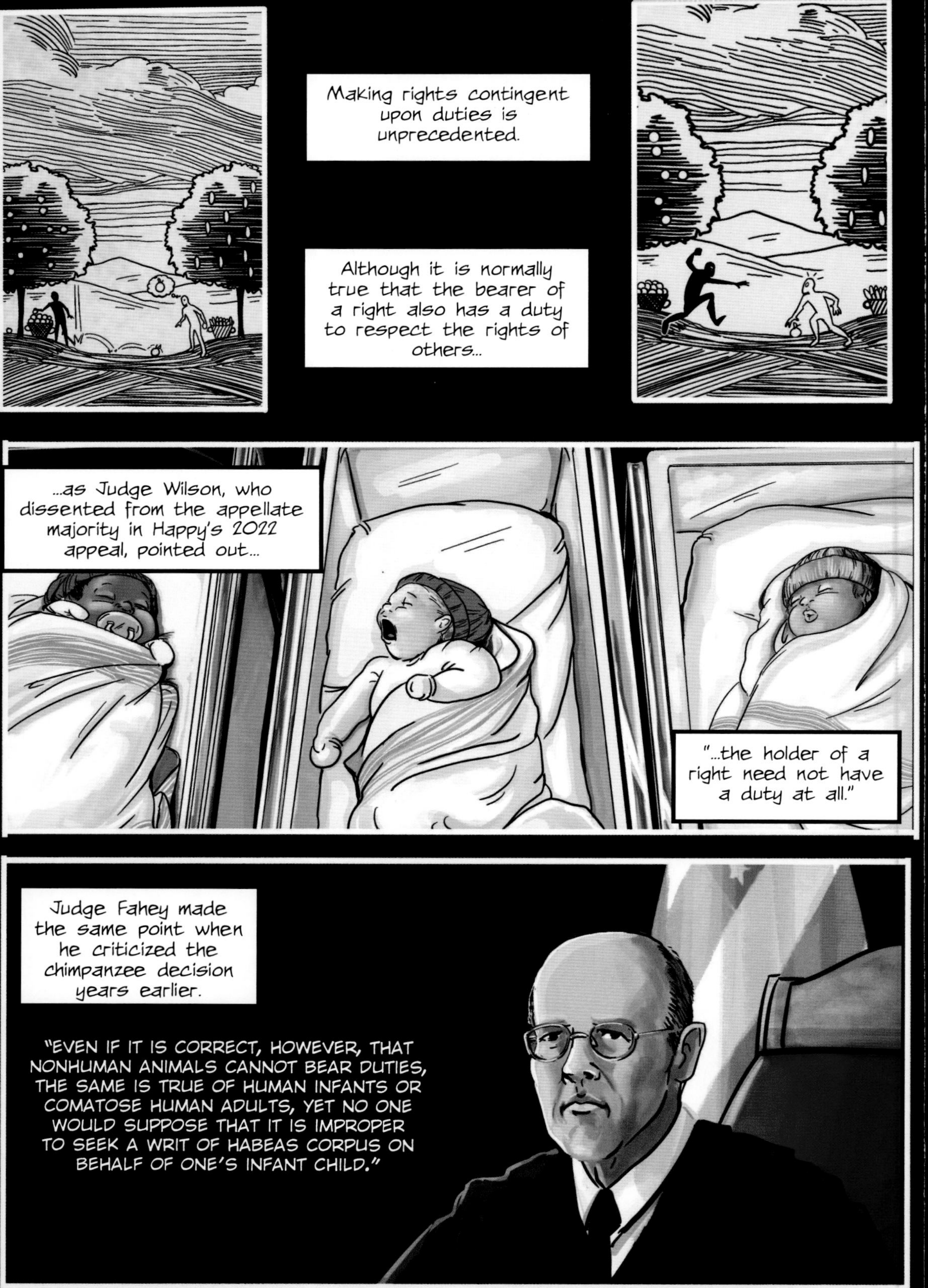
Making rights contingent upon duties is unprecedented.
Although it is normally true that the bearer of a right also has a duty to respect the rights of others...
...as Judge Wilson, who dissented from the appellate majority in Happy's 2022 appeal, pointed out...
"...the holder of a right need not have a duty at all."
Judge Fahey made the same point when he criticized the chimpanzee decision years earlier.
"EVEN IF IT IS CORRECT, HOWEVER, THAT NONHUMAN ANIMALS CANNOT BEAR DUTIES, THE SAME IS TRUE OF HUMAN INFANTS OR COMATOSE HUMAN ADULTS, YET NO ONE WOULD SUPPOSE THAT IT IS IMPROPER TO SEEK A WRIT OF HABEAS CORPUS ON BEHALF OF ONE'S INFANT CHILD."

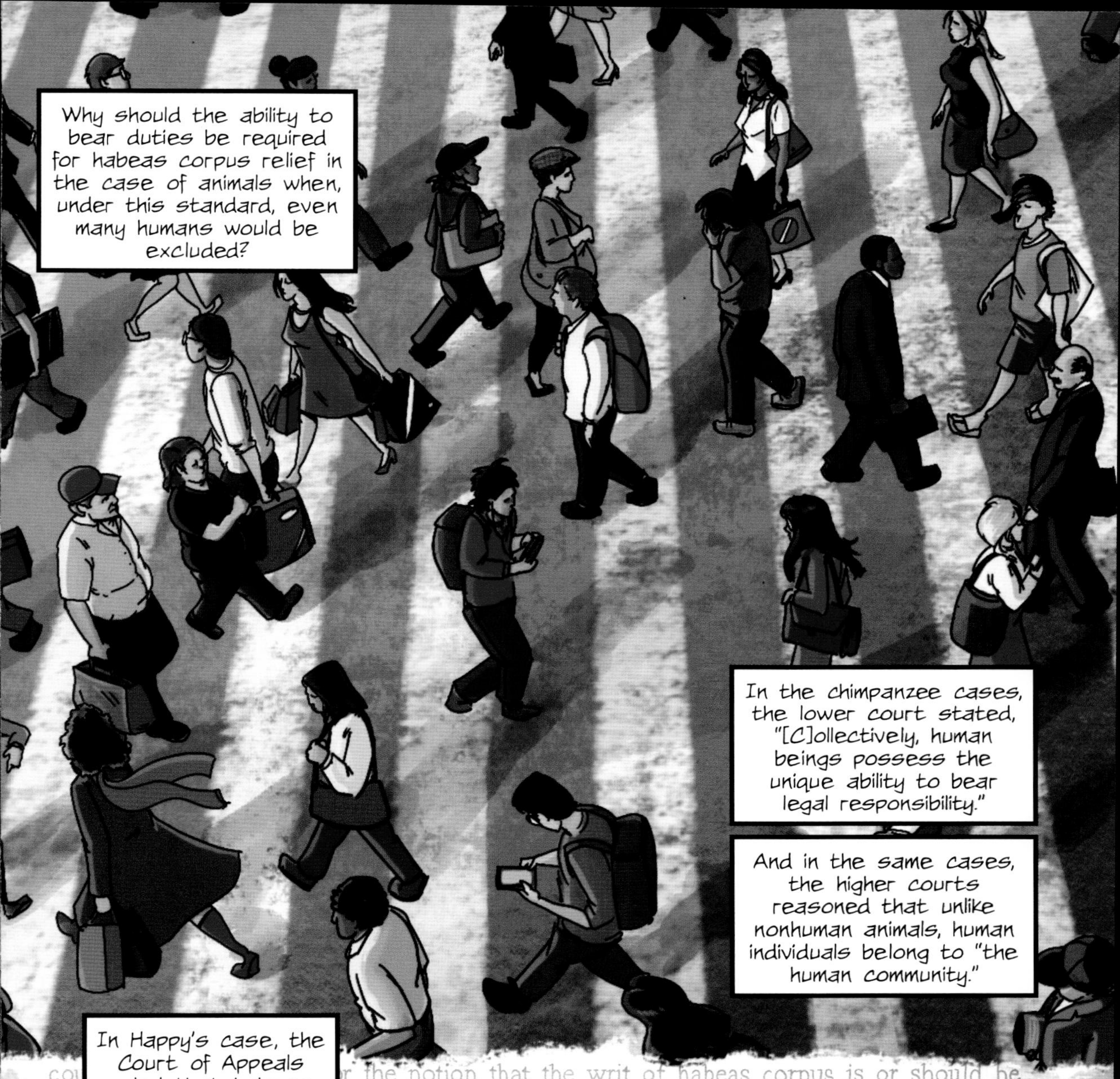
Why should the ability to bear duties be required for habeas corpus relief in the case of animals when, under this standard, even many humans would be excluded?
In the chimpanzee cases, the lower court stated, "[C]ollectively, human beings possess the unique ability to bear legal responsibility."
And in the same cases, the higher courts reasoned that unlike nonhuman animals, human individuals belong to "the human community."
In Happy's case, the Court of Appeals ruled that habeas corpus only protects humans.
The philosophers rejected these interpretations in their amicus brief.

cou… r the notion that the writ of habeas corpus is or should be app… imals. The selective capacity for autonomy, intelligence, and emotion of a particular nonhuman animal species is not a determinative factor in whether the writ is available as such factors are not what makes a person detained qualified to seek the writ. Rather, the great writ protects the right to liberty of humans *because* they are humans with certain fundamental liberty rights recognized by law (*see generally Preiser*, 411 US at 485; *Tweed*, 60 NY at 569; *Sisquoc Ranch Co. v Roth…*

[9th Cir 1946]). Nonhuman animals are not, and never have bee…

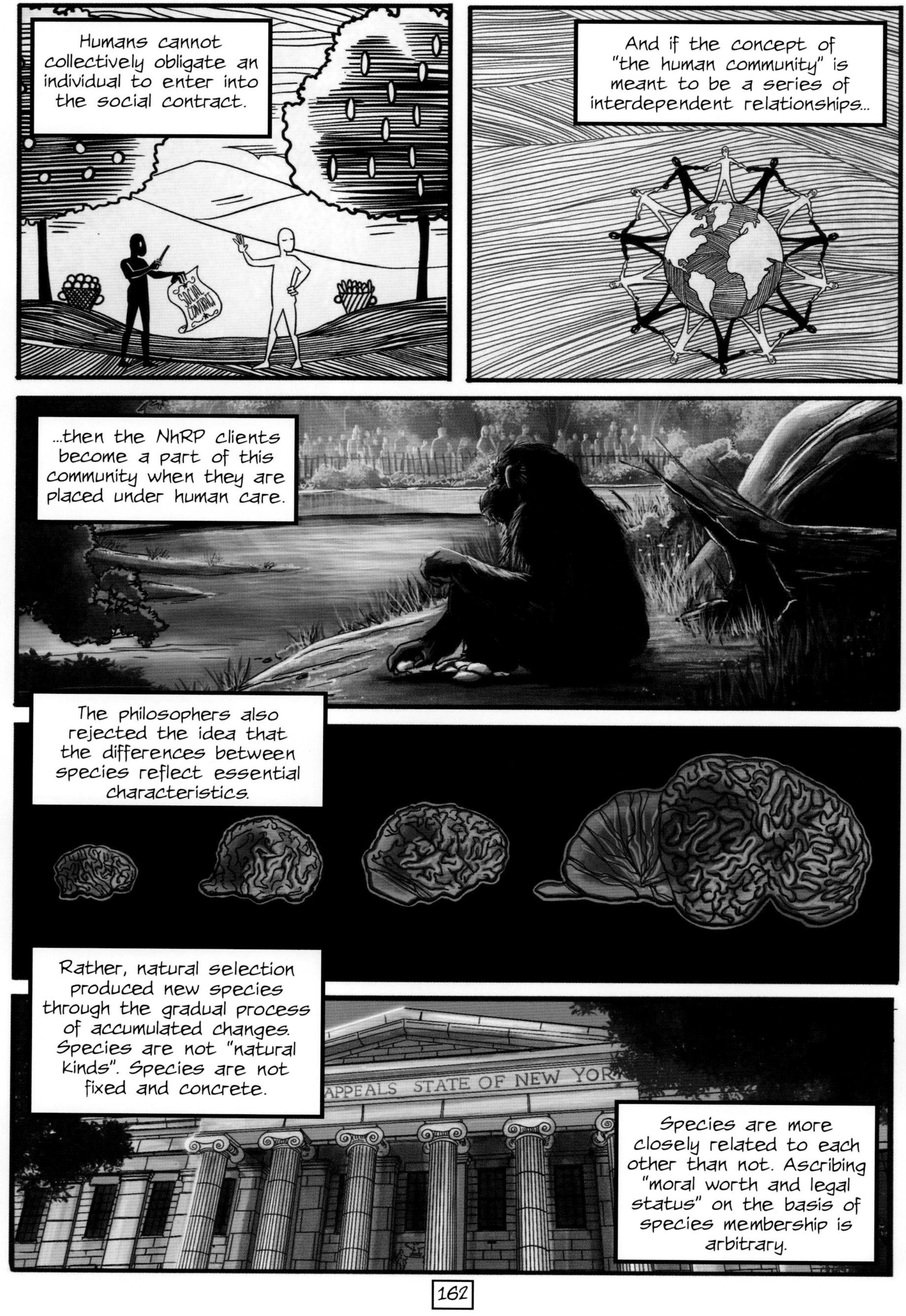
Humans cannot collectively obligate an individual to enter into the social contract.
SOCIAL CONTRACT
And if the concept of "the human community" is meant to be a series of interdependent relationships...
...then the NhRP clients become a part of this community when they are placed under human care.
The philosophers also rejected the idea that the differences between species reflect essential characteristics.
Rather, natural selection produced new species through the gradual process of accumulated changes. Species are not "natural kinds". Species are not fixed and concrete.
APPEALS STATE OF NEW YOR
Species are more closely related to each other than not. Ascribing "moral worth and legal status" on the basis of species membership is arbitrary.

Petitioner argues that the ability to acknowledge a legal duty or ... be determinative of entitlement to habeas relief ... for ... rehend that they owe duties or responsibilities and a comatose ... have legal rights." But, according to the First Department, the ... ese are still human beings, members of the human community." The NhRP fully supports the principle that human infants and comatose have, and should have, legal rights. However, that the very young and the comatose are "persons" who cannot bear duties and responsibilities, yet have the capacity to possess legal rights, explodes any claim that the capacity to bear duties and responsibilities is a necessary condition for personhood and legal rights. Instead of entering into the required mature weighing of public policy and moral principle that determines personhood in New York, *Byrn v. New York City Health and Hospitals Corp.*, 31 N.Y 2d. 194, 201 (1972), the First Department simply pronounced that humans, and only humans, can have legal rights, without providing any justification. This is merely a naked bias. We have seen such naked biases in other contexts. Before the United States Supreme Court in 1857, Dred Scott's lawyers "ignore[d] the fact" that he was not white. *Dred Scott v. Sandford*, 60 U.S. 393 (1857). The lawyers for the Native American, Chief Standing Bear, also "ignore(d) the fact" that Standing Bear was not white when, in 1879 the United States Attorney argued that a Native A... could not be a "person" for the purpose of habeas corpus after Standing Bear was ja... to his ancestral lands. *United States ex. rel Standing Bear v. Crook*, 25 F. Cas. ... D. Neb. 1879) (No. 1... The California Attorney General also "ign... a Chinese person ... when he insisted ... succe... nia Supreme Cou... person could te... man ... ell, 4 Cal. 399 (18... for Ms. Lavin... the fac... a man before th... me Court that, ... the righ... w because she was a... odell, 39 Wis. 2... Profess... ribe "ignore[d] the ... ael Hardwick w... unsucce... e United States Su... declare that the ... sodo... tional. *Bowers v. H... U.S. 186 (1986). Let... th... anzees are a... es corpus protects... uld b... should have th...

In Happy's case, this idea was critical. The judicial majority believed that recognizing Happy's right to liberty would lead to a slippery slope...

...it in a court of law" (*R.W. Commerford and Sons, Inc.*, 192.

...legal personhood to a nonhuman animal in such a manner

would have significant implications for the interactions of humans and animals in all facets

...risk ...beginning with Happy... property rights, the agricultural industry (among

...research efforts. Indeed, followed to its logical conclusion, such a

...all into question the very premises underlying pet ownership, the use

the enlist ...and progressing through the freeing of all pets... forms of work. With no clear

...with, of course, disastrous consequences for the agriculture industry.

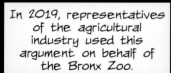

In 2019, representatives of the agricultural industry used this argument on behalf of the Bronx Zoo.

Judge Wilson called this the majority's "parade of horribles."

"THE WRIT IS A PROCEDURAL TOOL....THE MAJORITY COMPLAINS THAT GRANTING LEGAL PERSONHOOD AND LIBERTY RIGHTS TO HAPPY 'WOULD NOT BE AN INCREMENTAL STEP IN THE SLOW PROCESS OF DECISIONAL ACCRETION' REGARDING THE SCOPE AND FLEXIBILITY OF THE WRIT OF HABEAS CORPUS, BUT GRANTING A SINGLE ELEPHANT--NOT THE WHOLE ANIMAL KINGDOM--THE RIGHT TO A FULL HEARING....IS ABOUT AS INCREMENTAL AS ONE CAN GET."

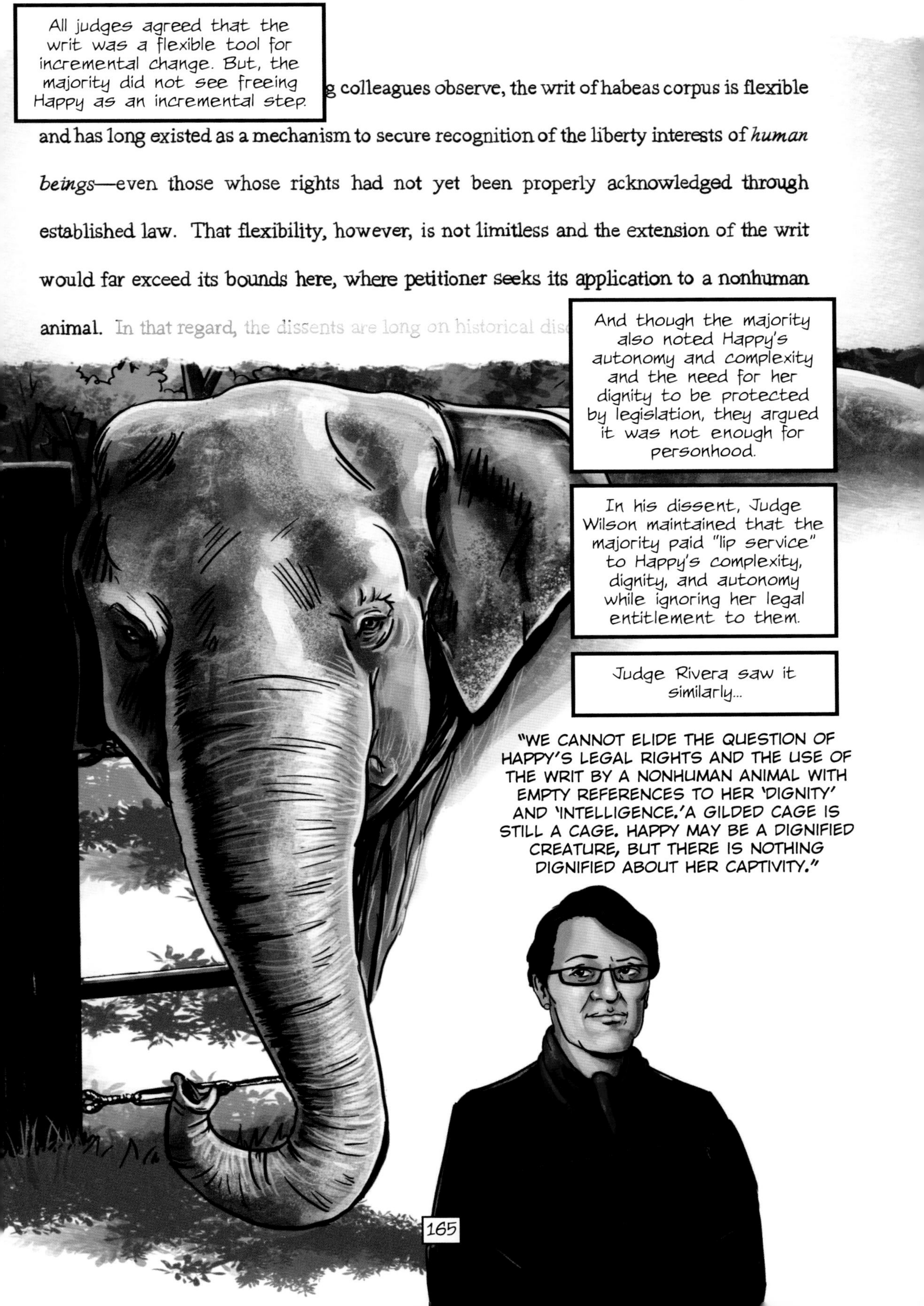
All judges agreed that the writ was a flexible tool for incremental change. But, the majority did not see freeing Happy as an incremental step.
g colleagues observe, the writ of habeas corpus is flexible and has long existed as a mechanism to secure recognition of the liberty interests of *human beings*—even those whose rights had not yet been properly acknowledged through established law. That flexibility, however, is not limitless and the extension of the writ would far exceed its bounds here, where petitioner seeks its application to a nonhuman animal. In that regard, the dissents are long on historical dis
And though the majority also noted Happy's autonomy and complexity and the need for her dignity to be protected by legislation, they argued it was not enough for personhood.
In his dissent, Judge Wilson maintained that the majority paid "lip service" to Happy's complexity, dignity, and autonomy while ignoring her legal entitlement to them.
Judge Rivera saw it similarly...
"WE CANNOT ELIDE THE QUESTION OF HAPPY'S LEGAL RIGHTS AND THE USE OF THE WRIT BY A NONHUMAN ANIMAL WITH EMPTY REFERENCES TO HER 'DIGNITY' AND 'INTELLIGENCE.' A GILDED CAGE IS STILL A CAGE. HAPPY MAY BE A DIGNIFIED CREATURE, BUT THERE IS NOTHING DIGNIFIED ABOUT HER CAPTIVITY."

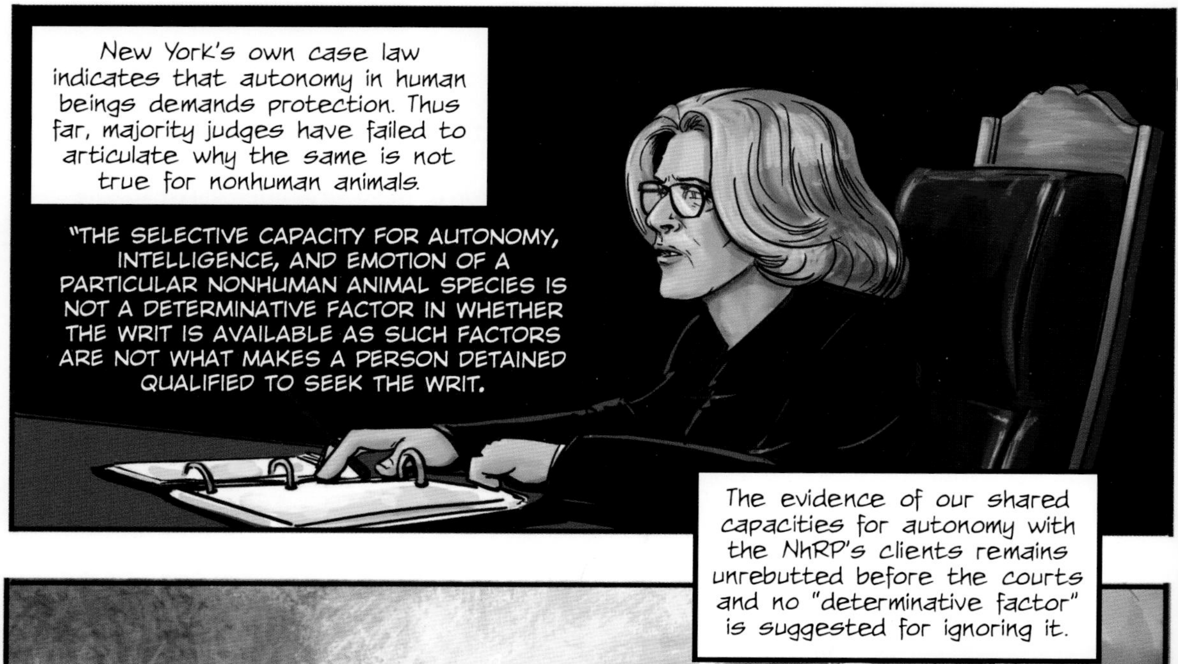
New York's own case law indicates that autonomy in human beings demands protection. Thus far, majority judges have failed to articulate why the same is not true for nonhuman animals.
"THE SELECTIVE CAPACITY FOR AUTONOMY, INTELLIGENCE, AND EMOTION OF A PARTICULAR NONHUMAN ANIMAL SPECIES IS NOT A DETERMINATIVE FACTOR IN WHETHER THE WRIT IS AVAILABLE AS SUCH FACTORS ARE NOT WHAT MAKES A PERSON DETAINED QUALIFIED TO SEEK THE WRIT.
The evidence of our shared capacities for autonomy with the NhRP's clients remains unrebutted before the courts and no "determinative factor" is suggested for ignoring it.

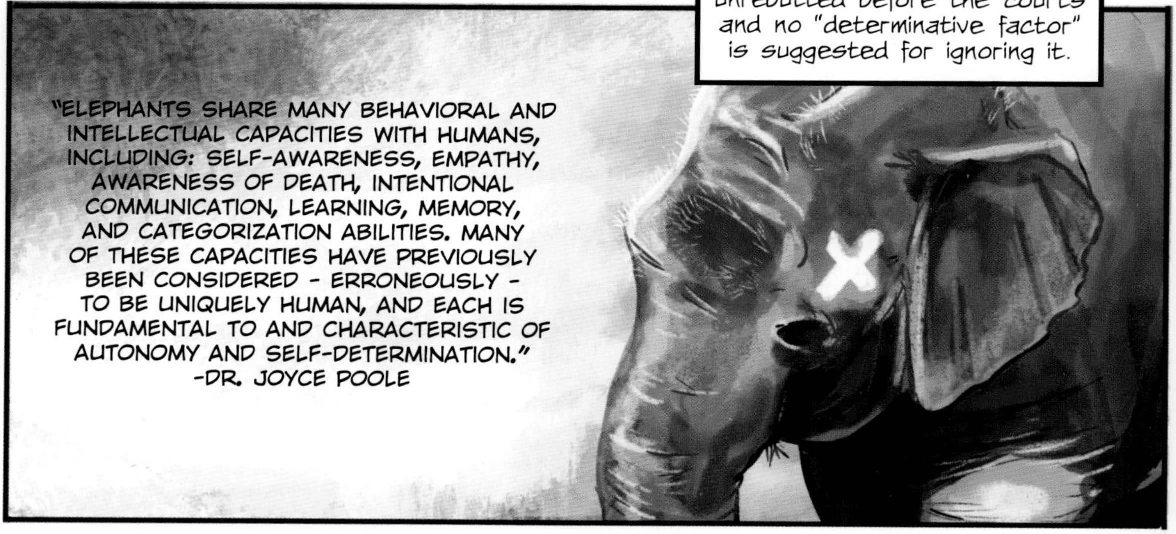
"ELEPHANTS SHARE MANY BEHAVIORAL AND INTELLECTUAL CAPACITIES WITH HUMANS, INCLUDING: SELF-AWARENESS, EMPATHY, AWARENESS OF DEATH, INTENTIONAL COMMUNICATION, LEARNING, MEMORY, AND CATEGORIZATION ABILITIES. MANY OF THESE CAPACITIES HAVE PREVIOUSLY BEEN CONSIDERED - ERRONEOUSLY - TO BE UNIQUELY HUMAN, AND EACH IS FUNDAMENTAL TO AND CHARACTERISTIC OF AUTONOMY AND SELF-DETERMINATION."
-DR. JOYCE POOLE

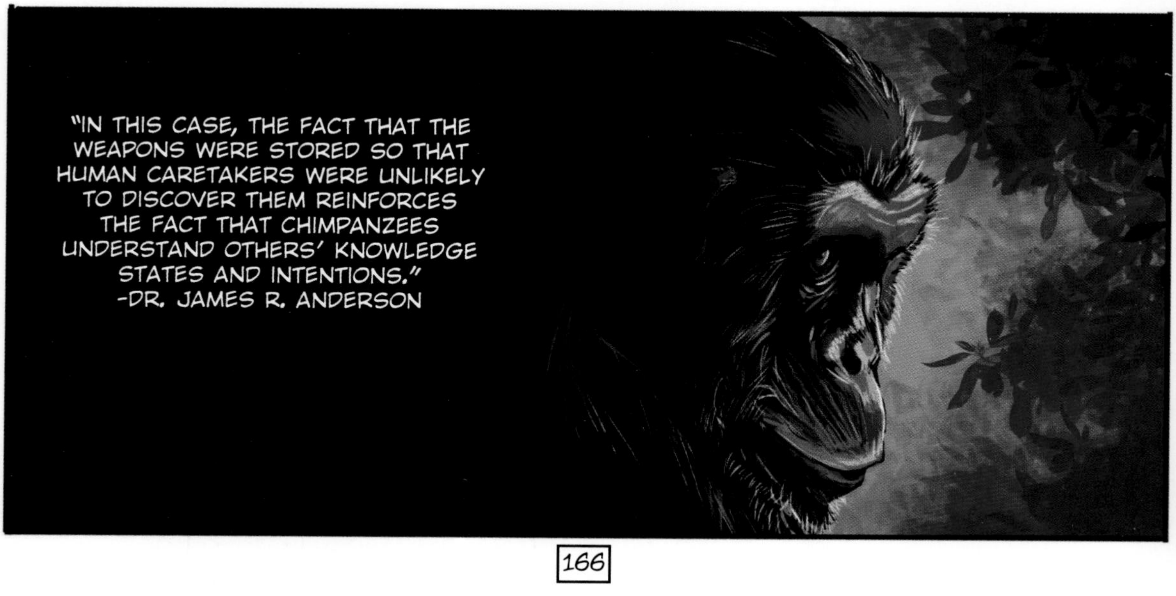
"IN THIS CASE, THE FACT THAT THE WEAPONS WERE STORED SO THAT HUMAN CARETAKERS WERE UNLIKELY TO DISCOVER THEM REINFORCES THE FACT THAT CHIMPANZEES UNDERSTAND OTHERS' KNOWLEDGE STATES AND INTENTIONS."
-DR. JAMES R. ANDERSON

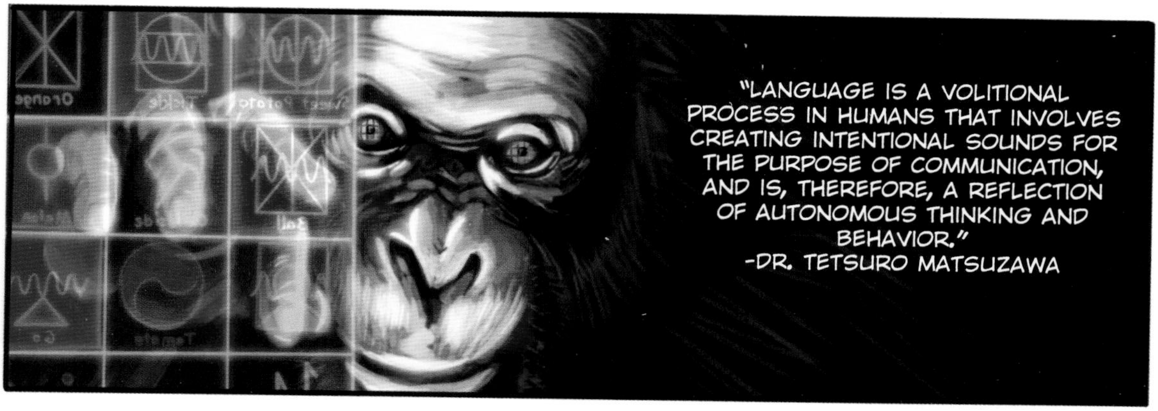
"LANGUAGE IS A VOLITIONAL PROCESS IN HUMANS THAT INVOLVES CREATING INTENTIONAL SOUNDS FOR THE PURPOSE OF COMMUNICATION, AND IS, THEREFORE, A REFLECTION OF AUTONOMOUS THINKING AND BEHAVIOR."
-DR. TETSURO MATSUZAWA

"MANY OF THE EXPRESSIONS IN CHIMPANZEES AND HUMANS ARE DISPLAYED IN SIMILAR CIRCUMSTANCES, SUGGESTING A COMMON FUNCTION OR MEANING."
-DR. JENNIFER MB FUGATE

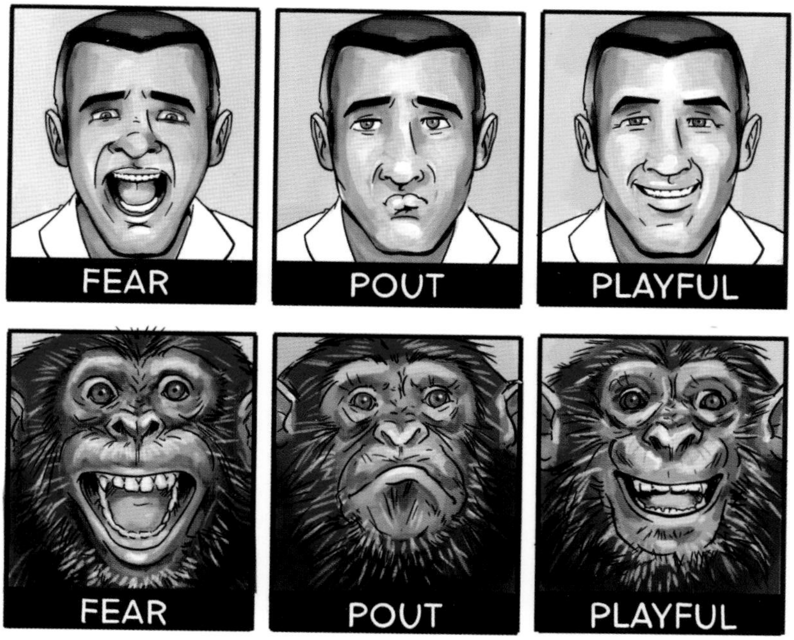
FEAR
POUT
PLAYFUL
FEAR
POUT
PLAYFUL

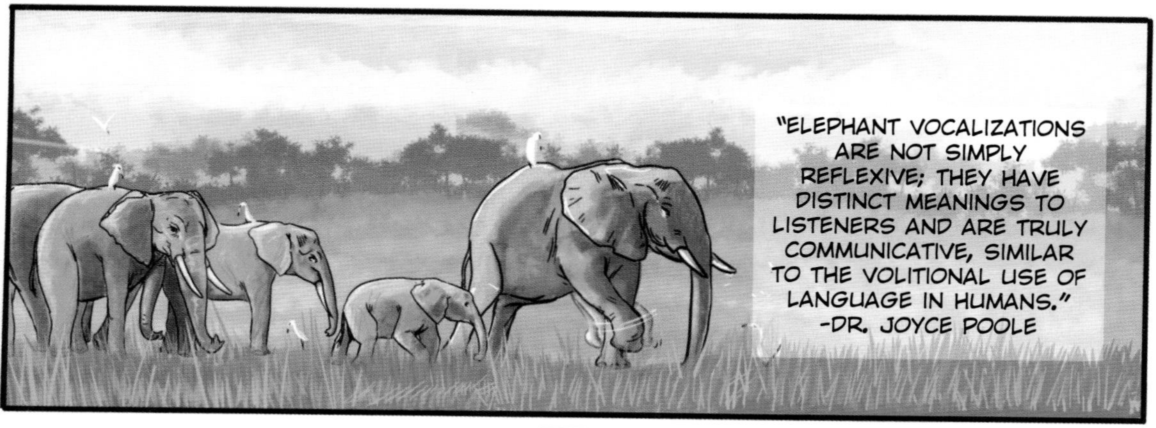
"ELEPHANT VOCALIZATIONS ARE NOT SIMPLY REFLEXIVE; THEY HAVE DISTINCT MEANINGS TO LISTENERS AND ARE TRULY COMMUNICATIVE, SIMILAR TO THE VOLITIONAL USE OF LANGUAGE IN HUMANS."
-DR. JOYCE POOLE

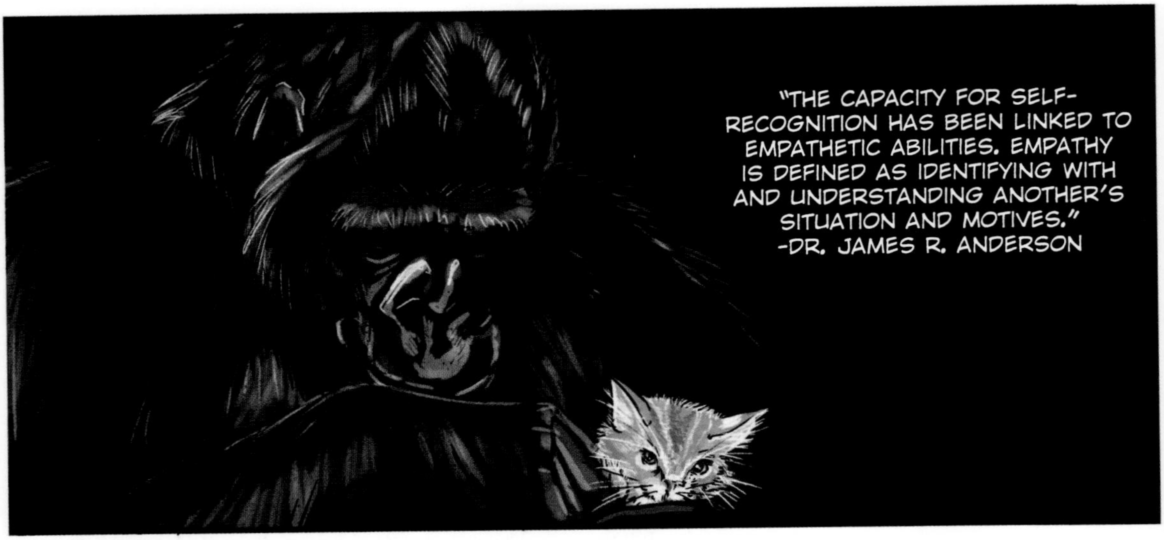
"THE CAPACITY FOR SELF-RECOGNITION HAS BEEN LINKED TO EMPATHETIC ABILITIES. EMPATHY IS DEFINED AS IDENTIFYING WITH AND UNDERSTANDING ANOTHER'S SITUATION AND MOTIVES."
-DR. JAMES R. ANDERSON

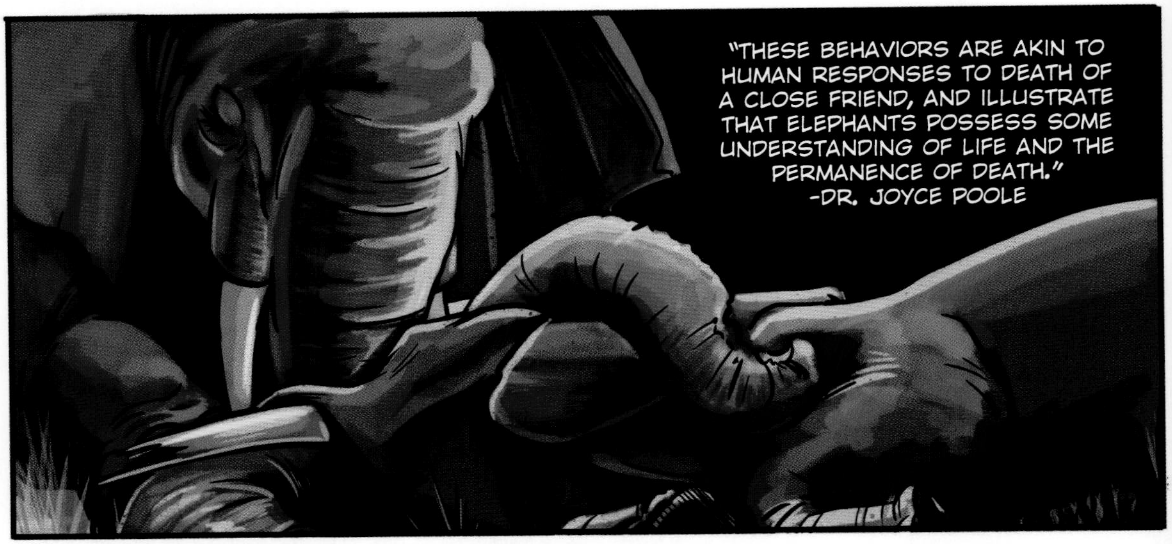
"THESE BEHAVIORS ARE AKIN TO HUMAN RESPONSES TO DEATH OF A CLOSE FRIEND, AND ILLUSTRATE THAT ELEPHANTS POSSESS SOME UNDERSTANDING OF LIFE AND THE PERMANENCE OF DEATH."
-DR. JOYCE POOLE

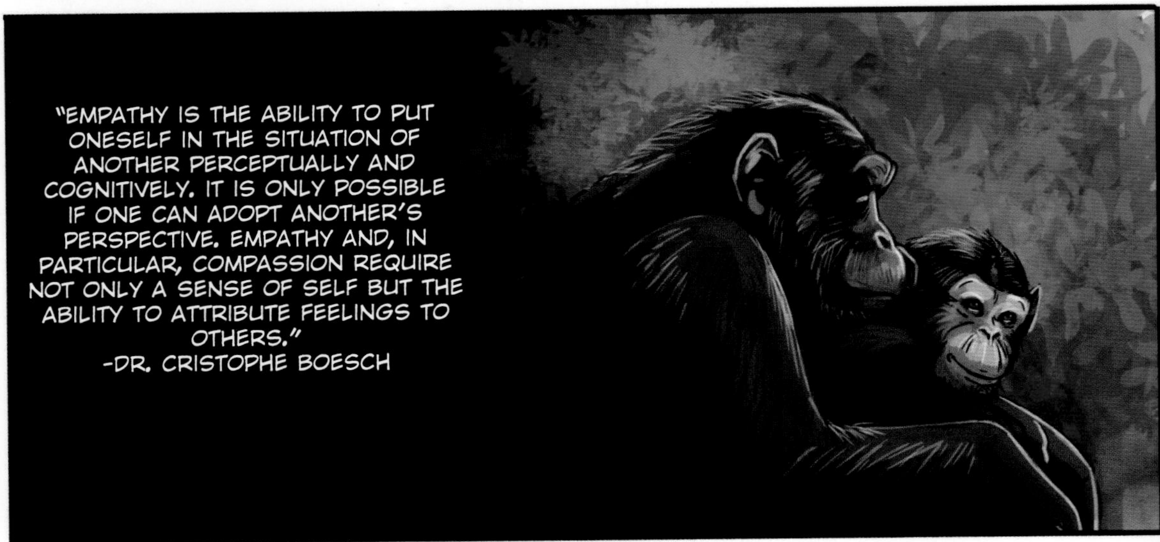
"EMPATHY IS THE ABILITY TO PUT ONESELF IN THE SITUATION OF ANOTHER PERCEPTUALLY AND COGNITIVELY. IT IS ONLY POSSIBLE IF ONE CAN ADOPT ANOTHER'S PERSPECTIVE. EMPATHY AND, IN PARTICULAR, COMPASSION REQUIRE NOT ONLY A SENSE OF SELF BUT THE ABILITY TO ATTRIBUTE FEELINGS TO OTHERS."
-DR. CRISTOPHE BOESCH

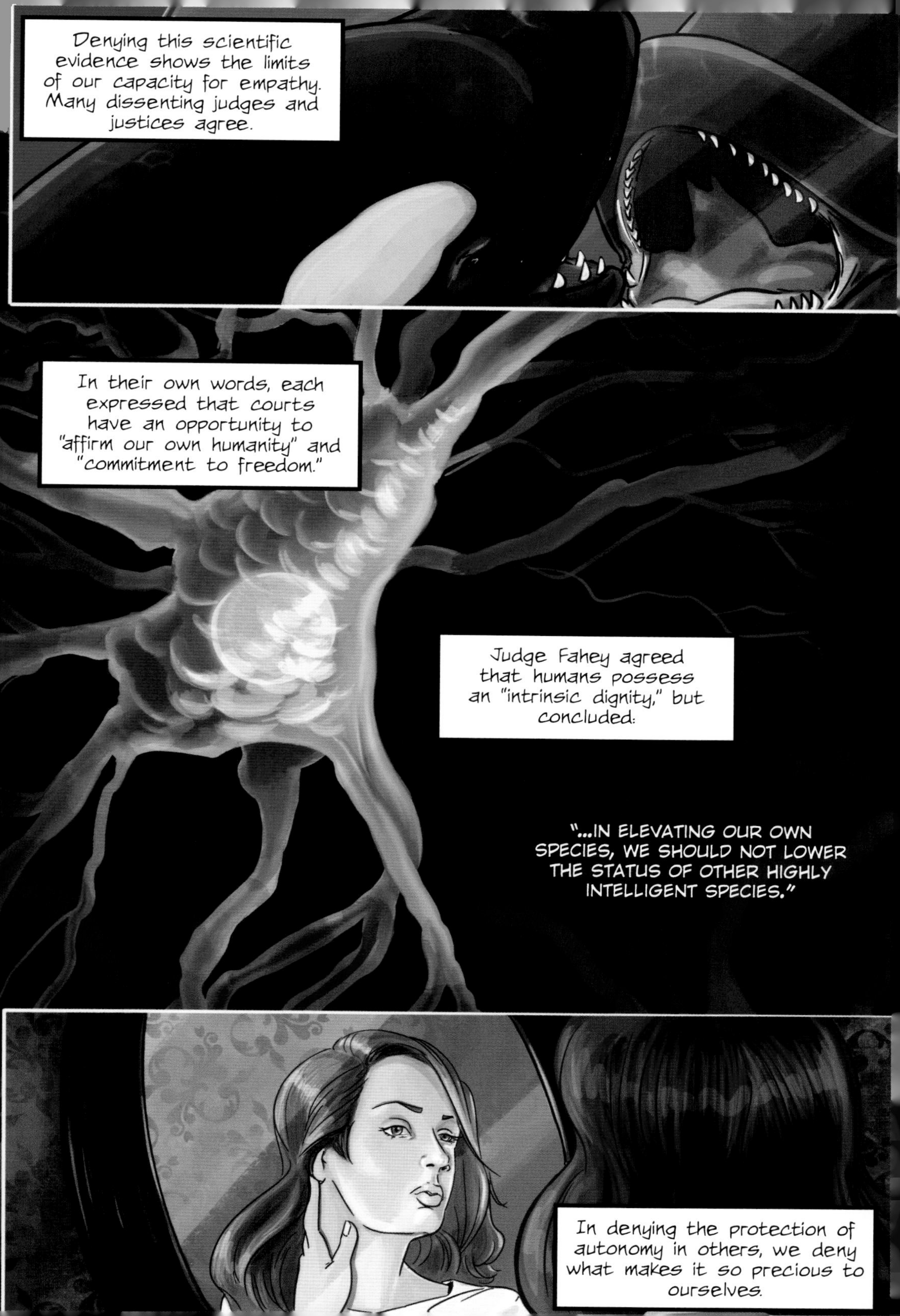
Denying this scientific evidence shows the limits of our capacity for empathy. Many dissenting judges and justices agree.
In their own words, each expressed that courts have an opportunity to "affirm our own humanity" and "commitment to freedom."
Judge Fahey agreed that humans possess an "intrinsic dignity," but concluded:
"...IN ELEVATING OUR OWN SPECIES, WE SHOULD NOT LOWER THE STATUS OF OTHER HIGHLY INTELLIGENT SPECIES."
In denying the protection of autonomy in others, we deny what makes it so precious to ourselves.

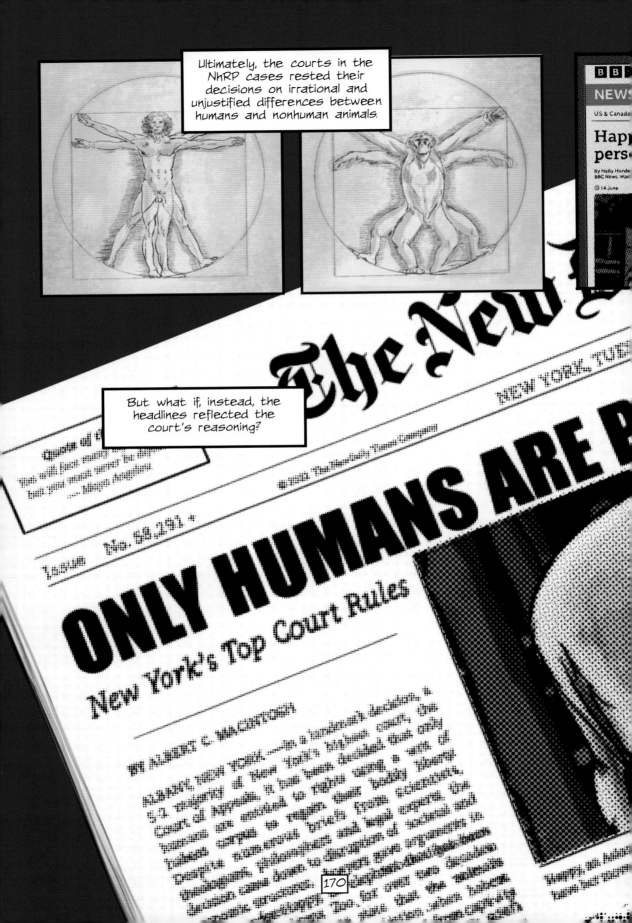

Ultimately, the courts in the NhRP cases rested their decisions on irrational and unjustified differences between humans and nonhuman animals.

But what if, instead, the headlines reflected the court's reasoning?

BBC NEWS

US & Canada

Happ
pers

By Holly Horde
BBC News, Was
14 June

The New

NEW YORK, TUE

ONLY HUMANS ARE P

New York's Top Court Rules

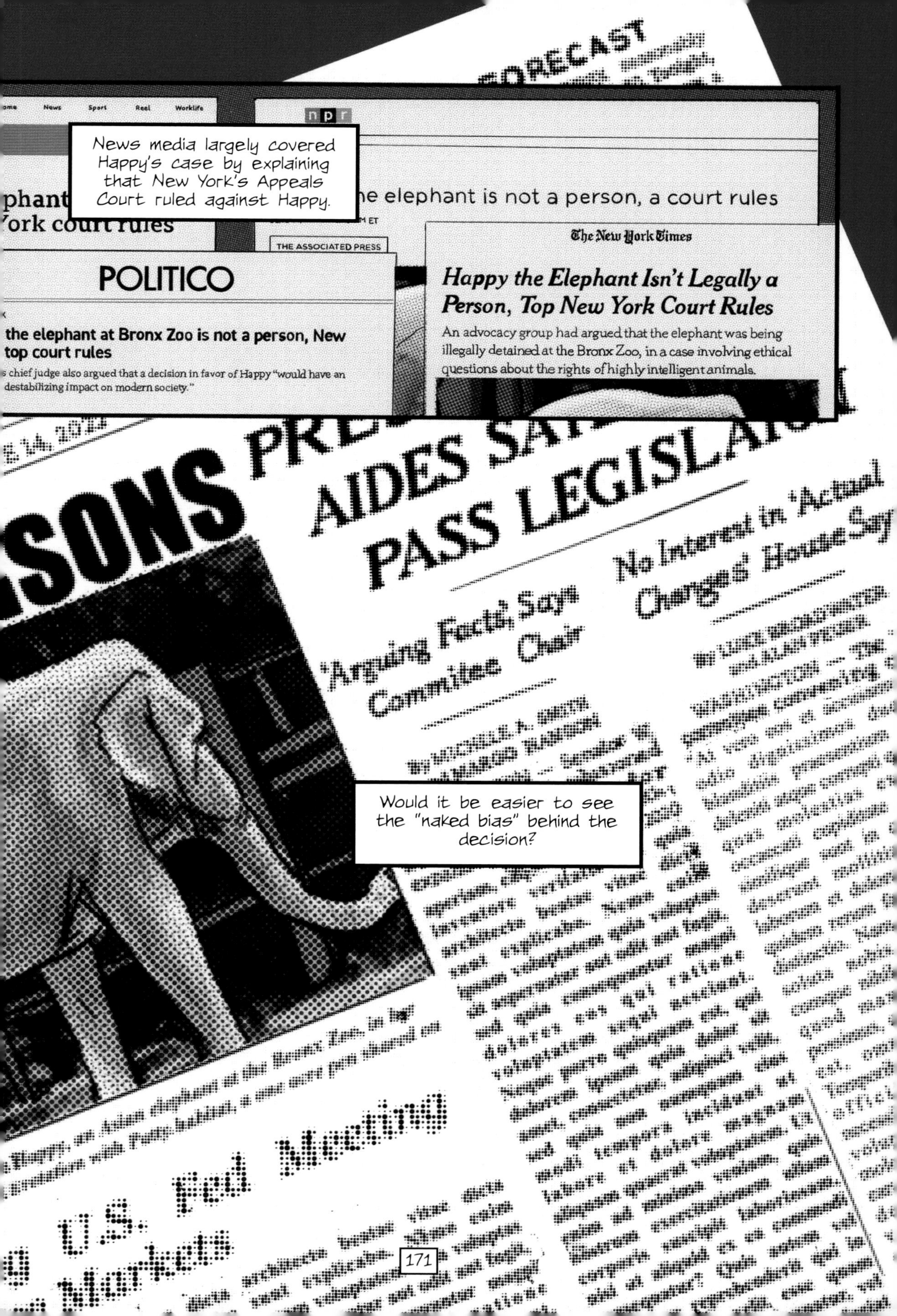

ORECAST
Home News Sport Reel Worklife
npr
News media largely covered Happy's case by explaining that New York's Appeals Court ruled against Happy.
phant
ork court rules
he elephant is not a person, a court rules
THE ASSOCIATED PRESS
POLITICO
the elephant at Bronx Zoo is not a person, New
top court rules
s chief judge also argued that a decision in favor of Happy "would have an
destabilizing impact on modern society."
The New York Times
Happy the Elephant Isn't Legally a Person, Top New York Court Rules
An advocacy group had argued that the elephant was being illegally detained at the Bronx Zoo, in a case involving ethical questions about the rights of highly intelligent animals.
SONS
AIDES
PASS LEGISL
No Interest in 'Actual Changes,' House Says
'Arguing Facts,' Says Commitee Chair
Would it be easier to see the "naked bias" behind the decision?
U.S.
Fed

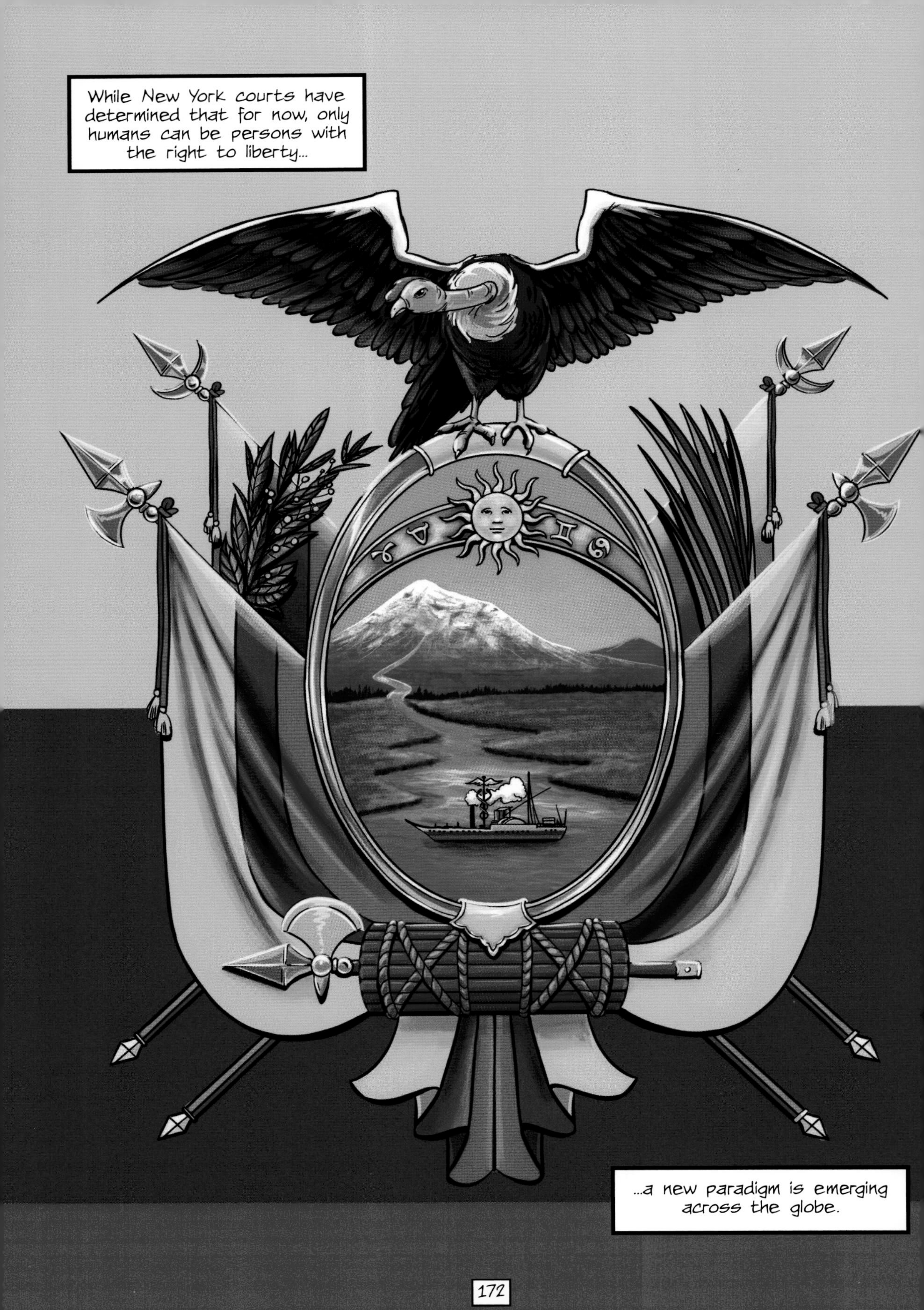
While New York courts have determined that for now, only humans can be persons with the right to liberty...
...a new paradigm is emerging across the globe.

CHAPTER 8: UNA PERSONA NO HUMANA

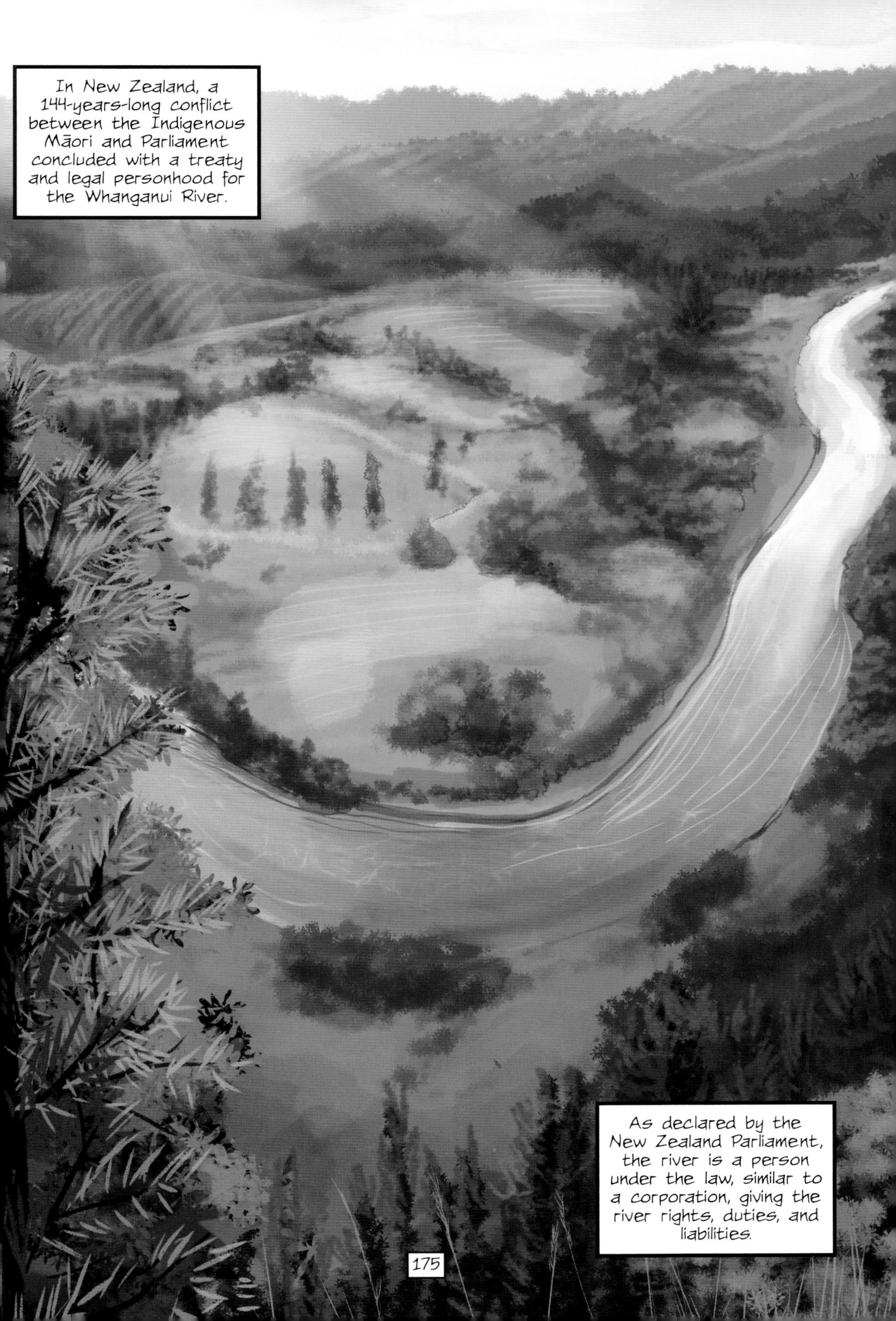
In New Zealand, a 144-years-long conflict between the Indigenous Māori and Parliament concluded with a treaty and legal personhood for the Whanganui River.
As declared by the New Zealand Parliament, the river is a person under the law, similar to a corporation, giving the river rights, duties, and liabilities.

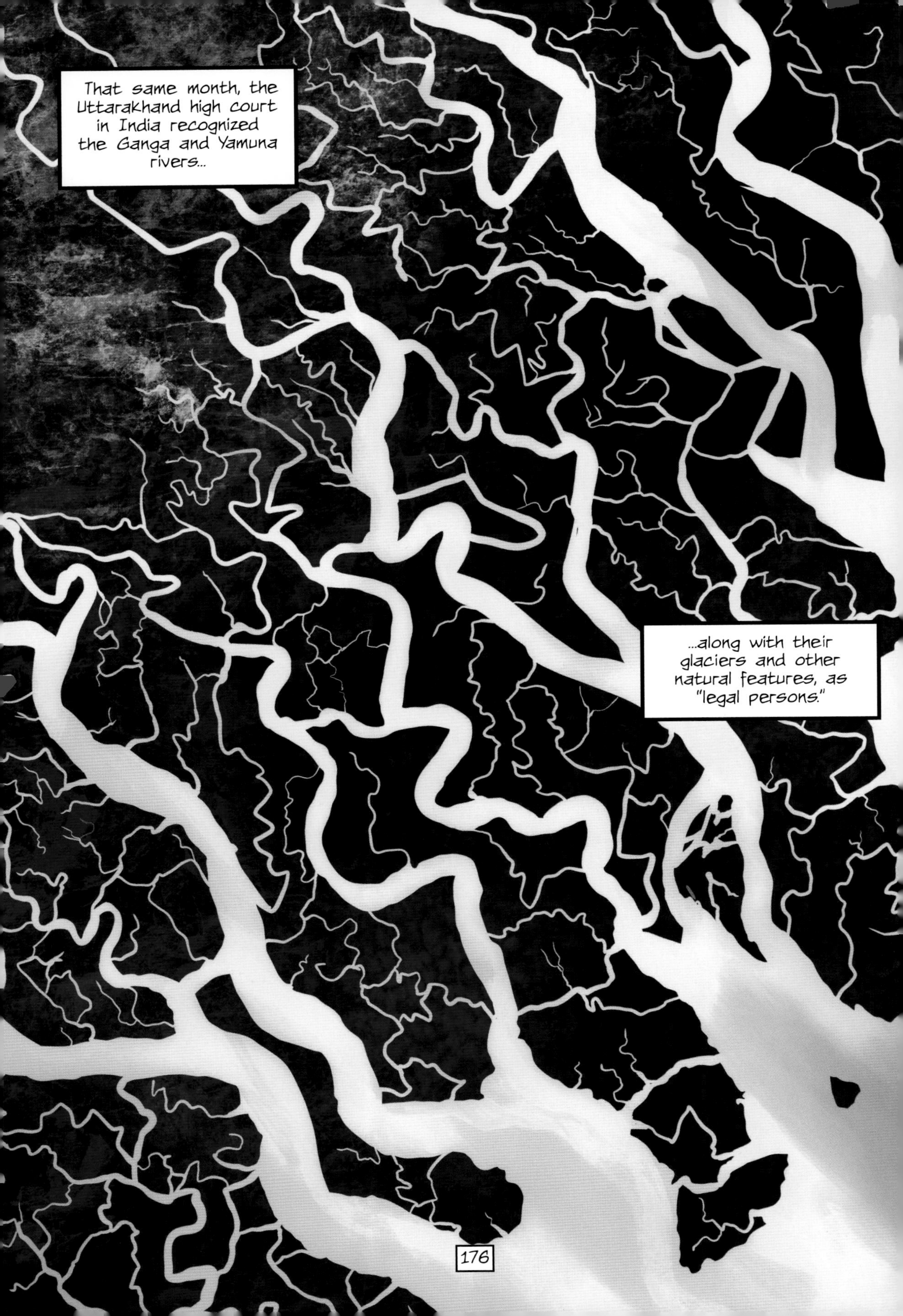
That same month, the Uttarakhand high court in India recognized the Ganga and Yamuna rivers...
...along with their glaciers and other natural features, as "legal persons."

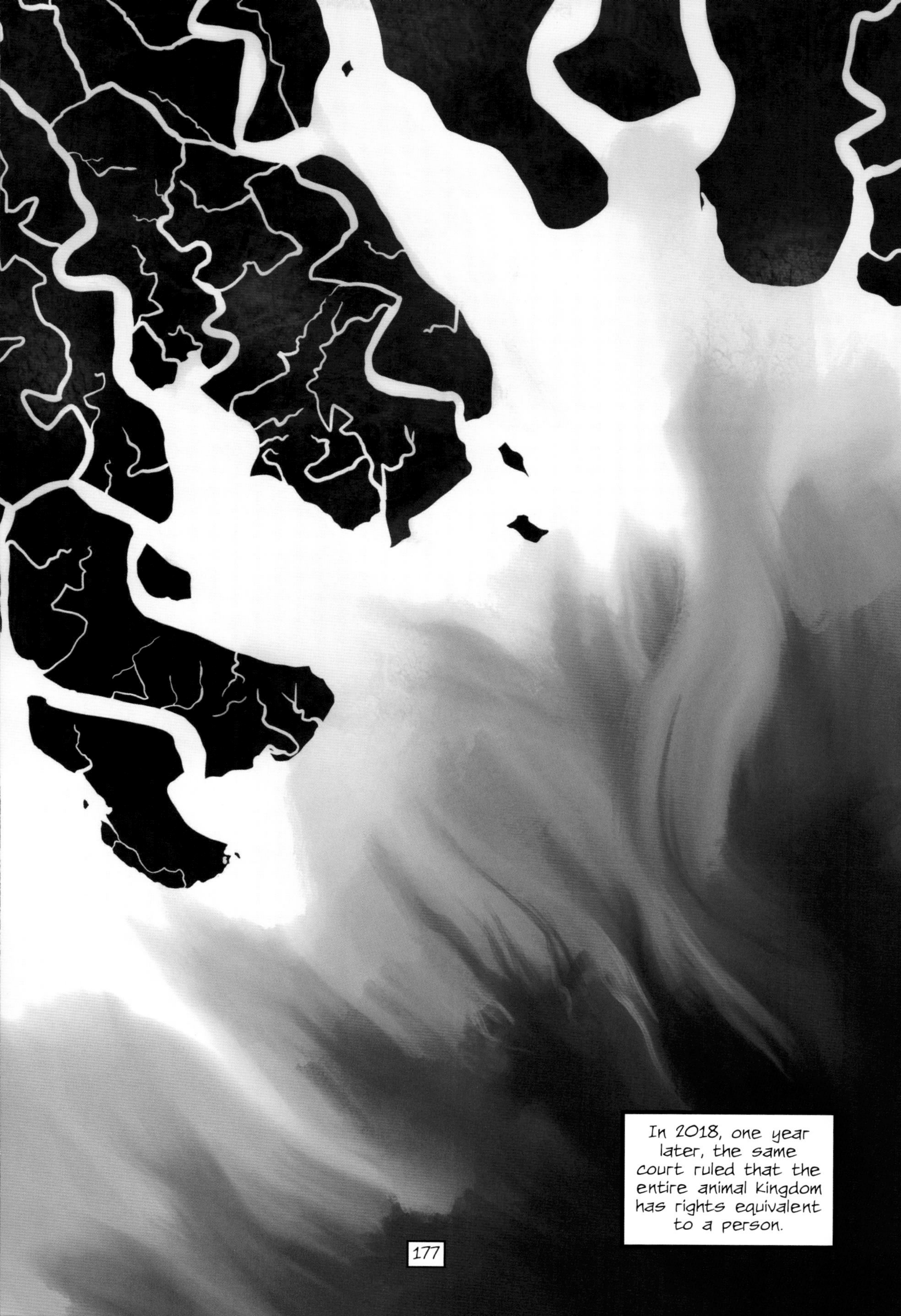
In 2018, one year later, the same court ruled that the entire animal kingdom has rights equivalent to a person.

The Uttarakhand High Court rulings were drawn from case law, the Animal Welfare Act, and ideals laid out in the Indian constitution.

In New Zealand, the extension of personhood to the Whanganui River was based on treaty negotiation...

Proclamation

WHEREAS, to say that animals are sentient is to state explicitly that they can experience both positive and negative emotions, including pain and distress, and;

WHEREAS, the 2013 Cambridge Declaration on Consciousness concluded that non-human animals have the neuroanatomical, neuroch... ...trates of conscious states, thus providing scien... ...lings in the same way that we do,

...or religions have principles of ani... ...lue and significance, and that p... ...f service, not in ex...

...to cruelty towards...

...animal welfare have been rapidly changing, and ...te for pets and farm stock are no longer acceptable

...rerning body has officially considered animals as...

...e first count... ...d the Governm... ...hat ALL animals...

...e sentience of... ...e judged by t...

STOP ANIMAL ABUSE AND CRUELTY

ANIMALS ARE HERE WITH US NOT FOR US

BE KIND

...and reflects the values of the Māori and a newly emerging legal framework.

However, New Zealand has yet to advance the idea of legal personhood for nonhuman animals...

...choosing instead to expand their animal welfare legislation to acknowledge that animals are sentient.

The acknowledgment provides no legal guidance concerning rights or impacts on animal welfare.

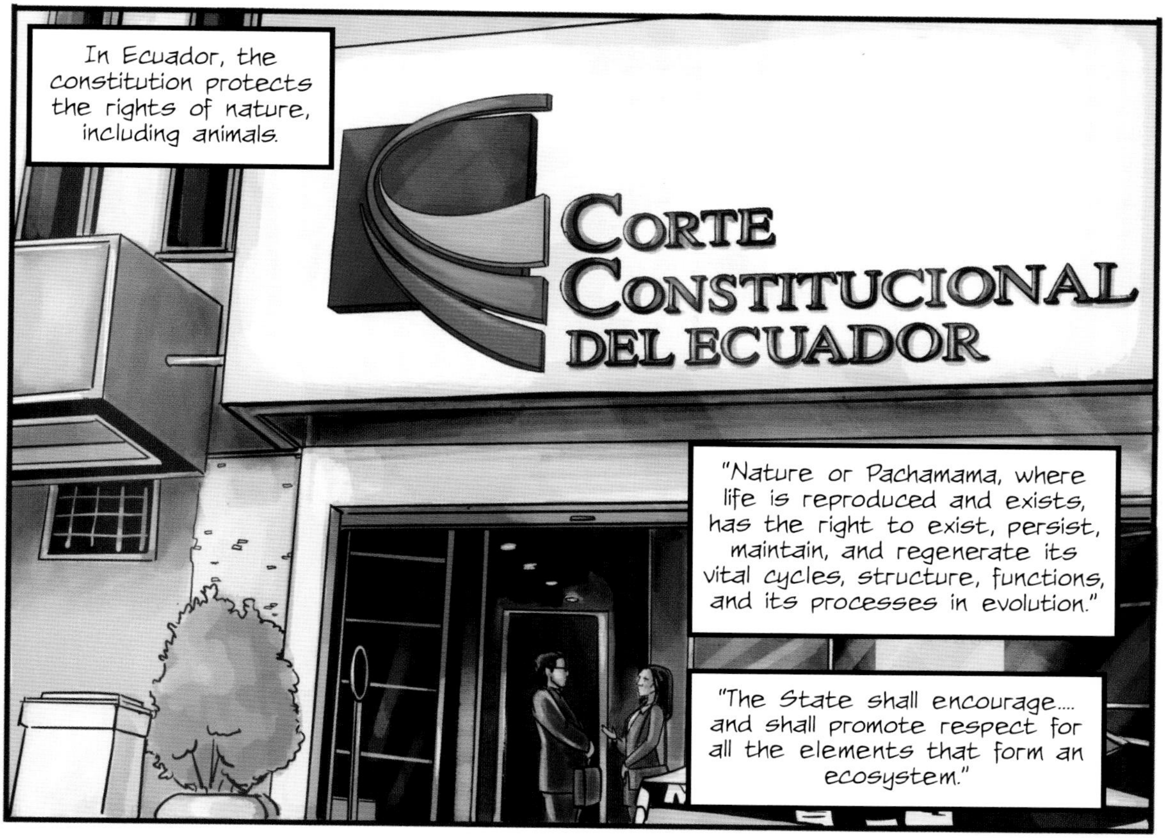
In Ecuador, the constitution protects the rights of nature, including animals.
CORTE CONSTITUCIONAL DEL ECUADOR
"Nature or Pachamama, where life is reproduced and exists, has the right to exist, persist, maintain, and regenerate its vital cycles, structure, functions, and its processes in evolution."
"The State shall encourage.... and shall promote respect for all the elements that form an ecosystem."

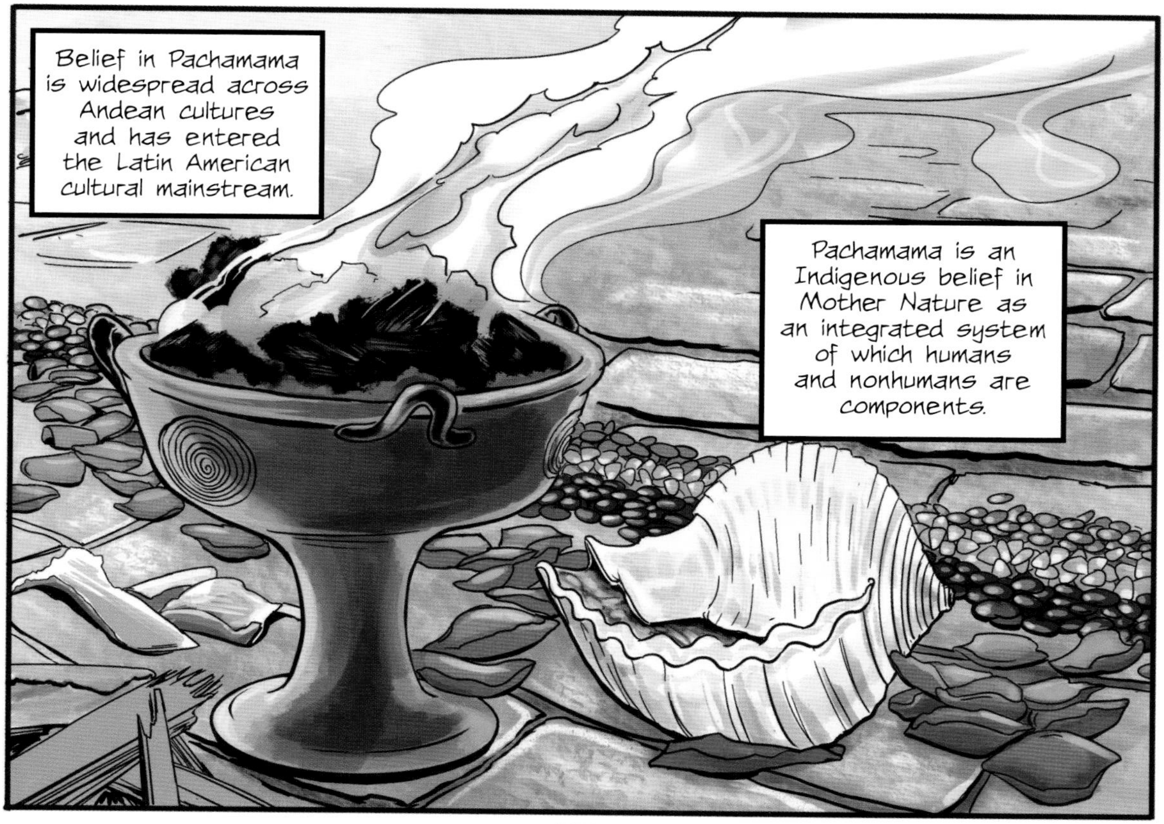
Belief in Pachamama is widespread across Andean cultures and has entered the Latin American cultural mainstream.
Pachamama is an Indigenous belief in Mother Nature as an integrated system of which humans and nonhumans are components.

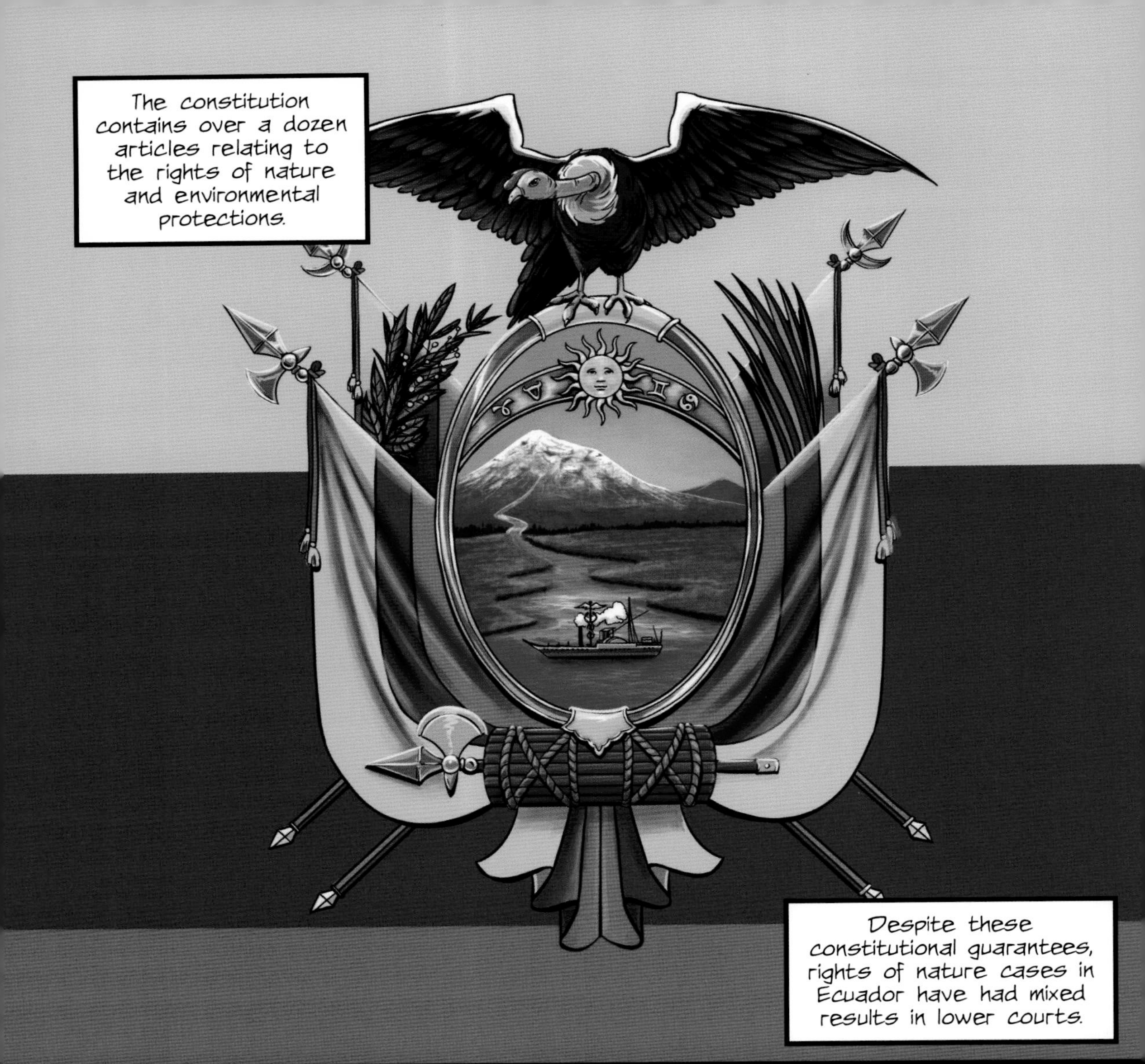
The constitution contains over a dozen articles relating to the rights of nature and environmental protections.
Despite these constitutional guarantees, rights of nature cases in Ecuador have had mixed results in lower courts.

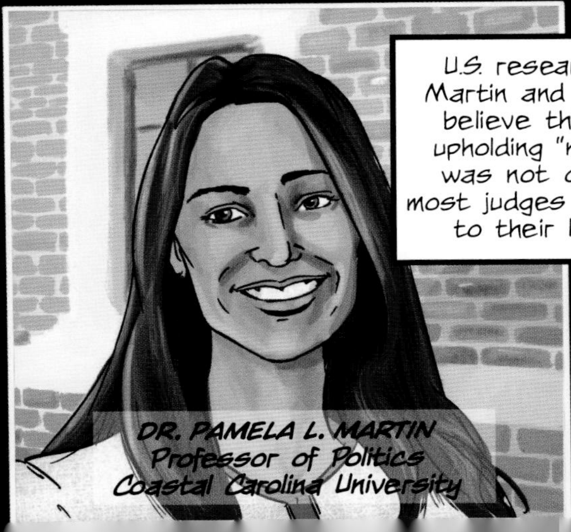
U.S. researchers Pamela Martin and Craig Kauffman believe this is because upholding "nature's rights was not only foreign to most judges but ran counter to their legal training."
DR. PAMELA L. MARTIN
Professor of Politics
Coastal Carolina University

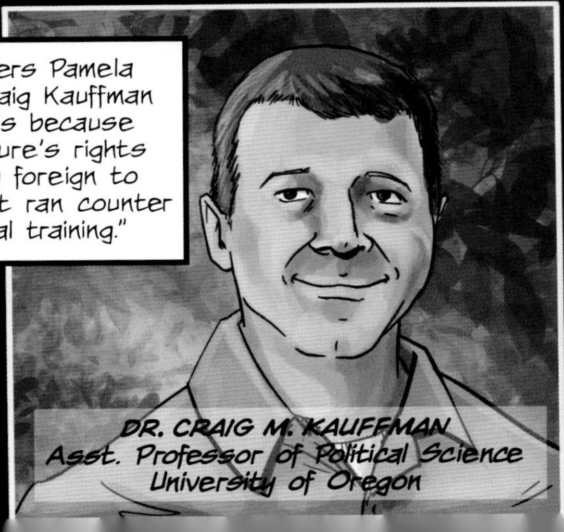
DR. CRAIG M. KAUFFMAN
Asst. Professor of Political Science
University of Oregon

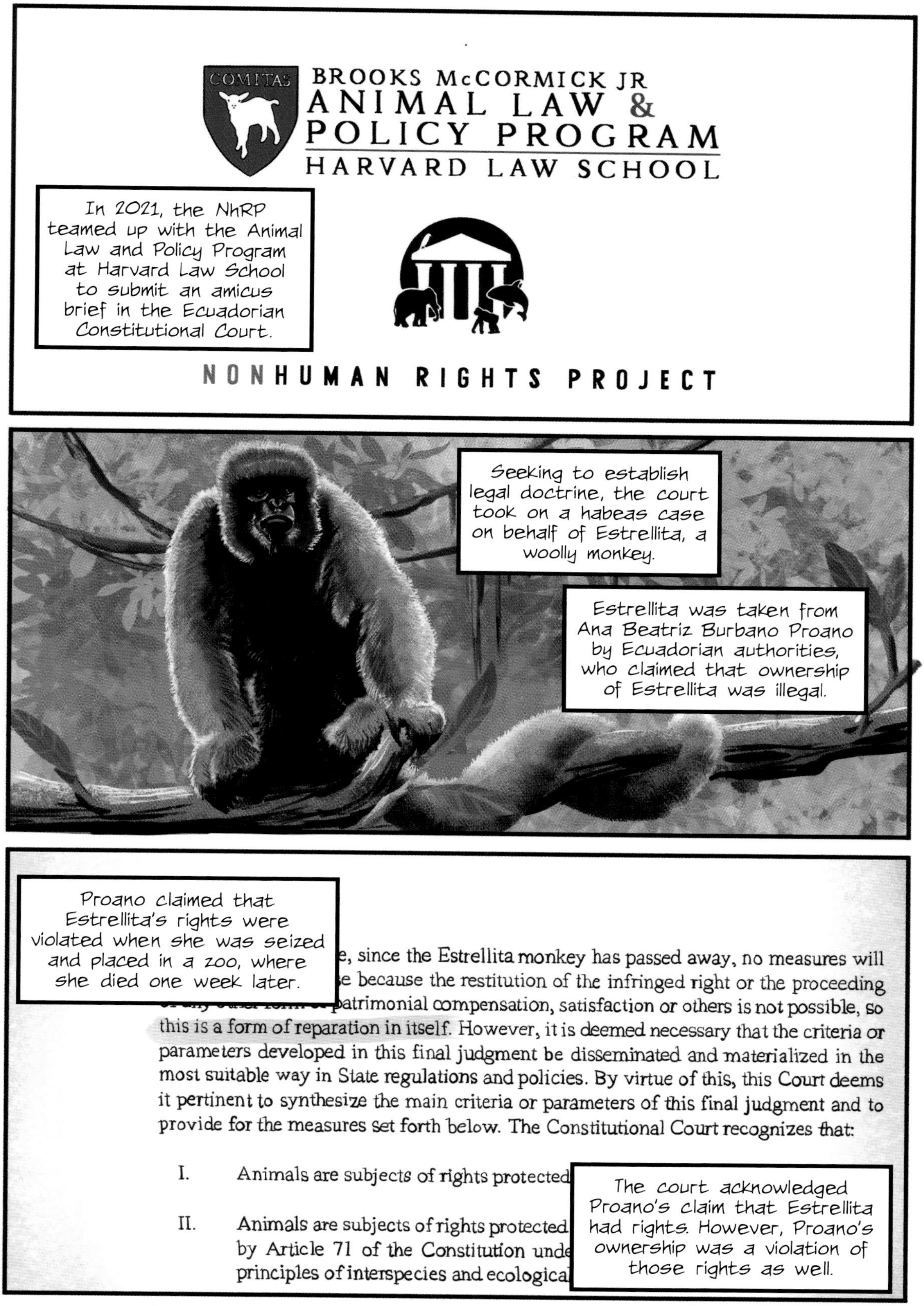
BROOKS McCORMICK JR
ANIMAL LAW &
POLICY PROGRAM
HARVARD LAW SCHOOL
COMITAS
In 2021, the NhRP teamed up with the Animal Law and Policy Program at Harvard Law School to submit an amicus brief in the Ecuadorian Constitutional Court.
NONHUMAN RIGHTS PROJECT
Seeking to establish legal doctrine, the court took on a habeas case on behalf of Estrellita, a woolly monkey.
Estrellita was taken from Ana Beatriz Burbano Proano by Ecuadorian authorities, who claimed that ownership of Estrellita was illegal.
Proano claimed that Estrellita's rights were violated when she was seized and placed in a zoo, where she died one week later.
e, since the Estrellita monkey has passed away, no measures will
e because the restitution of the infringed right or the proceeding
patrimonial compensation, satisfaction or others is not possible, so
this is a form of reparation in itself. However, it is deemed necessary that the criteria or
parameters developed in this final judgment be disseminated and materialized in the
most suitable way in State regulations and policies. By virtue of this, this Court deems
it pertinent to synthesize the main criteria or parameters of this final judgment and to
provide for the measures set forth below. The Constitutional Court recognizes that:
I. Animals are subjects of rights protected
II. Animals are subjects of rights protected
by Article 71 of the Constitution unde
principles of interspecies and ecologica
The court acknowledged Proano's claim that Estrellita had rights. However, Proano's ownership was a violation of those rights as well.

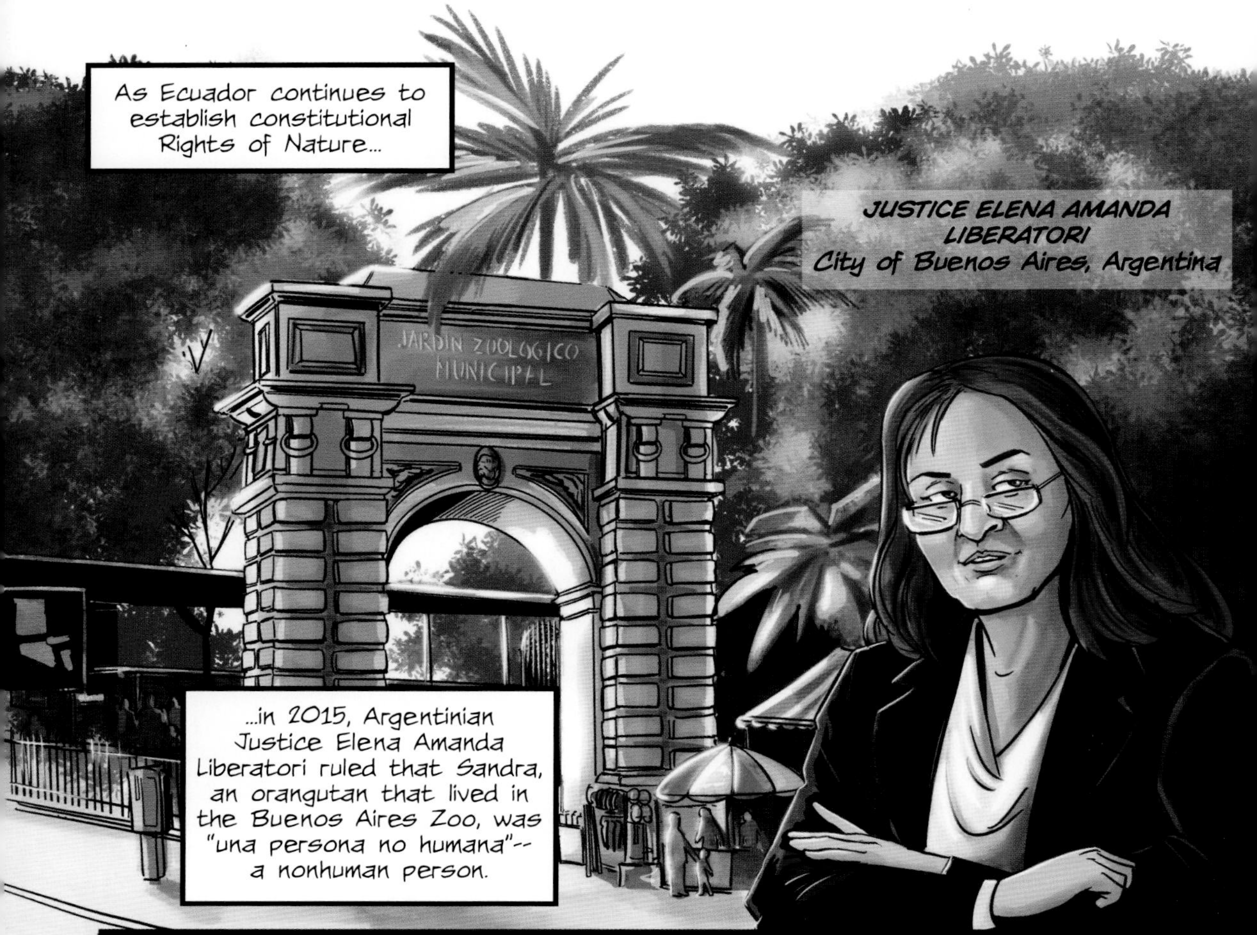
As Ecuador continues to establish constitutional Rights of Nature...
JUSTICE ELENA AMANDA LIBERATORI
City of Buenos Aires, Argentina
JARDIN ZOOLOGICO MUNICIPAL
...in 2015, Argentinian Justice Elena Amanda Liberatori ruled that Sandra, an orangutan that lived in the Buenos Aires Zoo, was "una persona no humana"--a nonhuman person.

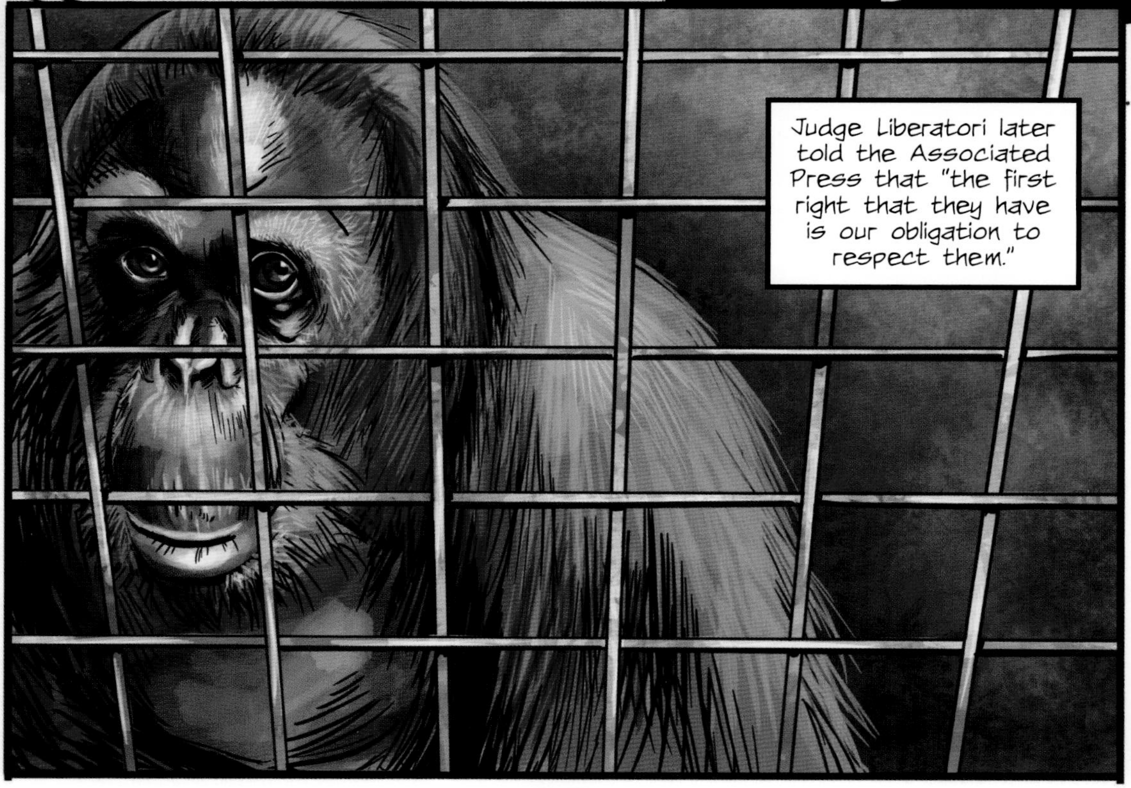
Judge Liberatori later told the Associated Press that "the first right that they have is our obligation to respect them."

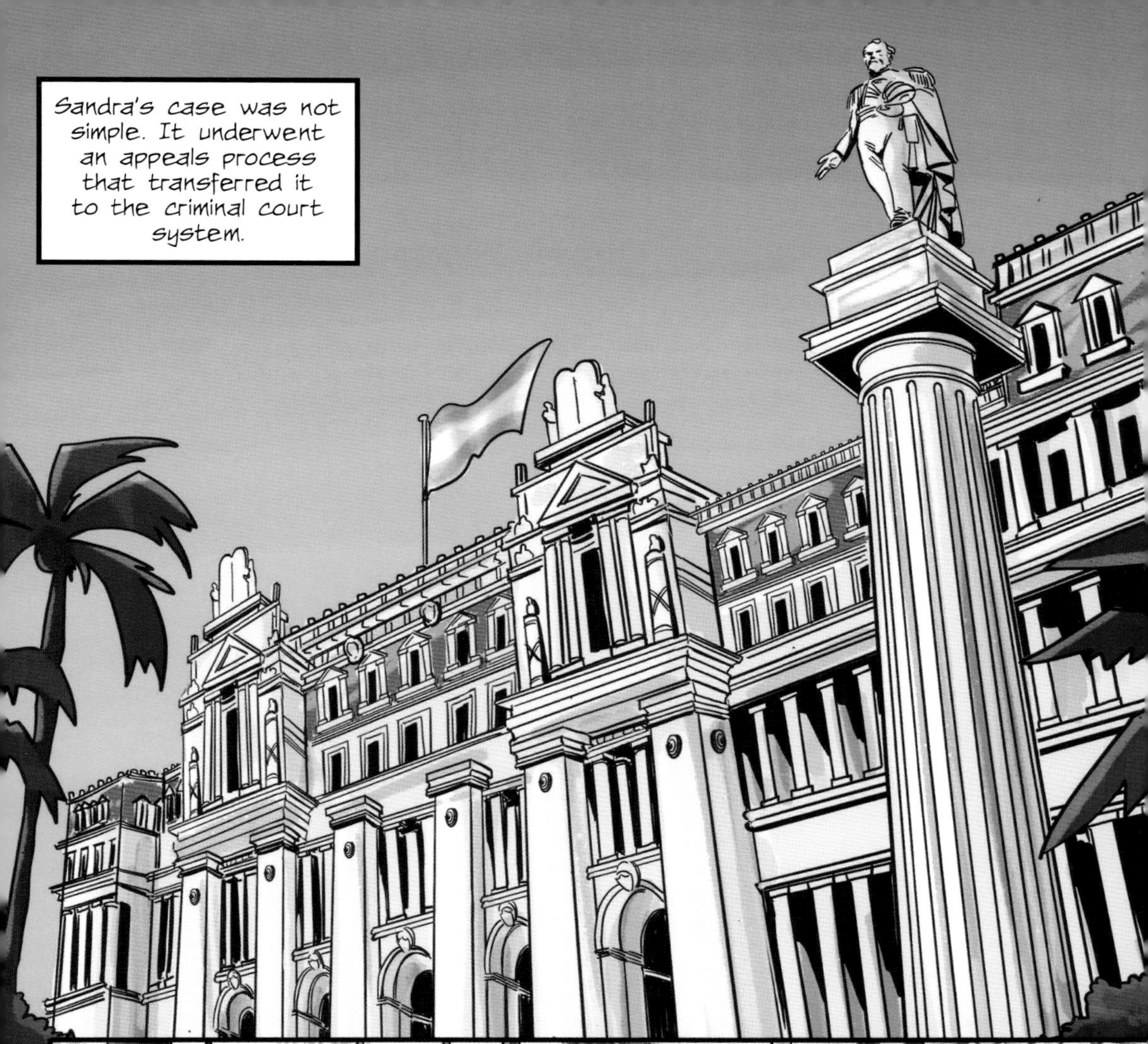
Sandra's case was not simple. It underwent an appeals process that transferred it to the criminal court system.

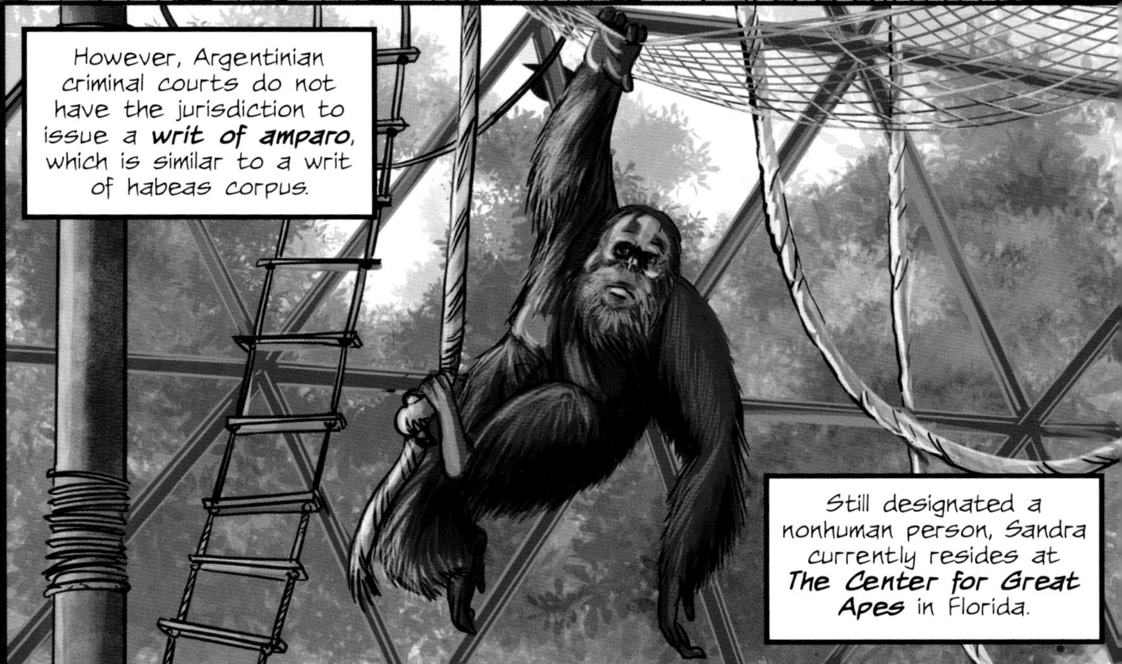
However, Argentinian criminal courts do not have the jurisdiction to issue a ***writ of amparo***, which is similar to a writ of habeas corpus.
Still designated a nonhuman person, Sandra currently resides at ***The Center for Great Apes*** in Florida.

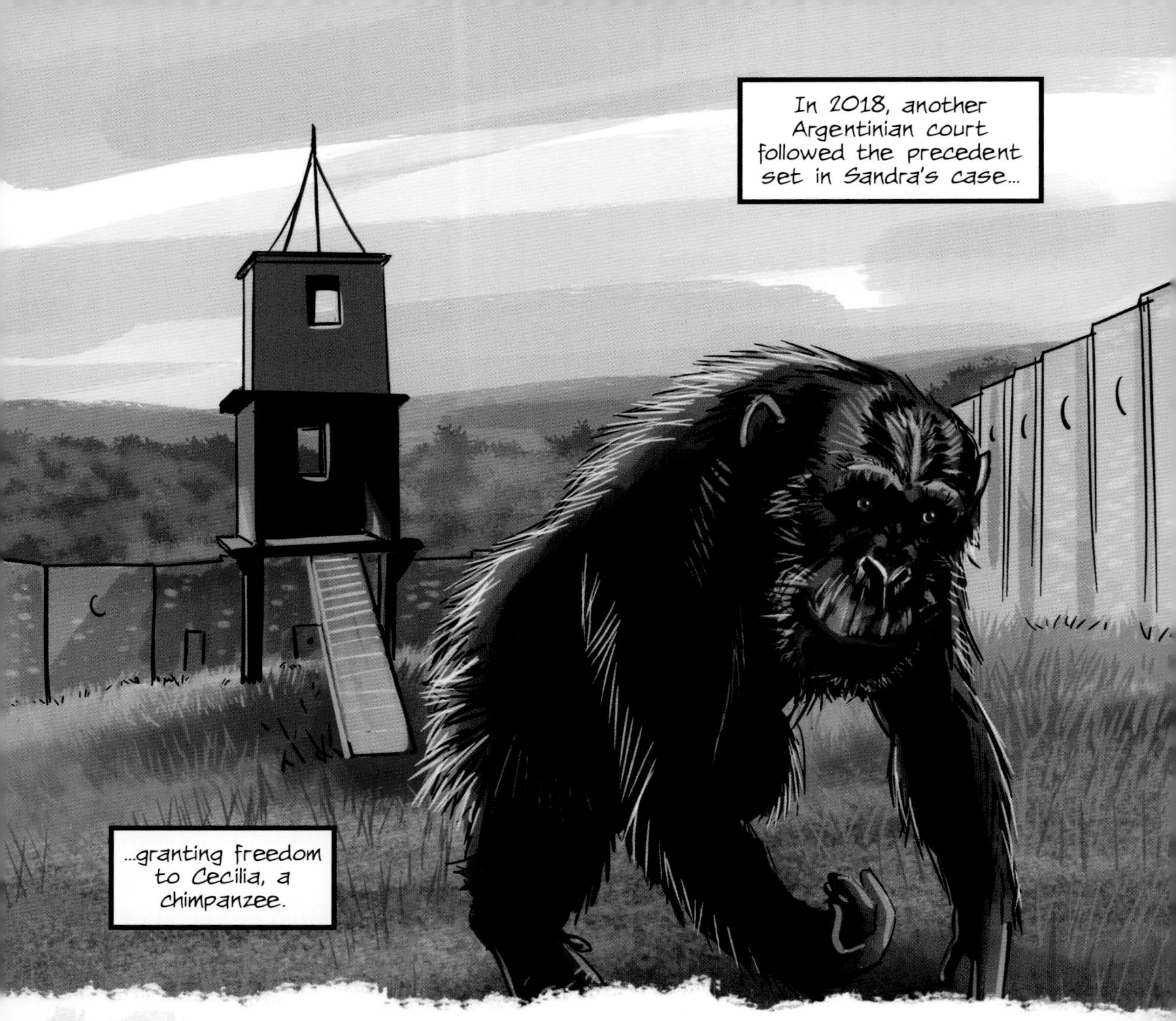
In 2018, another Argentinian court followed the precedent set in Sandra's case...
...granting freedom to Cecilia, a chimpanzee.

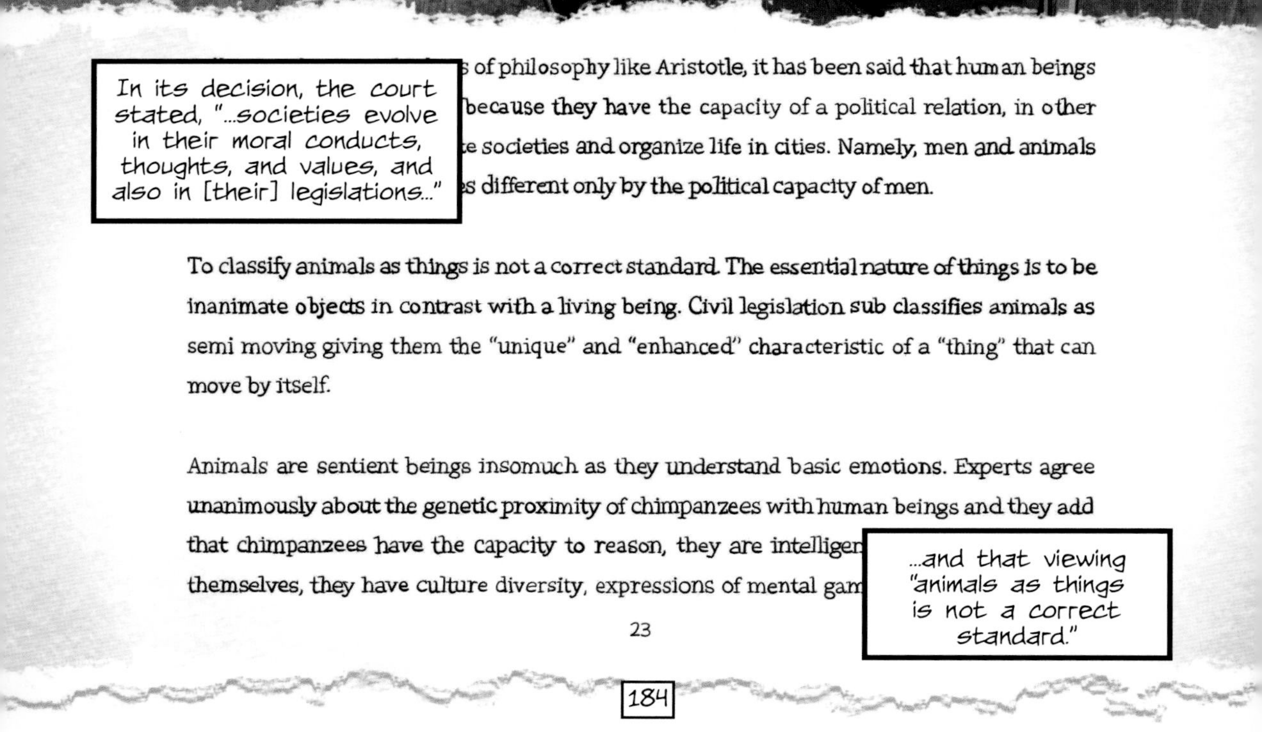
In its decision, the court stated, "...societies evolve in their moral conducts, thoughts, and values, and also in [their] legislations..."

s of philosophy like Aristotle, it has been said that human beings because they have the capacity of a political relation, in other e societies and organize life in cities. Namely, men and animals s different only by the political capacity of men.

To classify animals as things is not a correct standard. The essential nature of things is to be inanimate objects in contrast with a living being. Civil legislation sub classifies animals as semi moving giving them the "unique" and "enhanced" characteristic of a "thing" that can move by itself.

Animals are sentient beings insomuch as they understand basic emotions. Experts agree unanimously about the genetic proximity of chimpanzees with human beings and they add that chimpanzees have the capacity to reason, they are intelligen themselves, they have culture diversity, expressions of mental gam

23

...and that viewing "animals as things is not a correct standard."

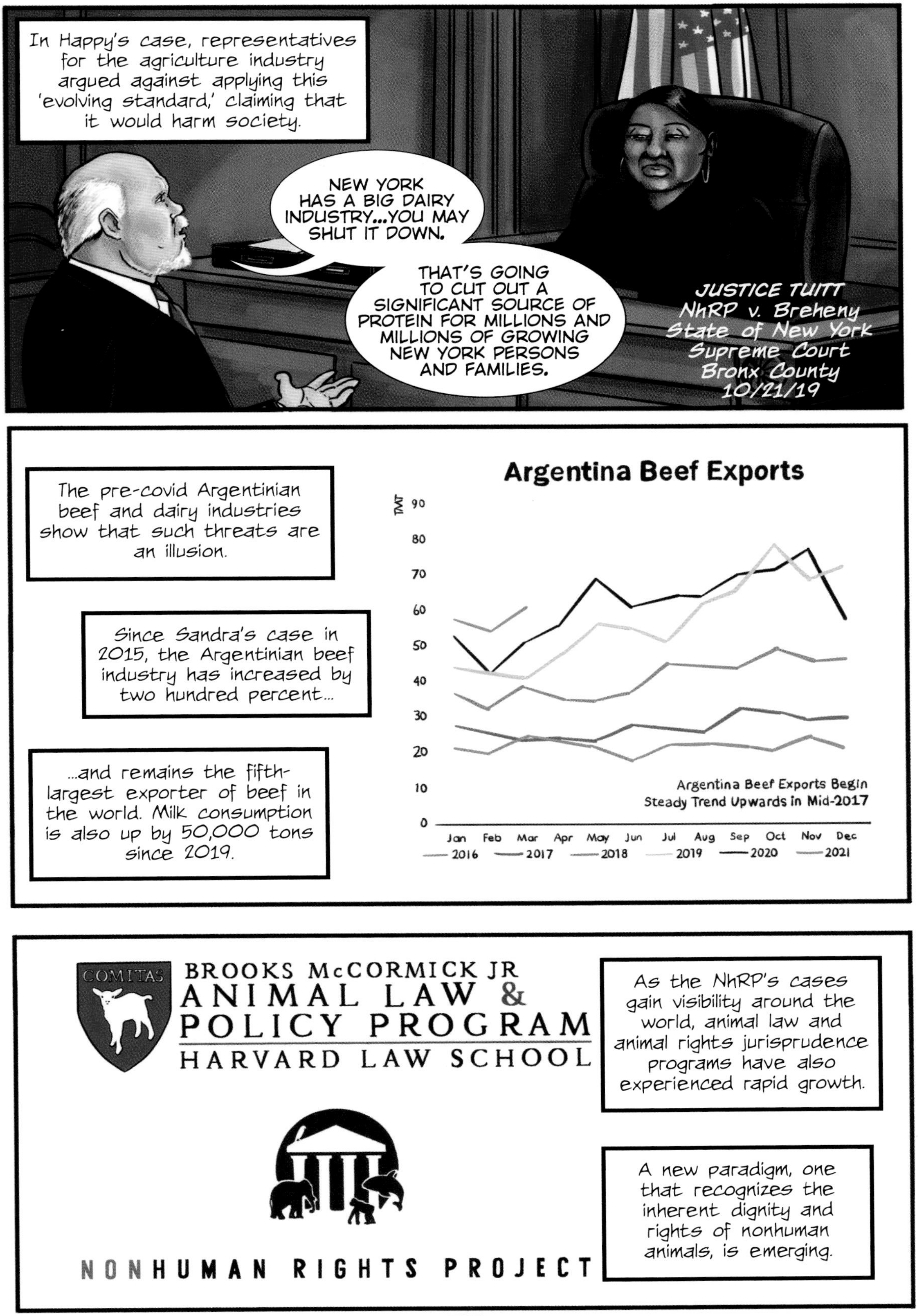
In Happy's case, representatives for the agriculture industry argued against applying this 'evolving standard,' claiming that it would harm society.
NEW YORK HAS A BIG DAIRY INDUSTRY...YOU MAY SHUT IT DOWN.
THAT'S GOING TO CUT OUT A SIGNIFICANT SOURCE OF PROTEIN FOR MILLIONS AND MILLIONS OF GROWING NEW YORK PERSONS AND FAMILIES.
JUSTICE TUITT NhRP v. Breheny State of New York Supreme Court Bronx County 10/21/19
The pre-covid Argentinian beef and dairy industries show that such threats are an illusion.
Since Sandra's case in 2015, the Argentinian beef industry has increased by two hundred percent...
...and remains the fifth-largest exporter of beef in the world. Milk consumption is also up by 50,000 tons since 2019.
Argentina Beef Exports
TMT
90
80
70
60
50
40
30
20
10
0
Jan Feb Mar Apr May Jun Jul Aug Sep Oct Nov Dec
2016 2017 2018 2019 2020 2021
Argentina Beef Exports Begin Steady Trend Upwards in Mid-2017
COMITAS
BROOKS McCORMICK JR ANIMAL LAW & POLICY PROGRAM HARVARD LAW SCHOOL
As the NhRP's cases gain visibility around the world, animal law and animal rights jurisprudence programs have also experienced rapid growth.
NONHUMAN RIGHTS PROJECT
A new paradigm, one that recognizes the inherent dignity and rights of nonhuman animals, is emerging.

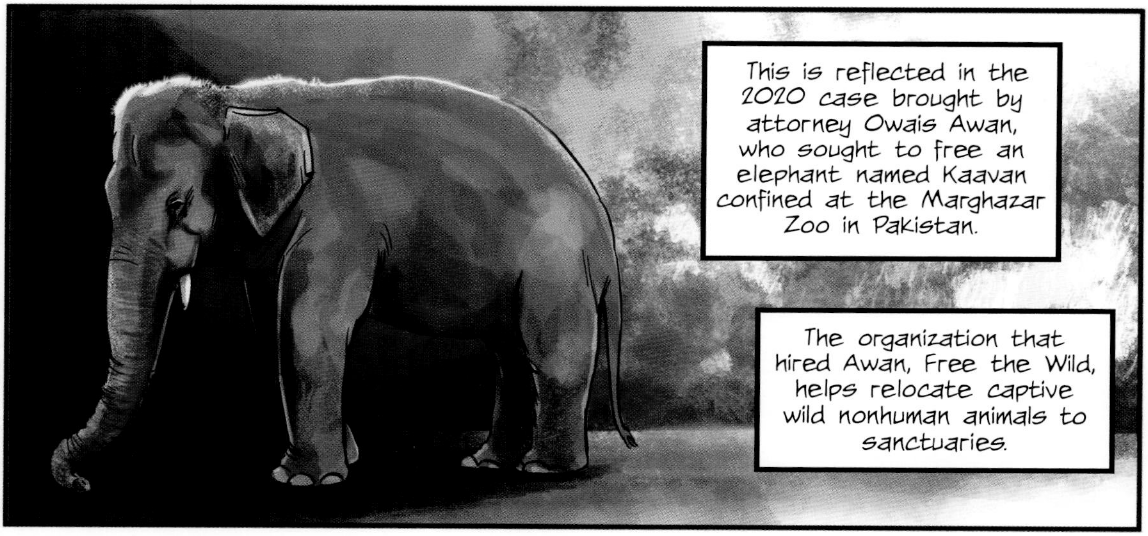
This is reflected in the 2020 case brought by attorney Owais Awan, who sought to free an elephant named Kaavan confined at the Marghazar Zoo in Pakistan.
The organization that hired Awan, Free the Wild, helps relocate captive wild nonhuman animals to sanctuaries.

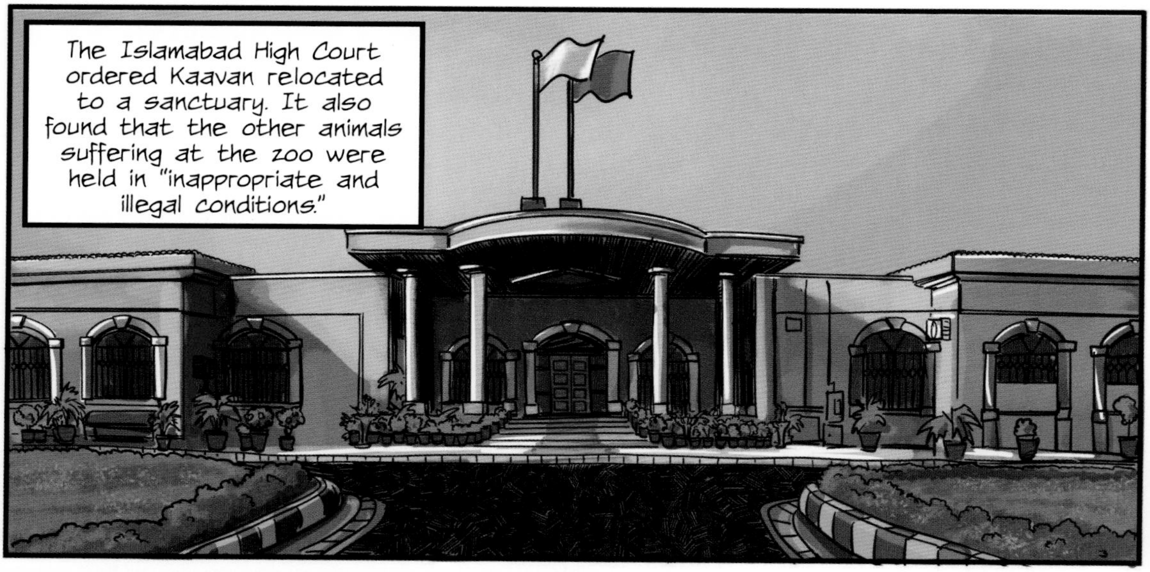
The Islamabad High Court ordered Kaavan relocated to a sanctuary. It also found that the other animals suffering at the zoo were held in "inappropriate and illegal conditions."

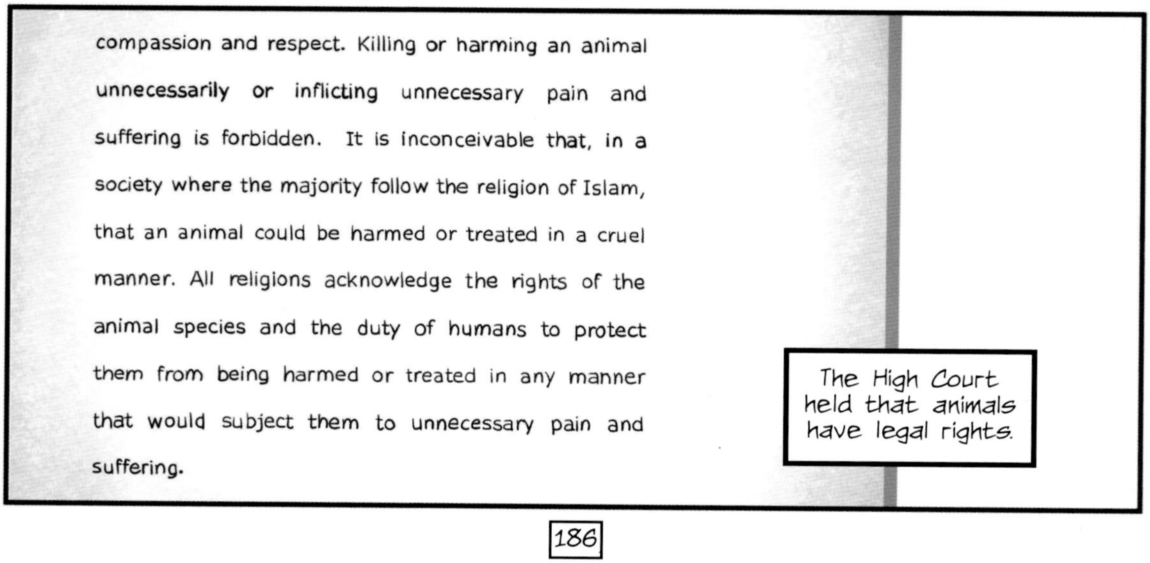
compassion and respect. Killing or harming an animal unnecessarily or inflicting unnecessary pain and suffering is forbidden. It is inconceivable that, in a society where the majority follow the religion of Islam, that an animal could be harmed or treated in a cruel manner. All religions acknowledge the rights of the animal species and the duty of humans to protect them from being harmed or treated in any manner that would subject them to unnecessary pain and suffering.
The High Court held that animals have legal rights.

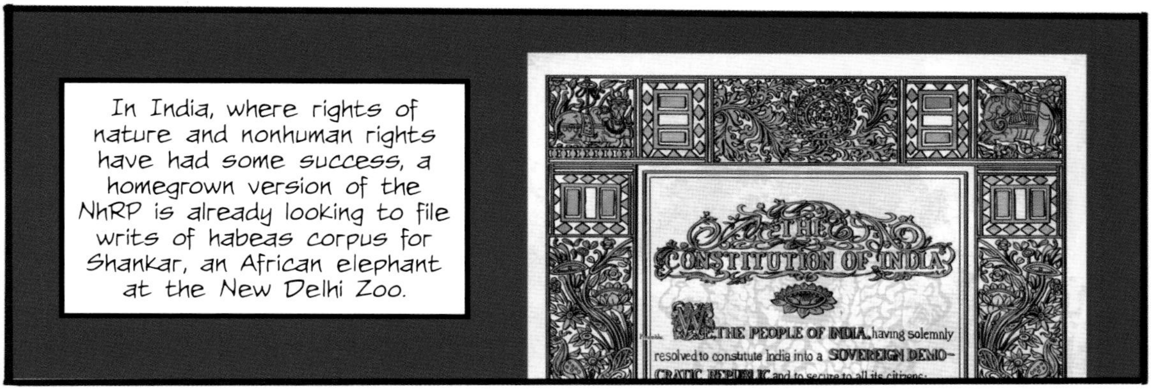
In India, where rights of nature and nonhuman rights have had some success, a homegrown version of the NhRP is already looking to file writs of habeas corpus for Shankar, an African elephant at the New Delhi Zoo.
THE CONSTITUTION OF INDIA
WE, THE PEOPLE OF INDIA, having solemnly
resolved to constitute India into a SOVEREIGN DEMO-

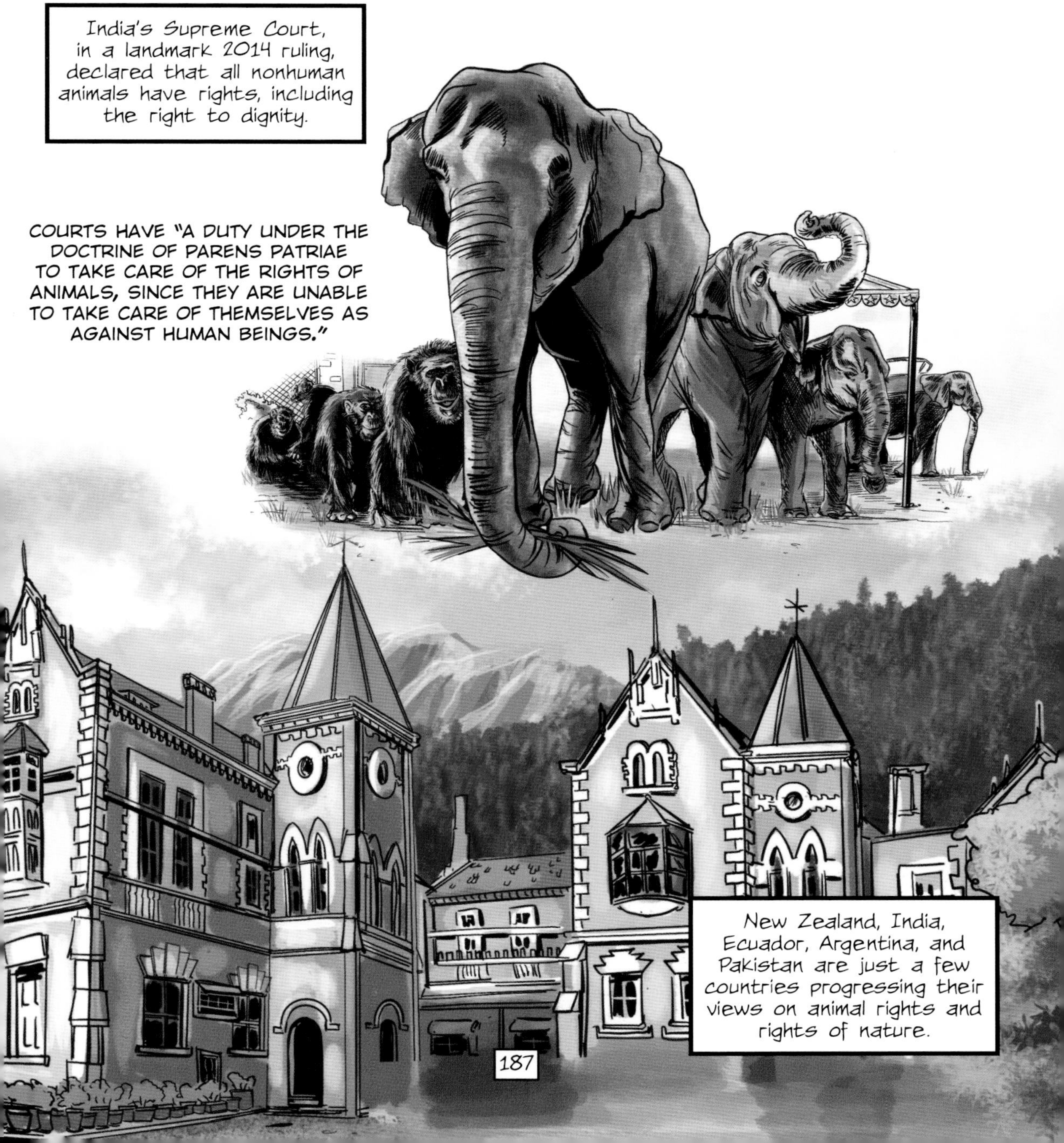
India's Supreme Court, in a landmark 2014 ruling, declared that all nonhuman animals have rights, including the right to dignity.
COURTS HAVE "A DUTY UNDER THE DOCTRINE OF PARENS PATRIAE TO TAKE CARE OF THE RIGHTS OF ANIMALS, SINCE THEY ARE UNABLE TO TAKE CARE OF THEMSELVES AS AGAINST HUMAN BEINGS."
New Zealand, India, Ecuador, Argentina, and Pakistan are just a few countries progressing their views on animal rights and rights of nature.

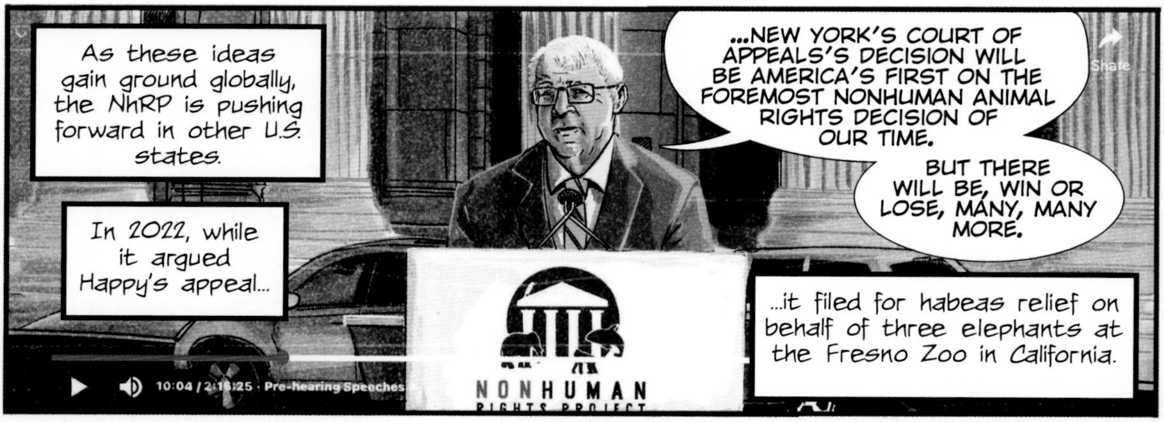
As these ideas gain ground globally, the NhRP is pushing forward in other U.S. states.
...NEW YORK'S COURT OF APPEALS'S DECISION WILL BE AMERICA'S FIRST ON THE FOREMOST NONHUMAN ANIMAL RIGHTS DECISION OF OUR TIME.
BUT THERE WILL BE, WIN OR LOSE, MANY, MANY MORE.
In 2022, while it argued Happy's appeal...
...it filed for habeas relief on behalf of three elephants at the Fresno Zoo in California.
10:04 / 2:18:25 · Pre-hearing Speeches
NONHUMAN

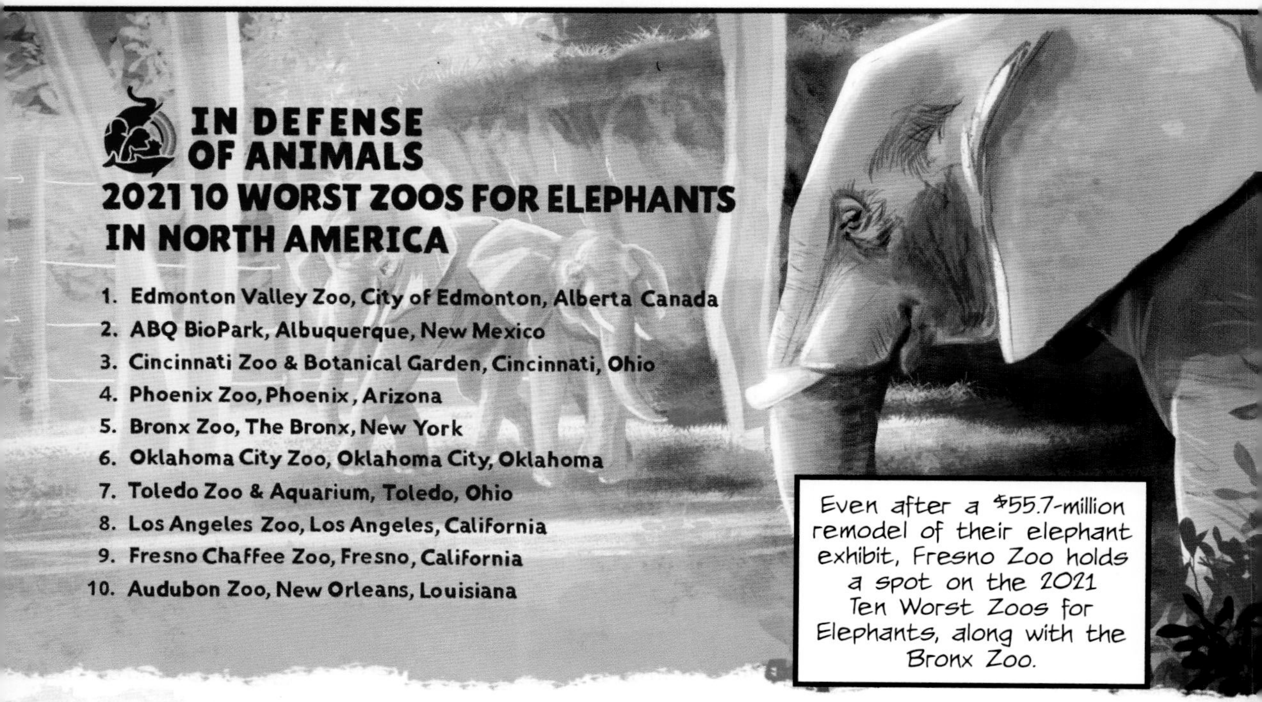
IN DEFENSE OF ANIMALS
2021 10 WORST ZOOS FOR ELEPHANTS IN NORTH AMERICA
1. Edmonton Valley Zoo, City of Edmonton, Alberta Canada
2. ABQ BioPark, Albuquerque, New Mexico
3. Cincinnati Zoo & Botanical Garden, Cincinnati, Ohio
4. Phoenix Zoo, Phoenix, Arizona
5. Bronx Zoo, The Bronx, New York
6. Oklahoma City Zoo, Oklahoma City, Oklahoma
7. Toledo Zoo & Aquarium, Toledo, Ohio
8. Los Angeles Zoo, Los Angeles, California
9. Fresno Chaffee Zoo, Fresno, California
10. Audubon Zoo, New Orleans, Louisiana
Even after a $55.7-million remodel of their elephant exhibit, Fresno Zoo holds a spot on the 2021 Ten Worst Zoos for Elephants, along with the Bronx Zoo.

c. **Progress of Society**

123. The California Supreme Court has long made clear that the common law is "constantly expanding and developing in keeping with advancing civilization and the new conditions and progress of society." *Rodriguez*, 12 Cal.3d at 394 (citation omitted). "The nature of the common law requires that each time a rule of law is applied, it be carefully scrutinized to make sure that the conditions and needs of the times have not so changed as to make further application of it the instrument of injustice." *Id.*[258]

that elephants

Citing California case law, the NhRP reminded the court that societies evolve in their values and conduct.

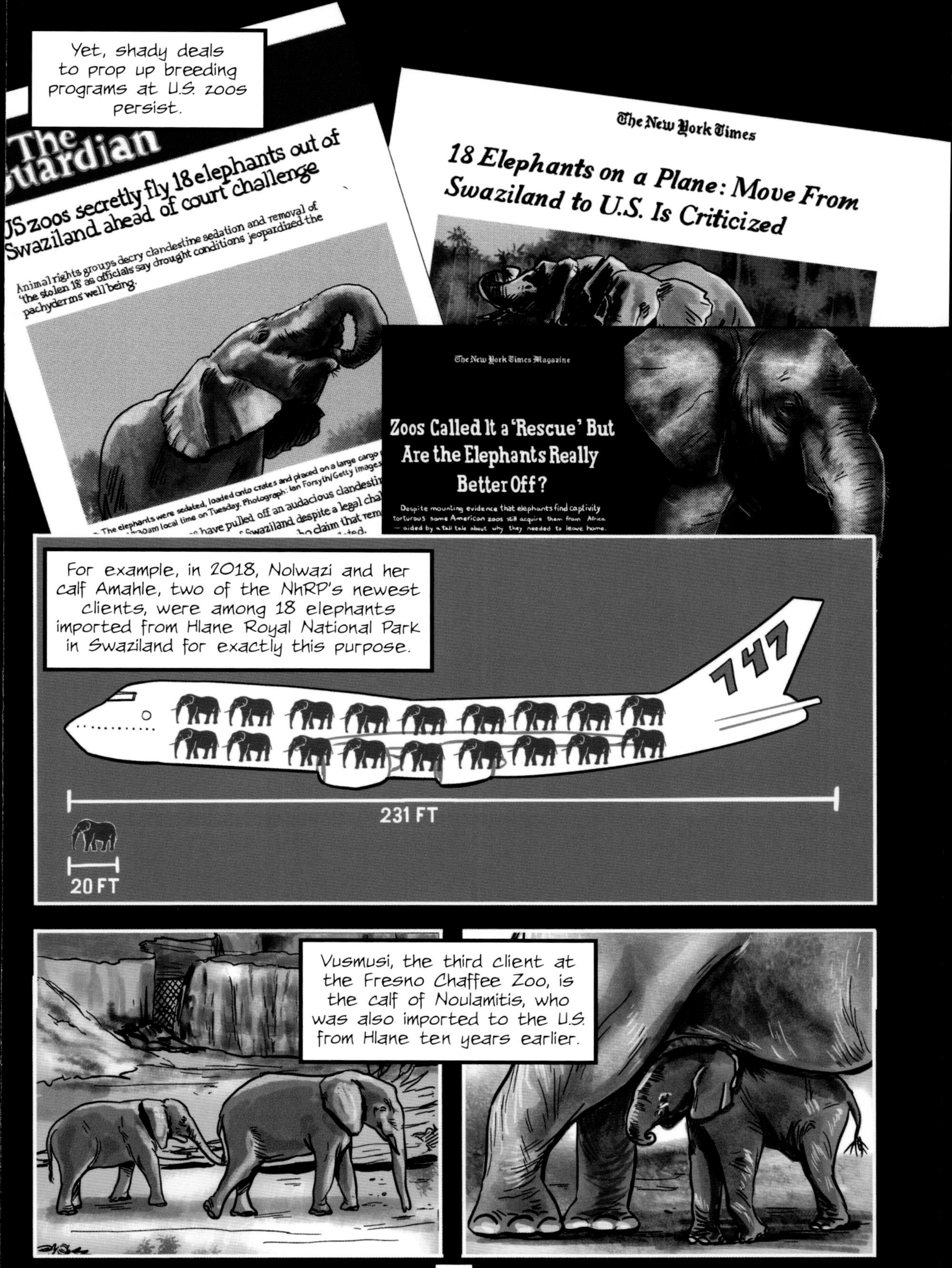

Yet, shady deals to prop up breeding programs at U.S. zoos persist.
The Guardian
US zoos secretly fly 18 elephants out of Swaziland ahead of court challenge
Animal rights groups decry clandestine sedation and removal of 'the stolen 18' as officials say drought conditions jeopardized the pachyderms' well being.
The New York Times
18 Elephants on a Plane: Move From Swaziland to U.S. Is Criticized
The New York Times Magazine
Zoos Called It a 'Rescue' But Are the Elephants Really Better Off?
Despite mounting evidence that elephants find captivity torturous some American zoos still acquire them from Africa — aided by a tall tale about why they needed to leave home.
For example, in 2018, Nolwazi and her calf Amahle, two of the NhRP's newest clients, were among 18 elephants imported from Hlane Royal National Park in Swaziland for exactly this purpose.
747
231 FT
20 FT
Vusmusi, the third client at the Fresno Chaffee Zoo, is the calf of Noulamitis, who was also imported to the U.S. from Hlane ten years earlier.

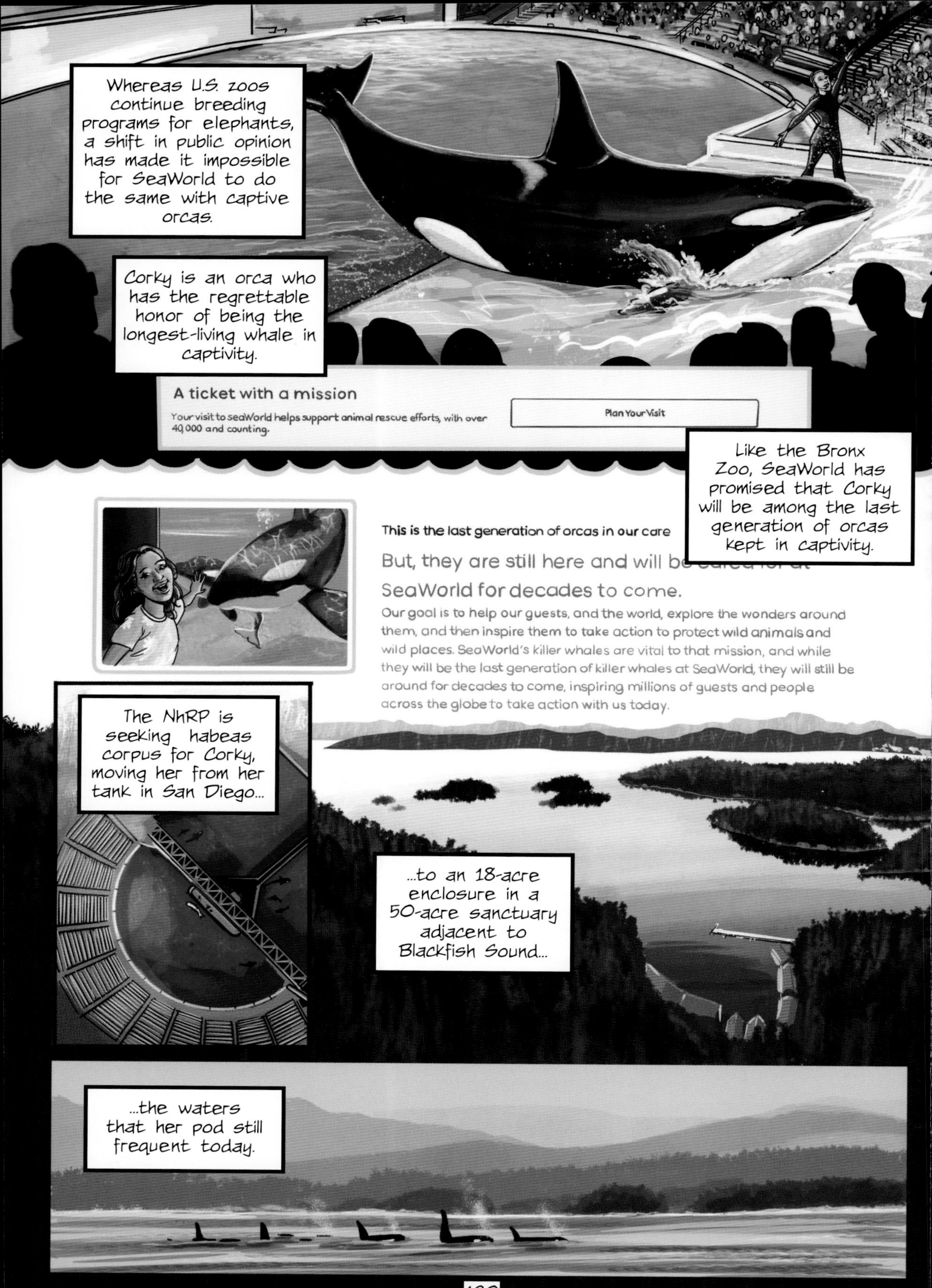
Whereas U.S. zoos continue breeding programs for elephants, a shift in public opinion has made it impossible for SeaWorld to do the same with captive orcas.
Corky is an orca who has the regrettable honor of being the longest-living whale in captivity.
A ticket with a mission
Your visit to seaWorld helps support animal rescue efforts, with over 40,000 and counting.
Plan Your Visit
Like the Bronx Zoo, SeaWorld has promised that Corky will be among the last generation of orcas kept in captivity.
This is the last generation of orcas in our care
But, they are still here and will be
SeaWorld for decades to come.
Our goal is to help our guests, and the world, explore the wonders around them, and then inspire them to take action to protect wild animals and wild places. SeaWorld's killer whales are vital to that mission, and while they will be the last generation of killer whales at SeaWorld, they will still be around for decades to come, inspiring millions of guests and people across the globe to take action with us today.
The NhRP is seeking habeas corpus for Corky, moving her from her tank in San Diego...
...to an 18-acre enclosure in a 50-acre sanctuary adjacent to Blackfish Sound...
...the waters that her pod still frequent today.

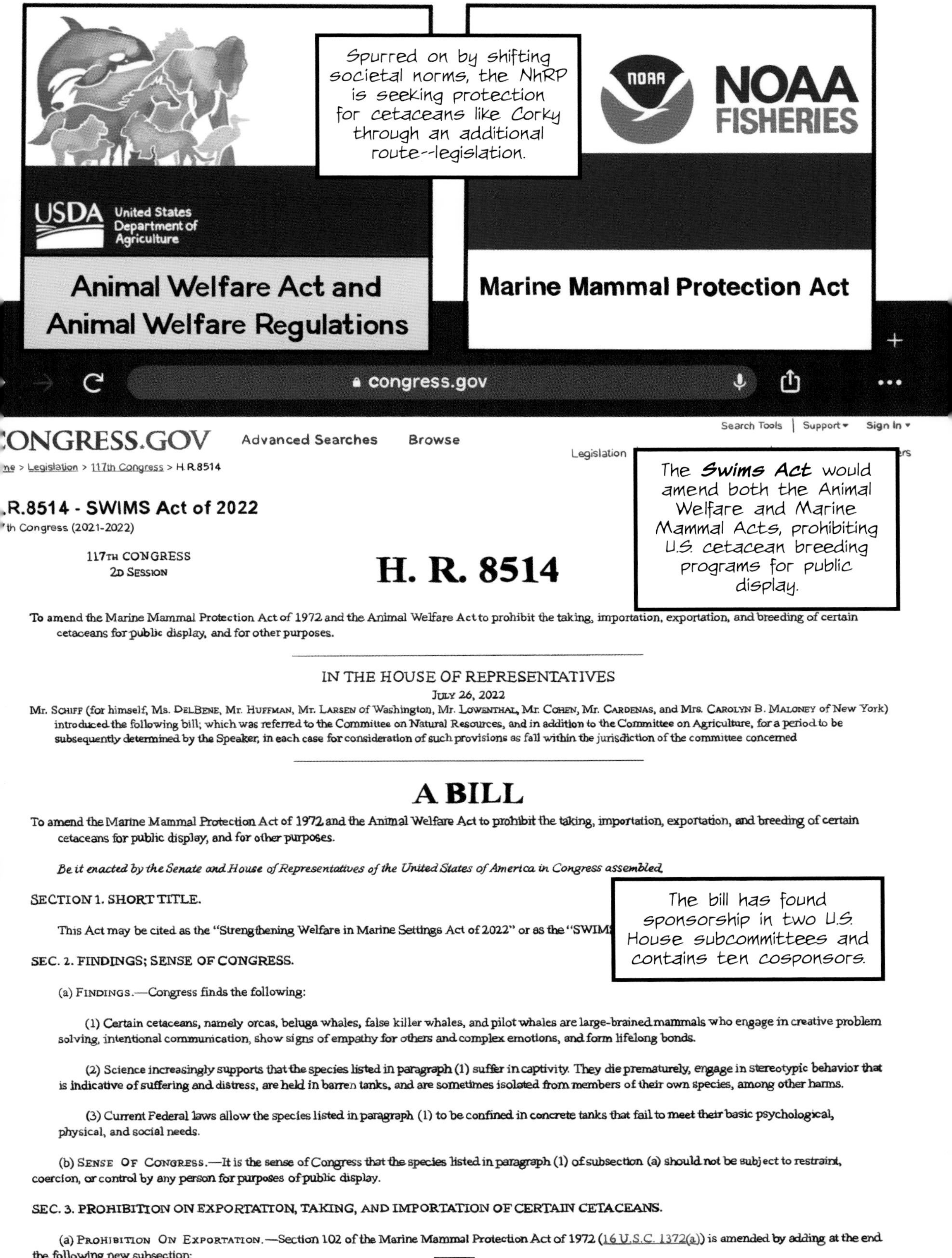

Spurred on by shifting societal norms, the NhRP is seeking protection for cetaceans like Corky through an additional route--legislation.
USDA United States Department of Agriculture
Animal Welfare Act and Animal Welfare Regulations
NOAA
NOAA FISHERIES
Marine Mammal Protection Act
congress.gov
ONGRESS.GOV
Advanced Searches
Browse
Search Tools | Support Sign In
Legislation
ne > Legislation > 117th Congress > H.R.8514
R.8514 - SWIMS Act of 2022
th Congress (2021-2022)
The Swims Act would amend both the Animal Welfare and Marine Mammal Acts, prohibiting U.S. cetacean breeding programs for public display.
117TH CONGRESS
2D SESSION
H. R. 8514
To amend the Marine Mammal Protection Act of 1972 and the Animal Welfare Act to prohibit the taking, importation, exportation, and breeding of certain cetaceans for public display, and for other purposes.
IN THE HOUSE OF REPRESENTATIVES
JULY 26, 2022
Mr. SCHIFF (for himself, Ms. DELBENE, Mr. HUFFMAN, Mr. LARSEN of Washington, Mr. LOWENTHAL, Mr. COHEN, Mr. CARDENAS, and Mrs. CAROLYN B. MALONEY of New York) introduced the following bill; which was referred to the Committee on Natural Resources, and in addition to the Committee on Agriculture, for a period to be subsequently determined by the Speaker, in each case for consideration of such provisions as fall within the jurisdiction of the committee concerned
A BILL
To amend the Marine Mammal Protection Act of 1972 and the Animal Welfare Act to prohibit the taking, importation, exportation, and breeding of certain cetaceans for public display, and for other purposes.
Be it enacted by the Senate and House of Representatives of the United States of America in Congress assembled,
SECTION 1. SHORT TITLE.
This Act may be cited as the "Strengthening Welfare in Marine Settings Act of 2022" or as the "SWIM
The bill has found sponsorship in two U.S. House subcommittees and contains ten cosponsors.
SEC. 2. FINDINGS; SENSE OF CONGRESS.
(a) FINDINGS.—Congress finds the following:
(1) Certain cetaceans, namely orcas, beluga whales, false killer whales, and pilot whales are large-brained mammals who engage in creative problem solving, intentional communication, show signs of empathy for others and complex emotions, and form lifelong bonds.
(2) Science increasingly supports that the species listed in paragraph (1) suffer in captivity. They die prematurely, engage in stereotypic behavior that is indicative of suffering and distress, are held in barren tanks, and are sometimes isolated from members of their own species, among other harms.
(3) Current Federal laws allow the species listed in paragraph (1) to be confined in concrete tanks that fail to meet their basic psychological, physical, and social needs.
(b) SENSE OF CONGRESS.—It is the sense of Congress that the species listed in paragraph (1) of subsection (a) should not be subject to restraint, coercion, or control by any person for purposes of public display.
SEC. 3. PROHIBITION ON EXPORTATION, TAKING, AND IMPORTATION OF CERTAIN CETACEANS.
(a) PROHIBITION ON EXPORTATION.—Section 102 of the Marine Mammal Protection Act of 1972 (16 U.S.C. 1372(a)) is amended by adding at the end the following new subsection:

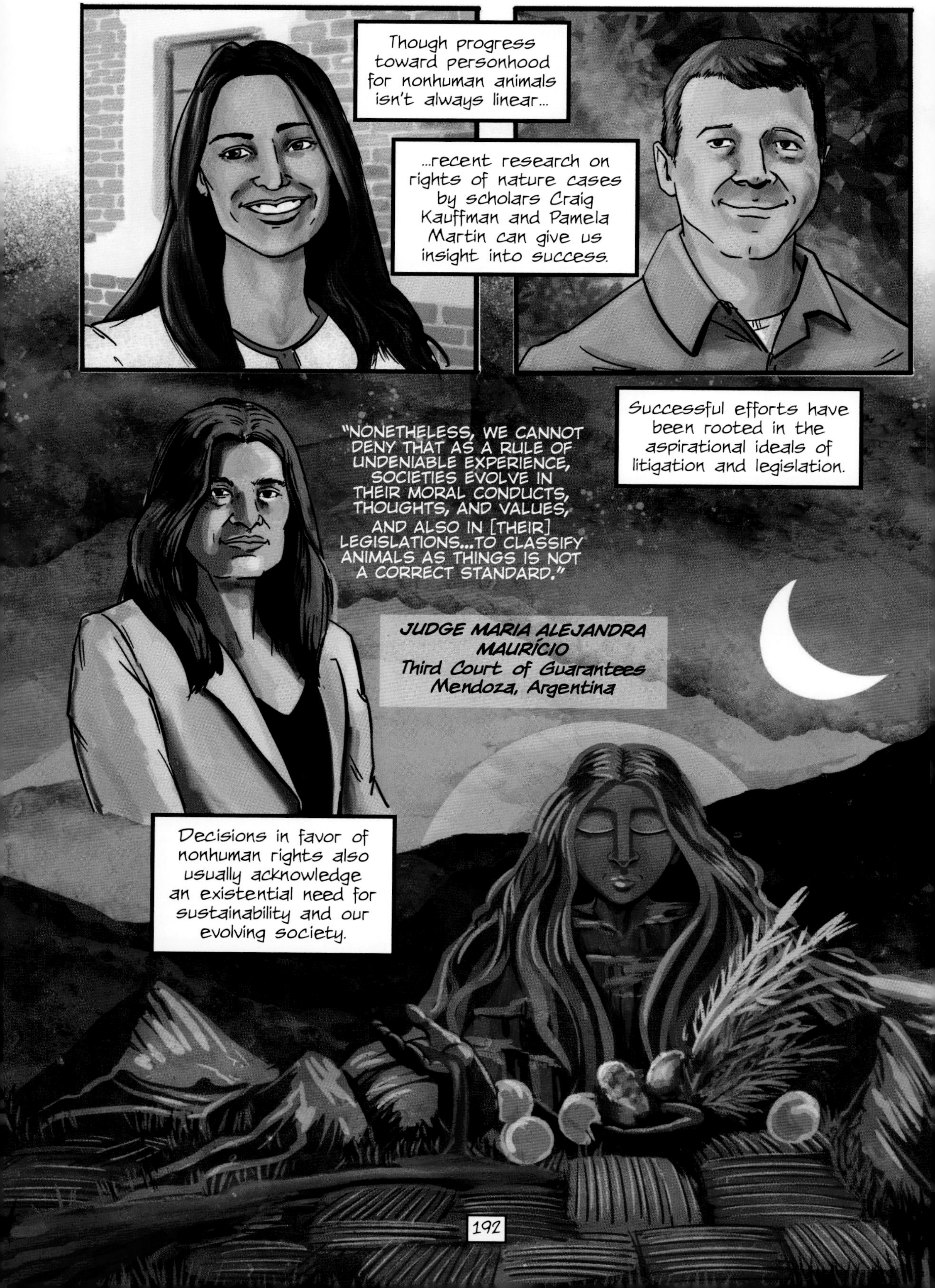
Though progress toward personhood for nonhuman animals isn't always linear...
...recent research on rights of nature cases by scholars Craig Kauffman and Pamela Martin can give us insight into success.
Successful efforts have been rooted in the aspirational ideals of litigation and legislation.
"NONETHELESS, WE CANNOT DENY THAT AS A RULE OF UNDENIABLE EXPERIENCE, SOCIETIES EVOLVE IN THEIR MORAL CONDUCTS, THOUGHTS, AND VALUES, AND ALSO IN [THEIR] LEGISLATIONS...TO CLASSIFY ANIMALS AS THINGS IS NOT A CORRECT STANDARD."
JUDGE MARIA ALEJANDRA MAURÍCIO
Third Court of Guarantees
Mendoza, Argentina
Decisions in favor of nonhuman rights also usually acknowledge an existential need for sustainability and our evolving society.

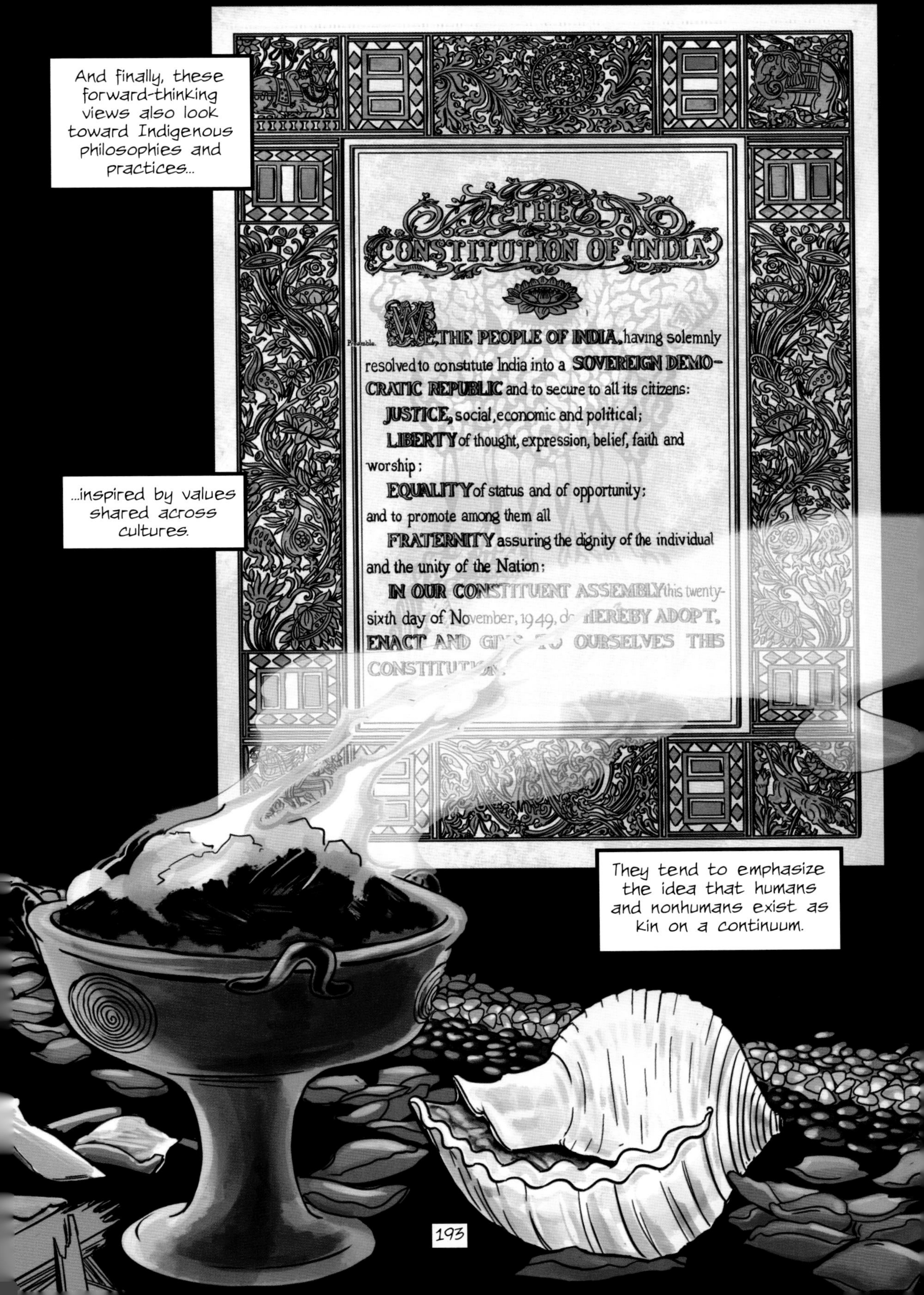

And finally, these forward-thinking views also look toward Indigenous philosophies and practices...
THE CONSTITUTION OF INDIA
Preamble.
WE, THE PEOPLE OF INDIA, having solemnly resolved to constitute India into a SOVEREIGN DEMO-CRATIC REPUBLIC and to secure to all its citizens:
JUSTICE, social, economic and political;
LIBERTY of thought, expression, belief, faith and worship;
EQUALITY of status and of opportunity;
and to promote among them all
FRATERNITY assuring the dignity of the individual and the unity of the Nation;
IN OUR CONSTITUENT ASSEMBLY this twenty-sixth day of November, 1949,
HEREBY ADOPT,
ENACT AND
TO OURSELVES THIS
...inspired by values shared across cultures.
They tend to emphasize the idea that humans and nonhumans exist as kin on a continuum.

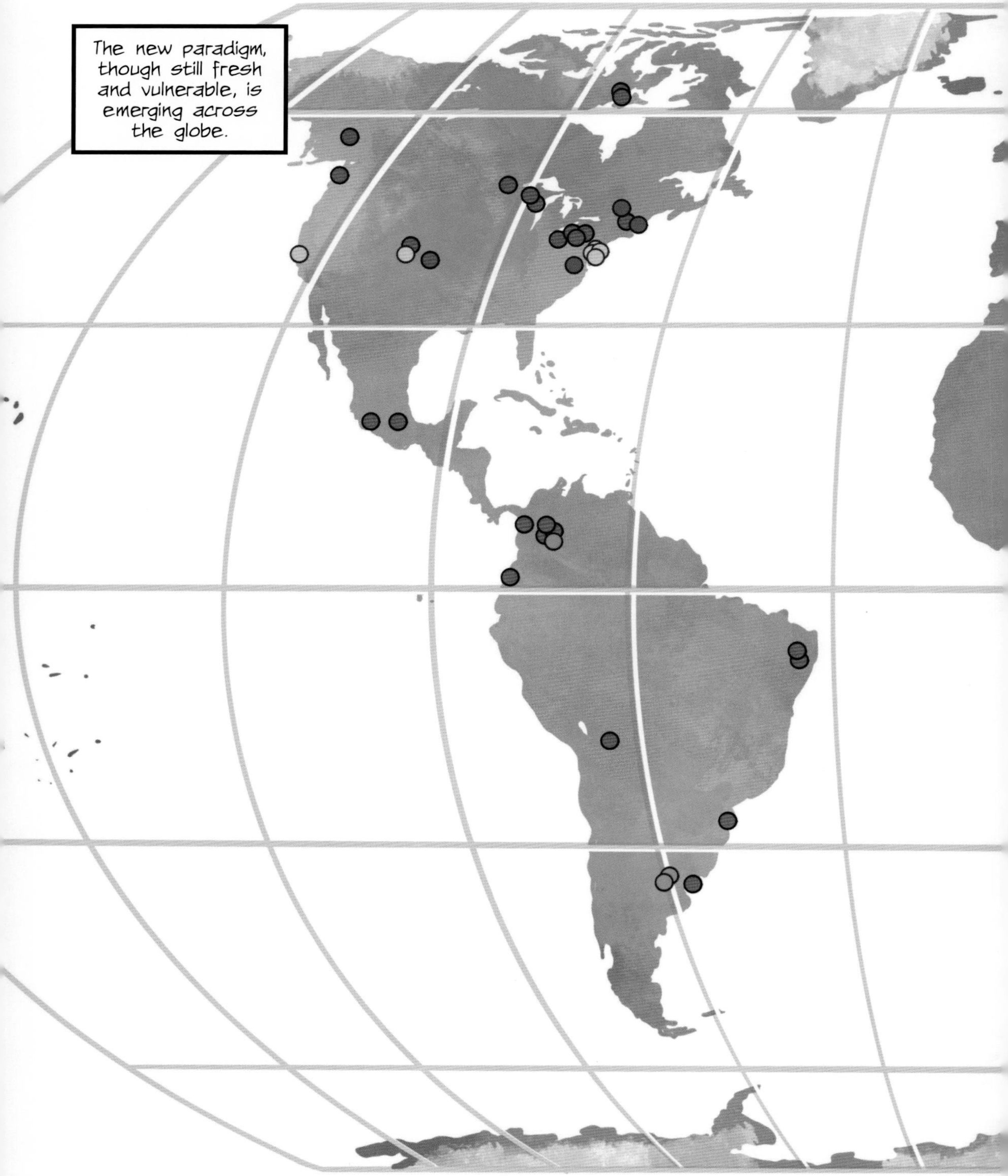
The new paradigm, though still fresh and vulnerable, is emerging across the globe.

Rights of Nature by law or statute

Nonhuman Personhood and Rights by law or statute

Nonhuman Rights cases

Animal Rights Jurisprudence programs

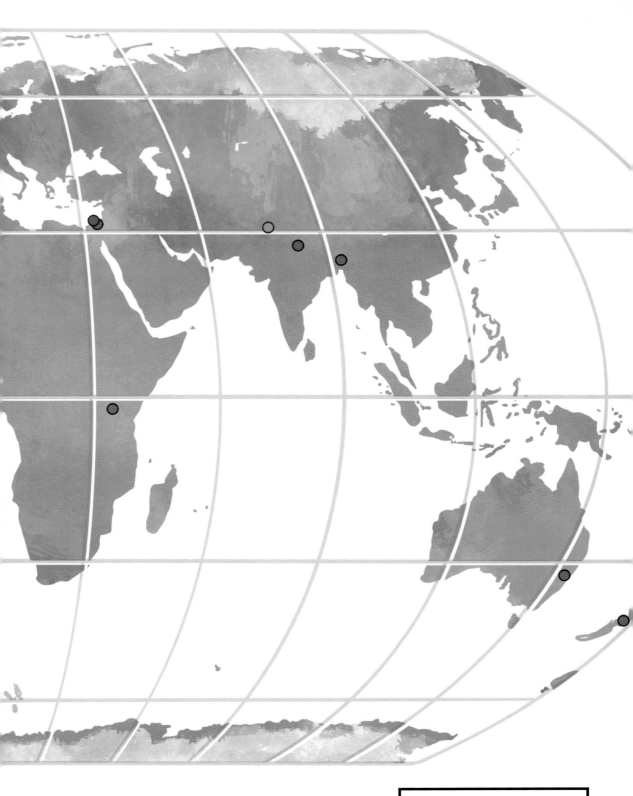

It recognizes our kinship with nonhuman animals and our duty to protect their dignity.

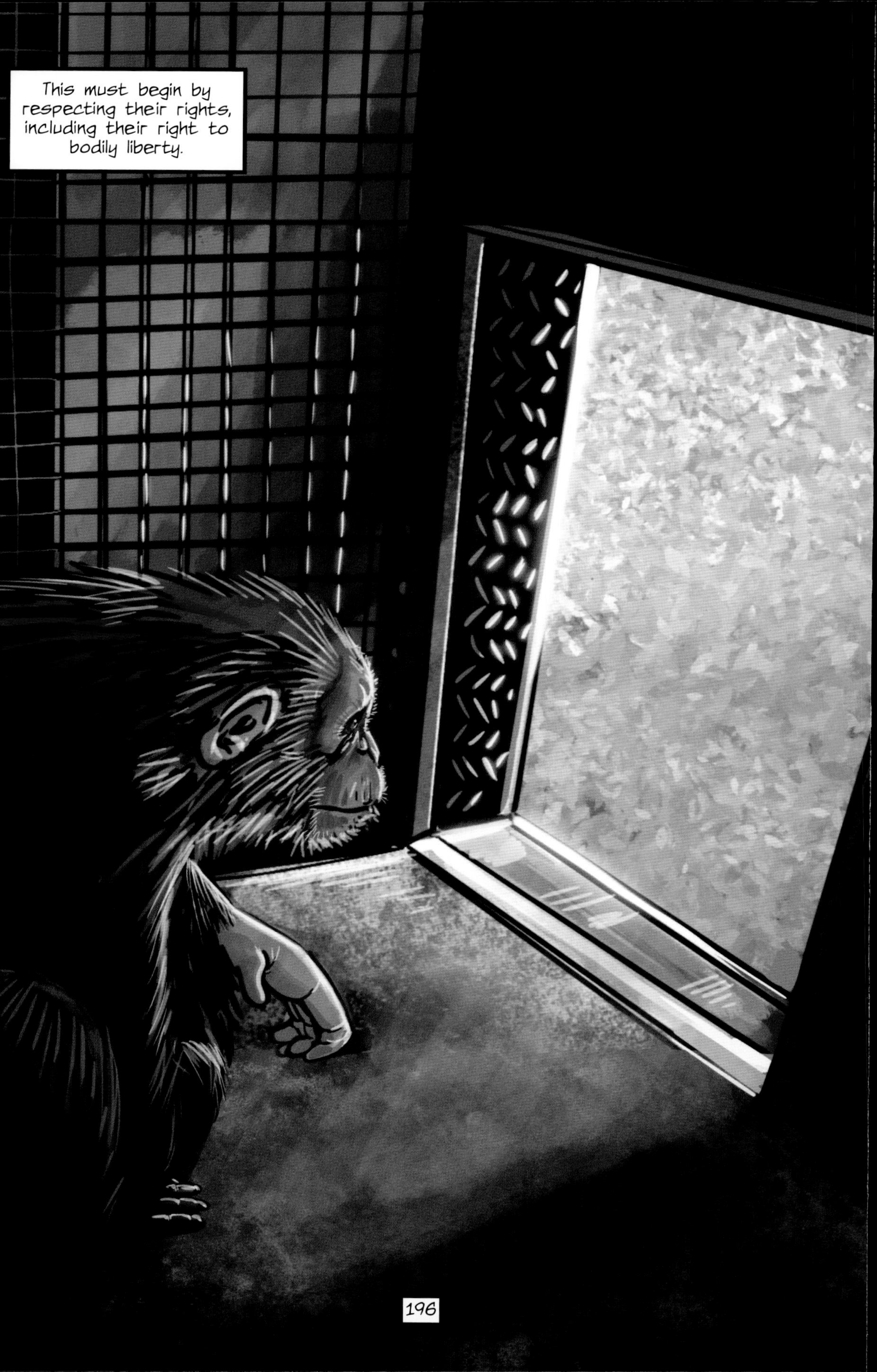
This must begin by respecting their rights, including their right to bodily liberty.

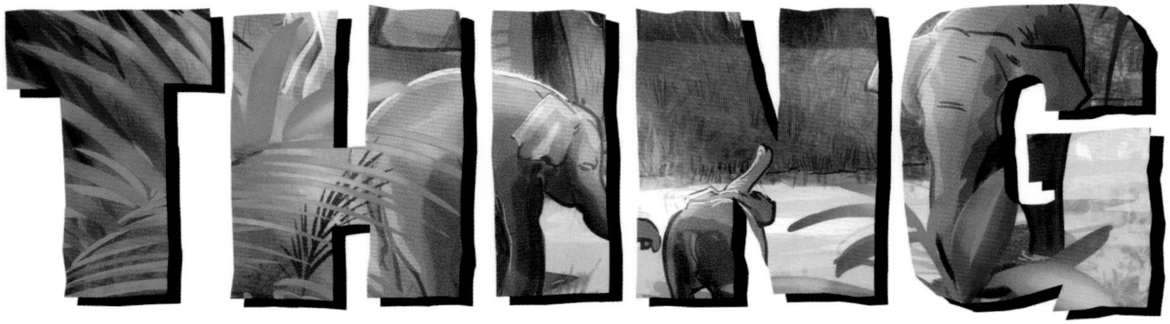
THING

CHAPTER 9: SANCTUARY

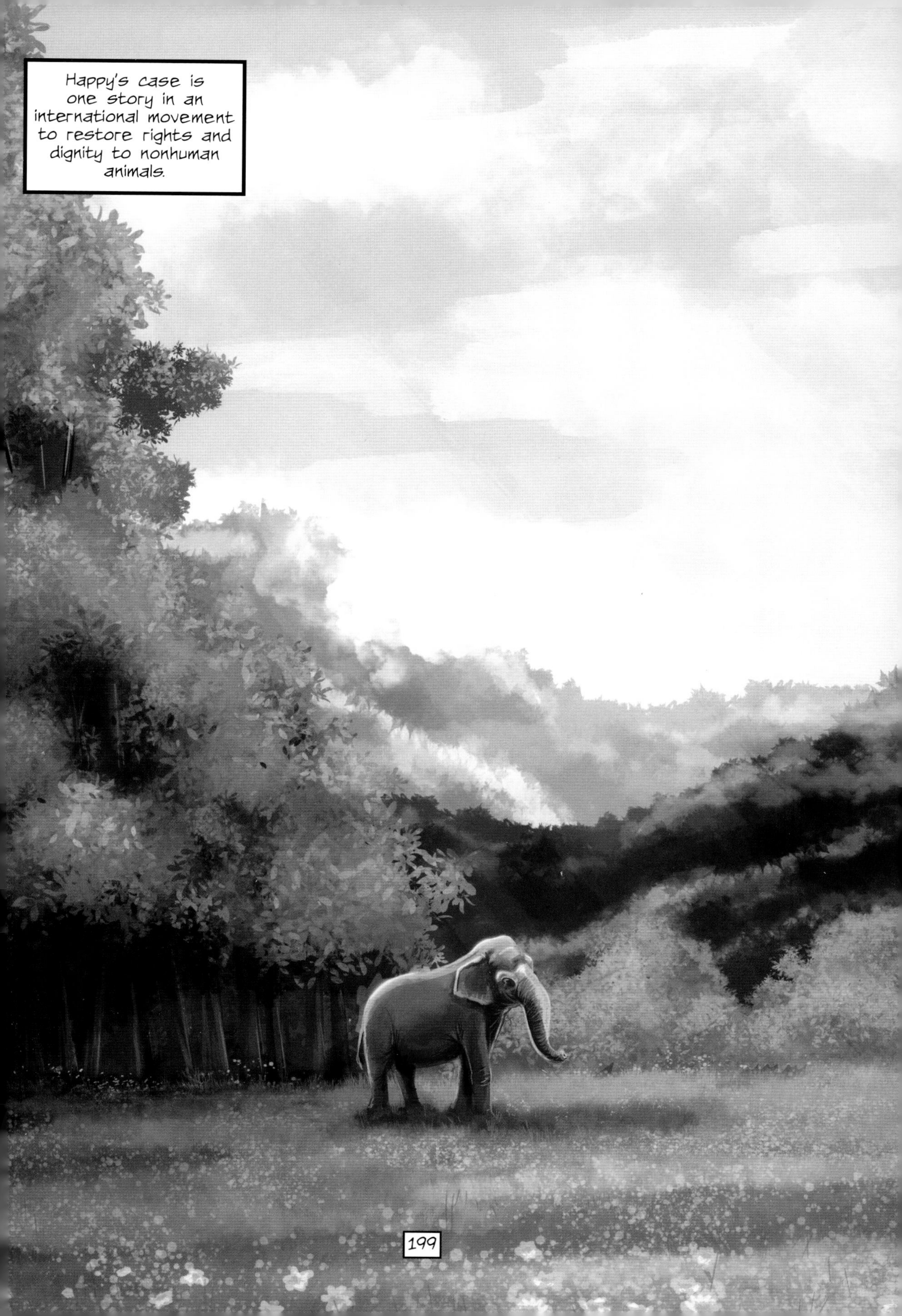
Happy's case is one story in an international movement to restore rights and dignity to nonhuman animals.

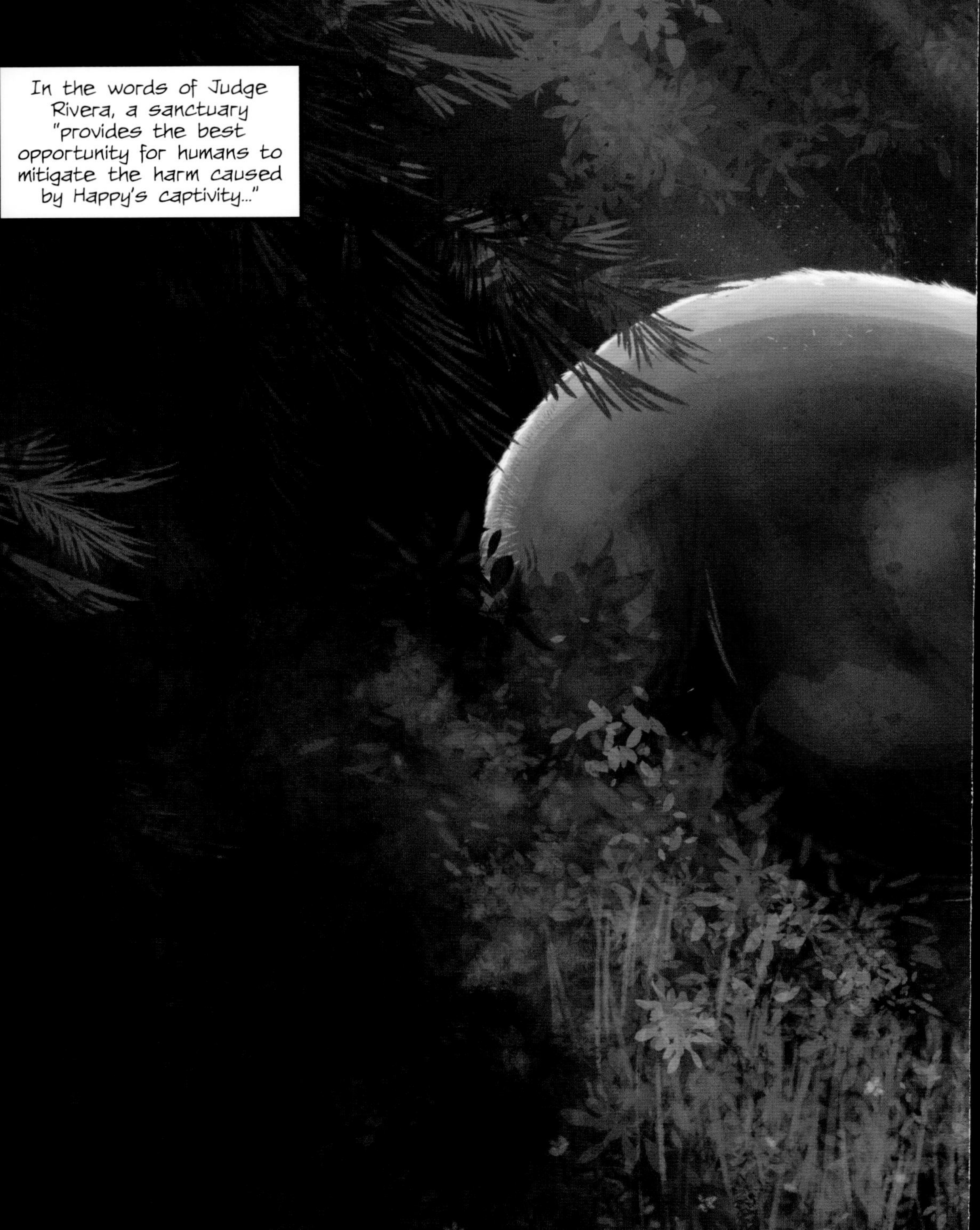
In the words of Judge Rivera, a sanctuary "provides the best opportunity for humans to mitigate the harm caused by Happy's captivity..."

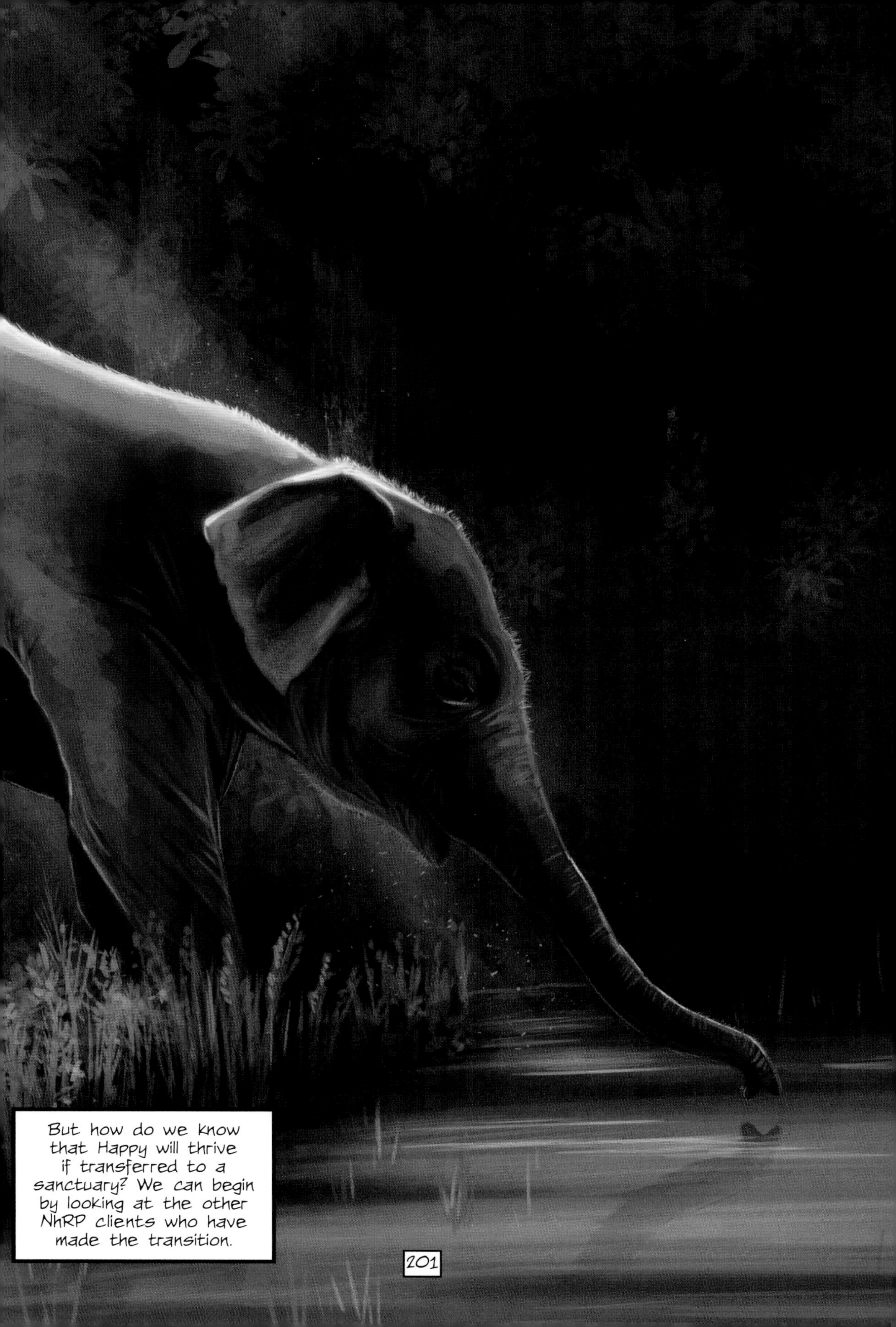
But how do we know that Happy will thrive if transferred to a sanctuary? We can begin by looking at the other NhRP clients who have made the transition.

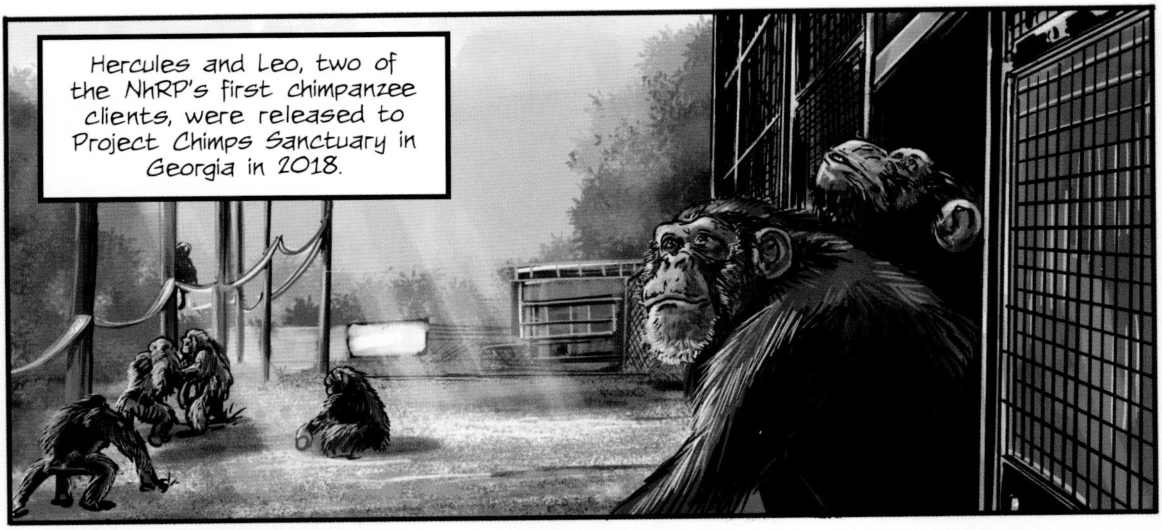
Hercules and Leo, two of the NhRP's first chimpanzee clients, were released to Project Chimps Sanctuary in Georgia in 2018.

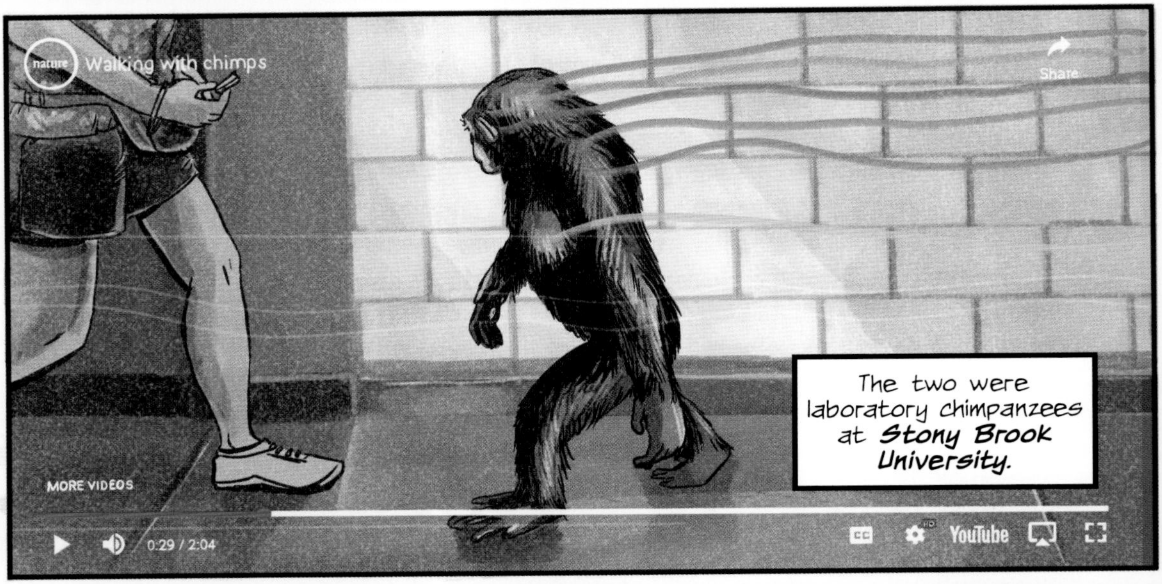
nature
Walking with chimps
Share
MORE VIDEOS
0:29 / 2:04
YouTube
The two were laboratory chimpanzees at **Stony Brook University.**

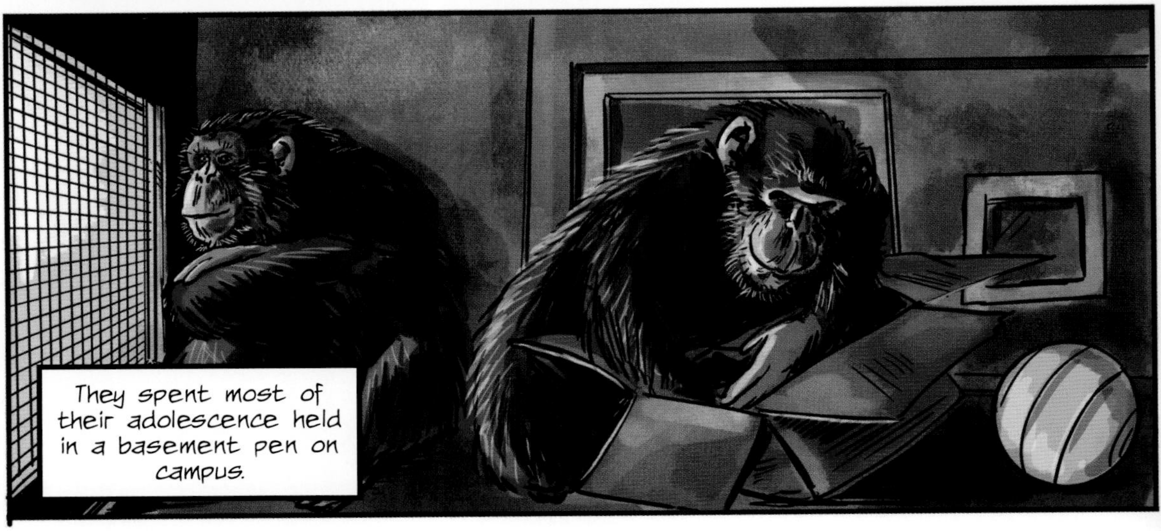
They spent most of their adolescence held in a basement pen on campus.

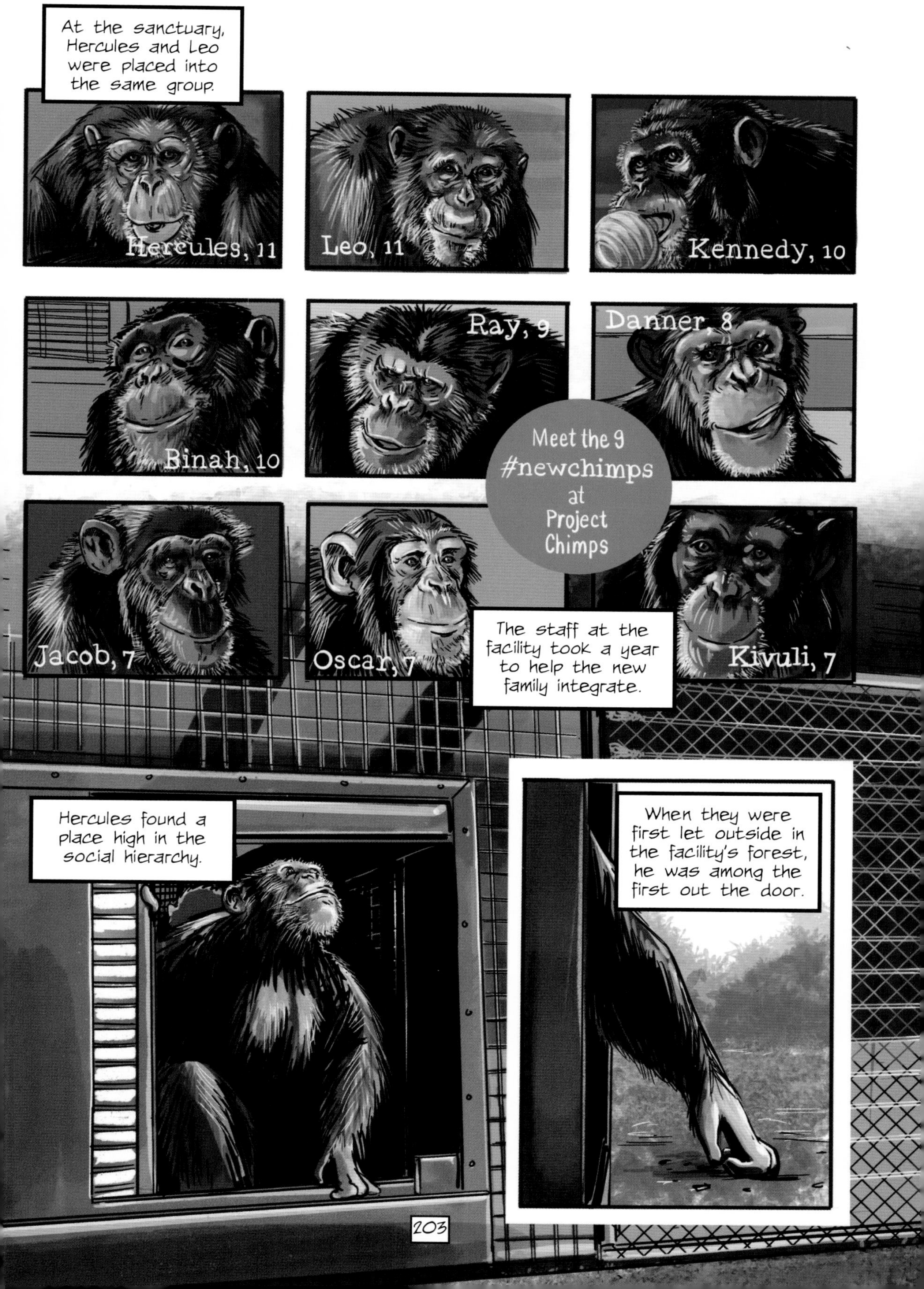
At the sanctuary, Hercules and Leo were placed into the same group.
Hercules, 11
Leo, 11
Kennedy, 10
Binah, 10
Ray, 9
Danner, 8
Meet the 9 #newchimps at Project Chimps
Jacob, 7
Oscar, 7
Kivuli, 7
The staff at the facility took a year to help the new family integrate.
Hercules found a place high in the social hierarchy.
When they were first let outside in the facility's forest, he was among the first out the door.

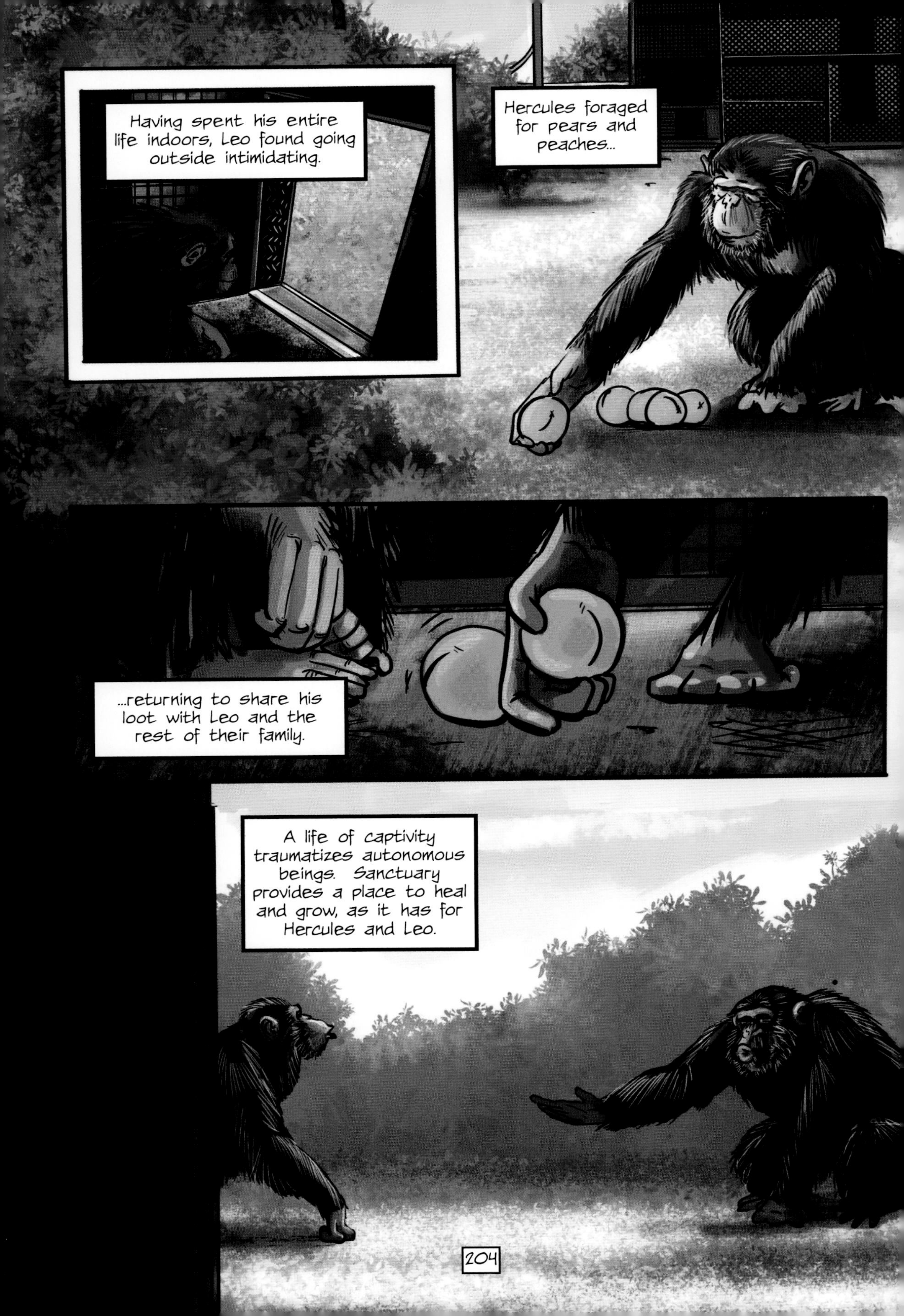
Having spent his entire life indoors, Leo found going outside intimidating.
Hercules foraged for pears and peaches...
...returning to share his loot with Leo and the rest of their family.
A life of captivity traumatizes autonomous beings. Sanctuary provides a place to heal and grow, as it has for Hercules and Leo.

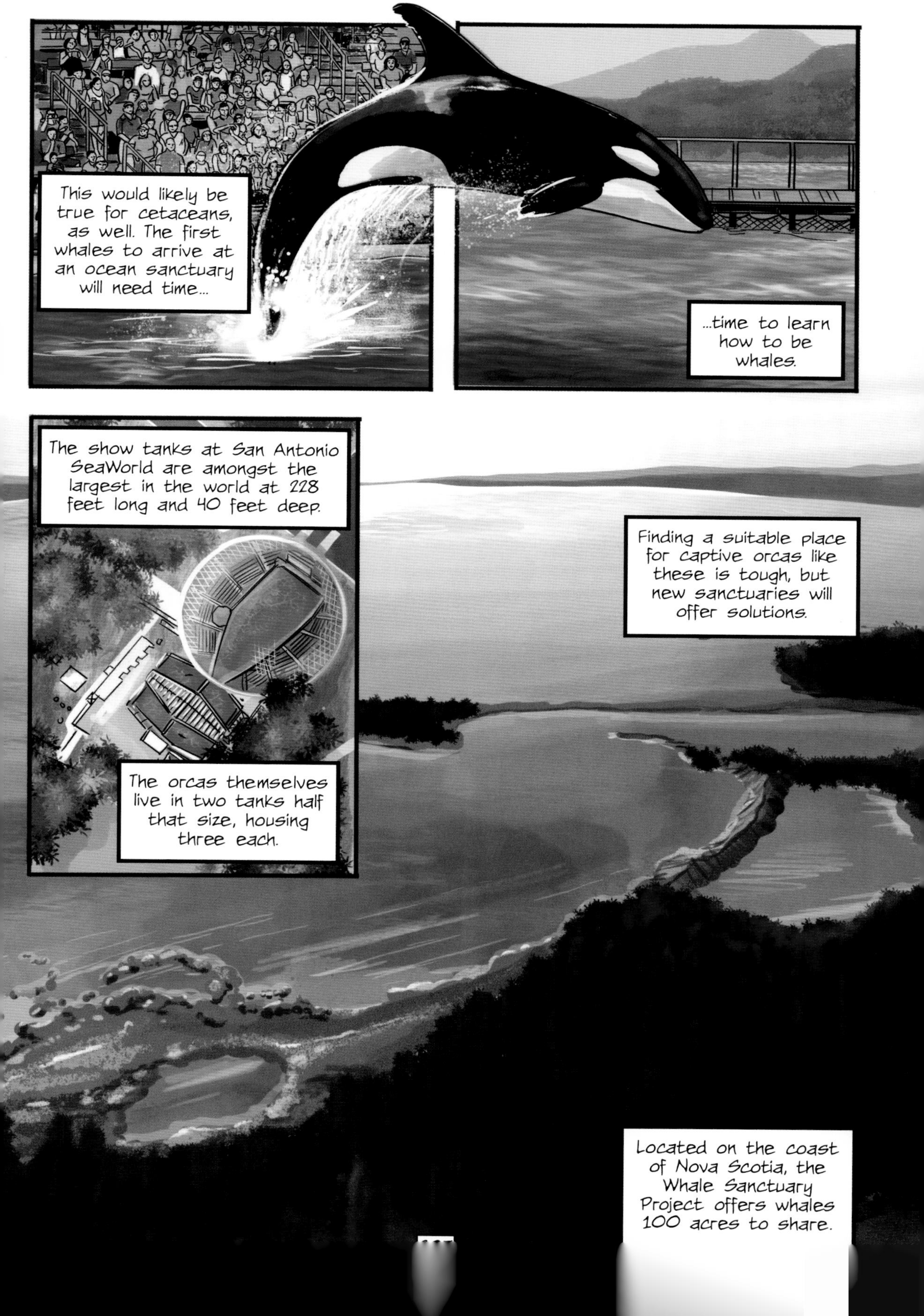
This would likely be true for cetaceans, as well. The first whales to arrive at an ocean sanctuary will need time...
...time to learn how to be whales.
The show tanks at San Antonio SeaWorld are amongst the largest in the world at 228 feet long and 40 feet deep.
The orcas themselves live in two tanks half that size, housing three each.
Finding a suitable place for captive orcas like these is tough, but new sanctuaries will offer solutions.
Located on the coast of Nova Scotia, the Whale Sanctuary Project offers whales 100 acres to share.

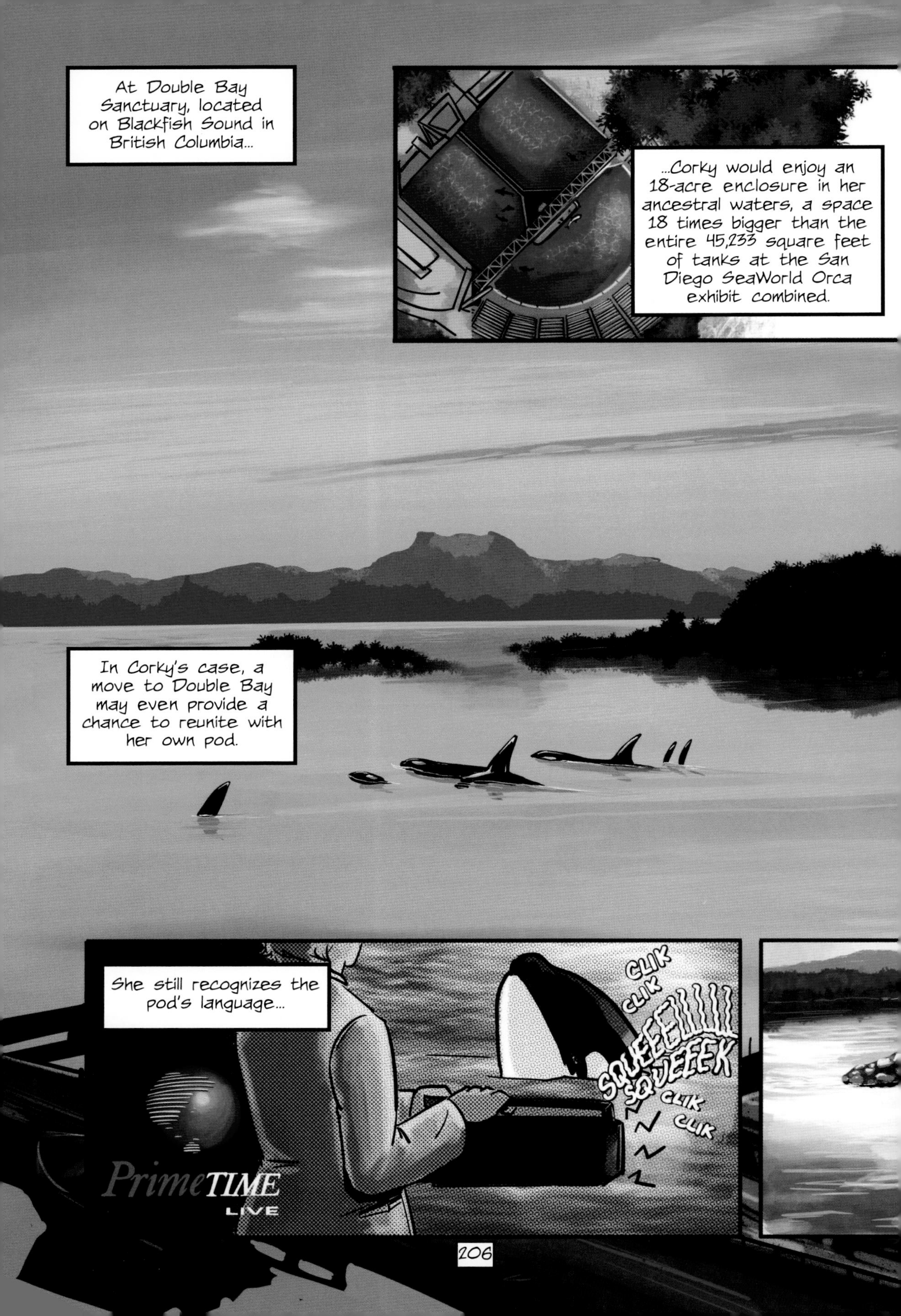
At Double Bay Sanctuary, located on Blackfish Sound in British Columbia...
...Corky would enjoy an 18-acre enclosure in her ancestral waters, a space 18 times bigger than the entire 45,233 square feet of tanks at the San Diego SeaWorld Orca exhibit combined.
In Corky's case, a move to Double Bay may even provide a chance to reunite with her own pod.
She still recognizes the pod's language...
PrimeTIME LIVE
CLIK CLIK
SQUEEEIIIIII SQUEEEEK
CLIK CLIK

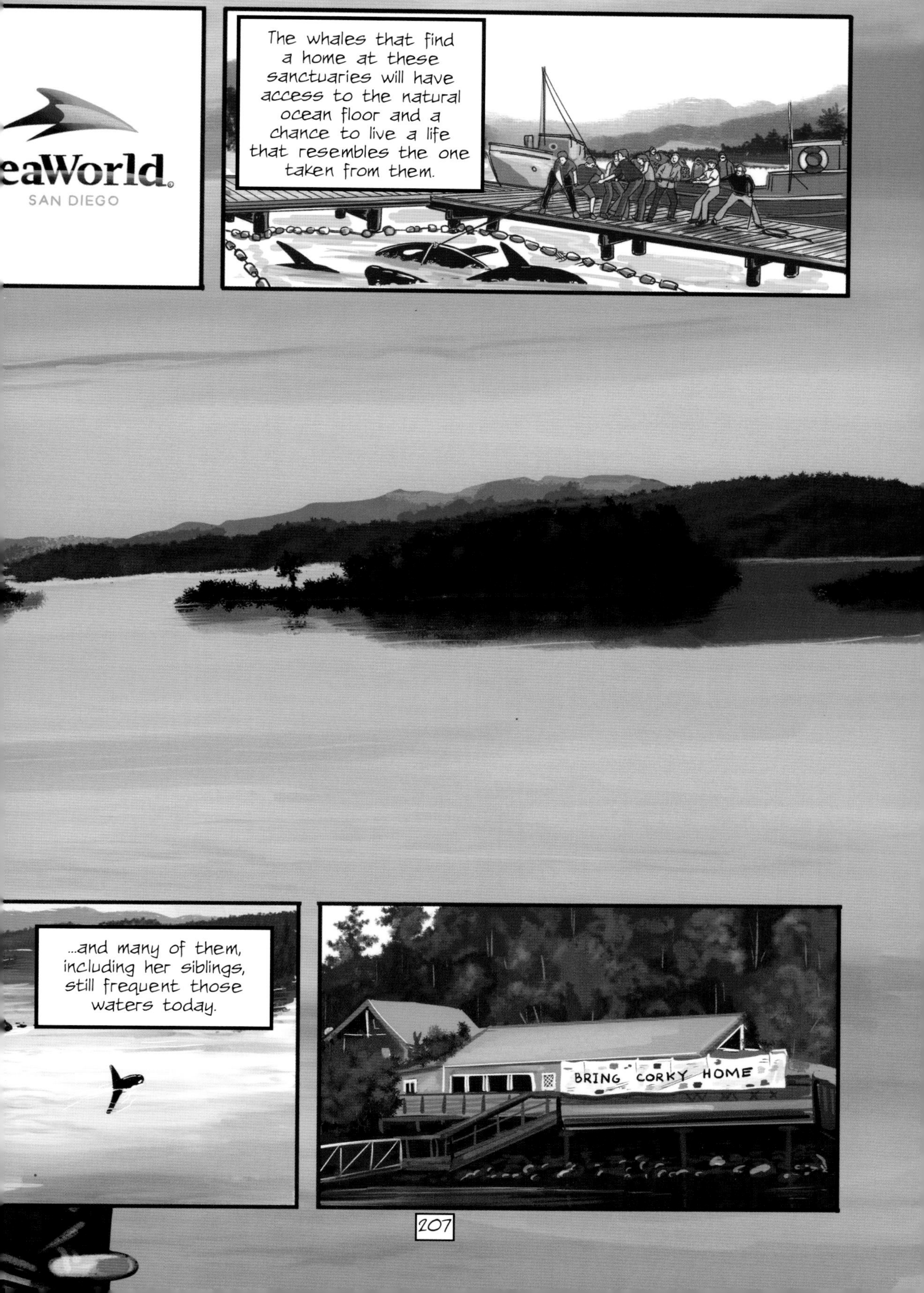
eaWorld
SAN DIEGO
The whales that find a home at these sanctuaries will have access to the natural ocean floor and a chance to live a life that resembles the one taken from them.
...and many of them, including her siblings, still frequent those waters today.
BRING CORKY HOME

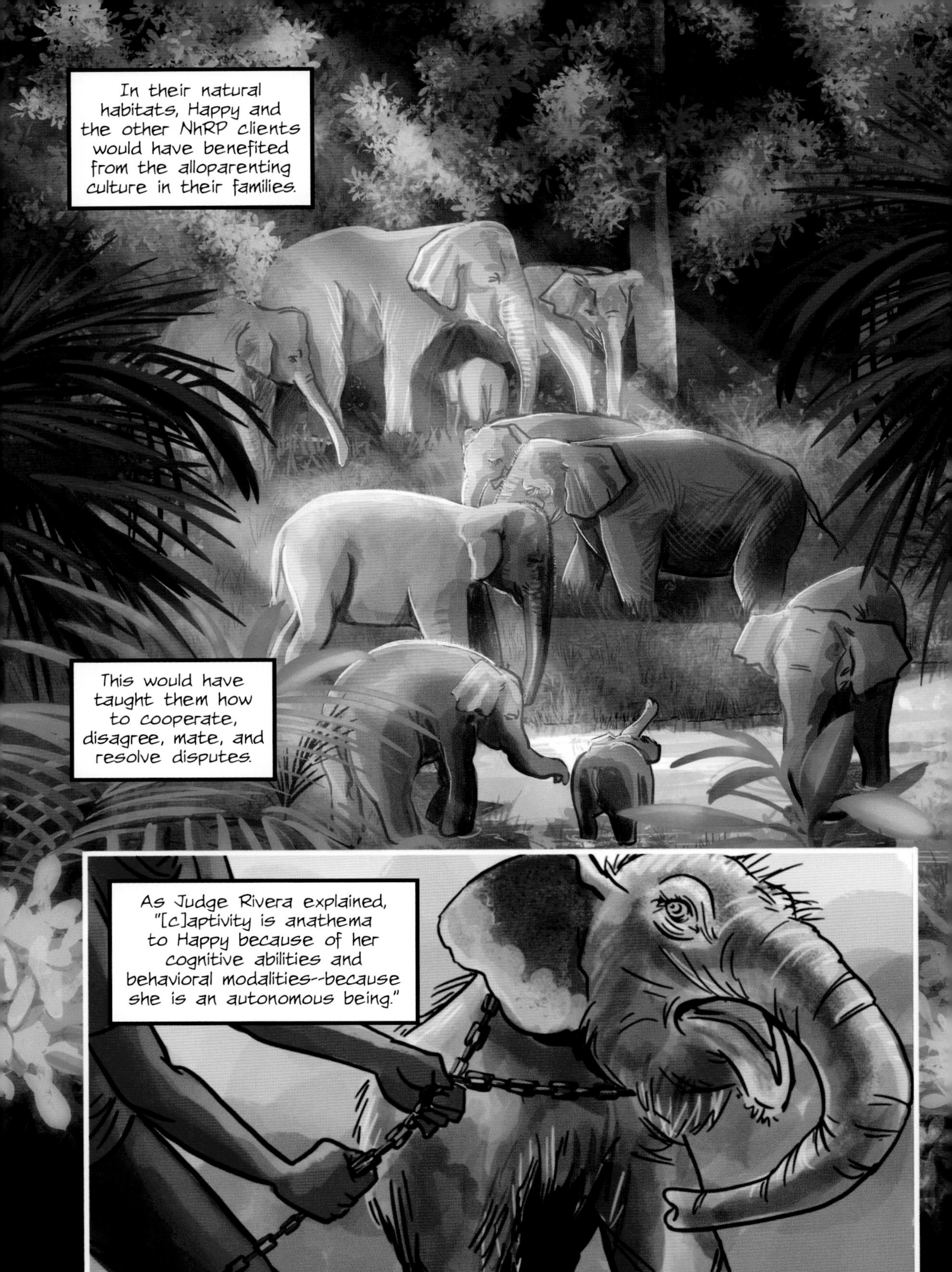
In their natural habitats, Happy and the other NhRP clients would have benefited from the alloparenting culture in their families.
This would have taught them how to cooperate, disagree, mate, and resolve disputes.
As Judge Rivera explained, "[c]aptivity is anathema to Happy because of her cognitive abilities and behavioral modalities--because she is an autonomous being."

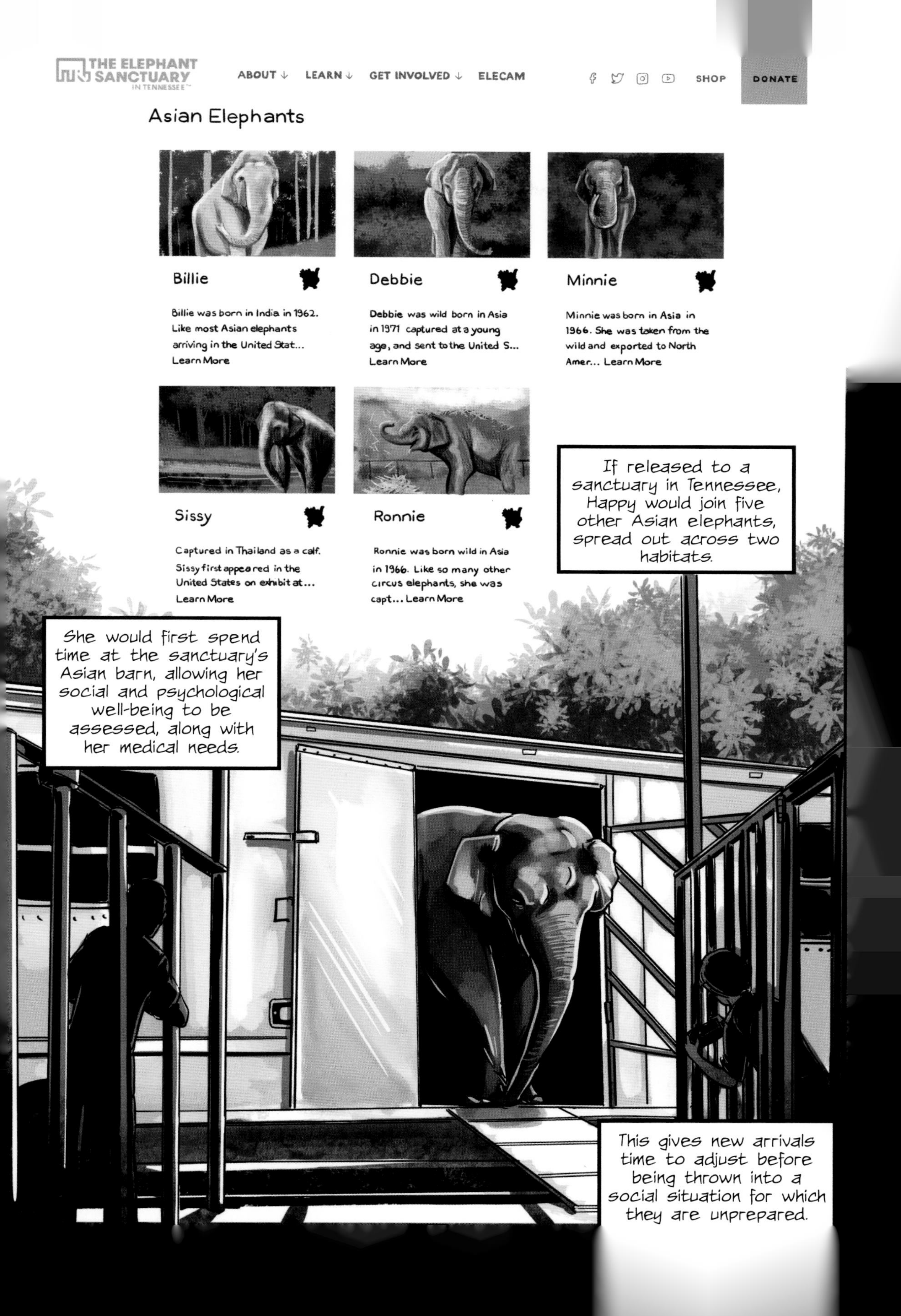
THE ELEPHANT SANCTUARY IN TENNESSEE
ABOUT
LEARN
GET INVOLVED
ELECAM
SHOP
DONATE
Asian Elephants
Billie
Billie was born in India in 1962. Like most Asian elephants arriving in the United Stat... Learn More
Debbie
Debbie was wild born in Asia in 1971 captured at a young age, and sent to the United S... Learn More
Minnie
Minnie was born in Asia in 1966. She was taken from the wild and exported to North Amer... Learn More
Sissy
Captured in Thailand as a calf. Sissy first appeared in the United States on exhibit at... Learn More
Ronnie
Ronnie was born wild in Asia in 1966. Like so many other circus elephants, she was capt... Learn More
If released to a sanctuary in Tennessee, Happy would join five other Asian elephants, spread out across two habitats.
She would first spend time at the sanctuary's Asian barn, allowing her social and psychological well-being to be assessed, along with her medical needs.
This gives new arrivals time to adjust before being thrown into a social situation for which they are unprepared.

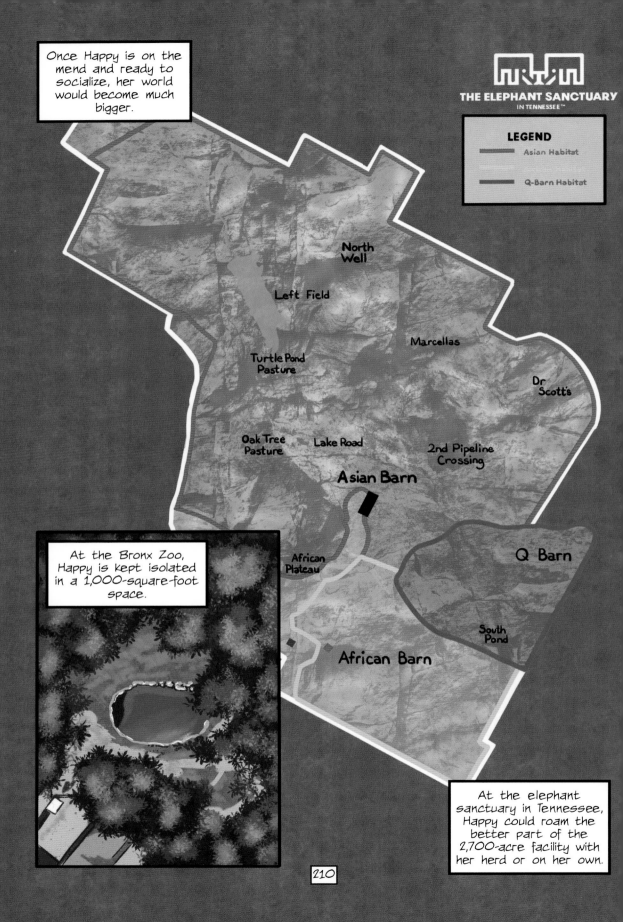

Once Happy is on the mend and ready to socialize, her world would become much bigger.

THE ELEPHANT SANCTUARY
IN TENNESSEE™

LEGEND
———— Asian Habitat
———— Q-Barn Habitat

North Well

Left Field

Marcellas

Turtle Pond Pasture

Dr Scott's

Oak Tree Pasture

Lake Road

2nd Pipeline Crossing

Asian Barn

Q Barn

African Plateau

South Pond

African Barn

At the Bronx Zoo, Happy is kept isolated in a 1,000-square-foot space.

At the elephant sanctuary in Tennessee, Happy could roam the better part of the 2,700-acre facility with her herd or on her own.

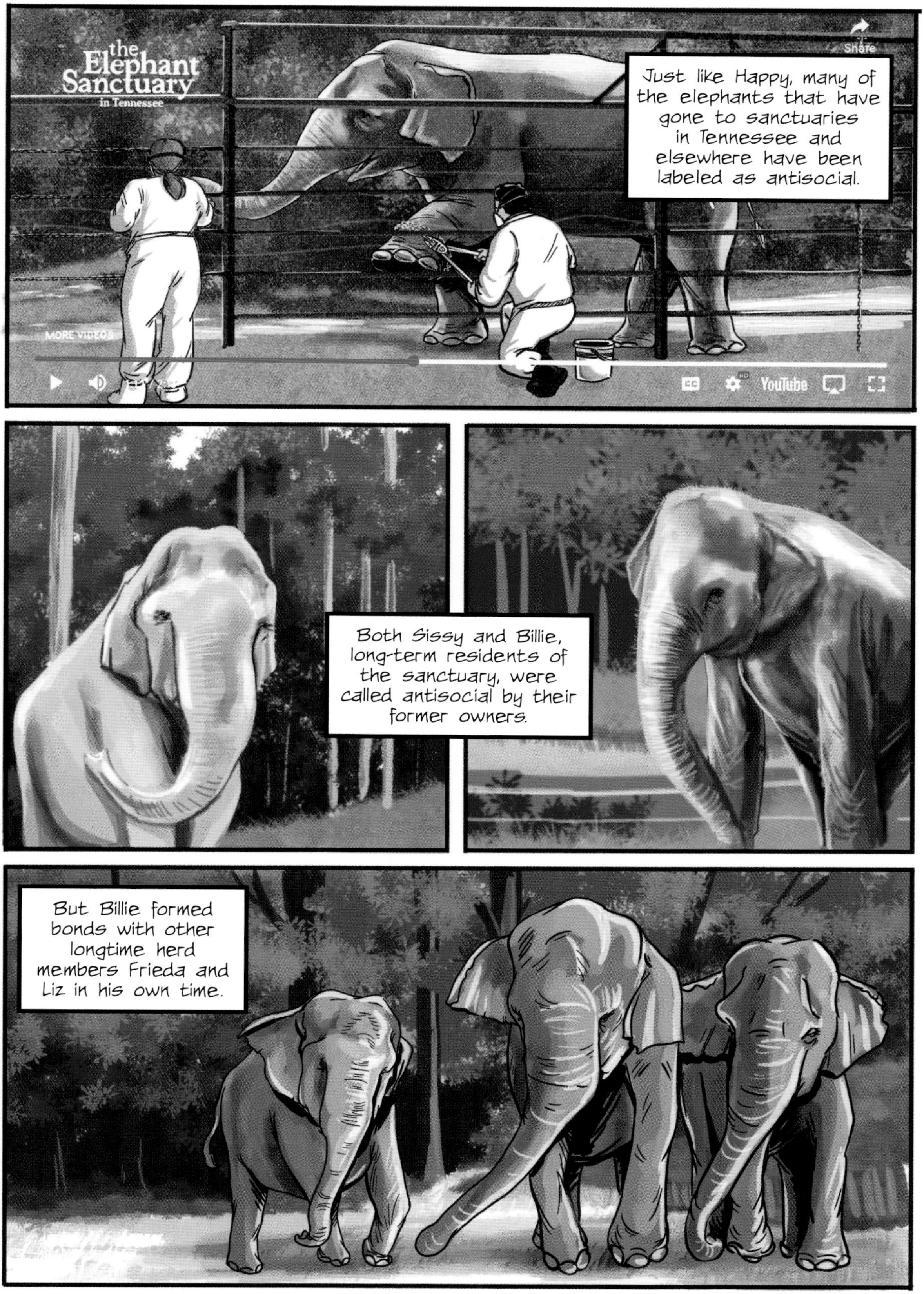
the Elephant Sanctuary in Tennessee
Share
Just like Happy, many of the elephants that have gone to sanctuaries in Tennessee and elsewhere have been labeled as antisocial.
MORE VIDEOS
YouTube
Both Sissy and Billie, long-term residents of the sanctuary, were called antisocial by their former owners.
But Billie formed bonds with other longtime herd members Frieda and Liz in his own time.

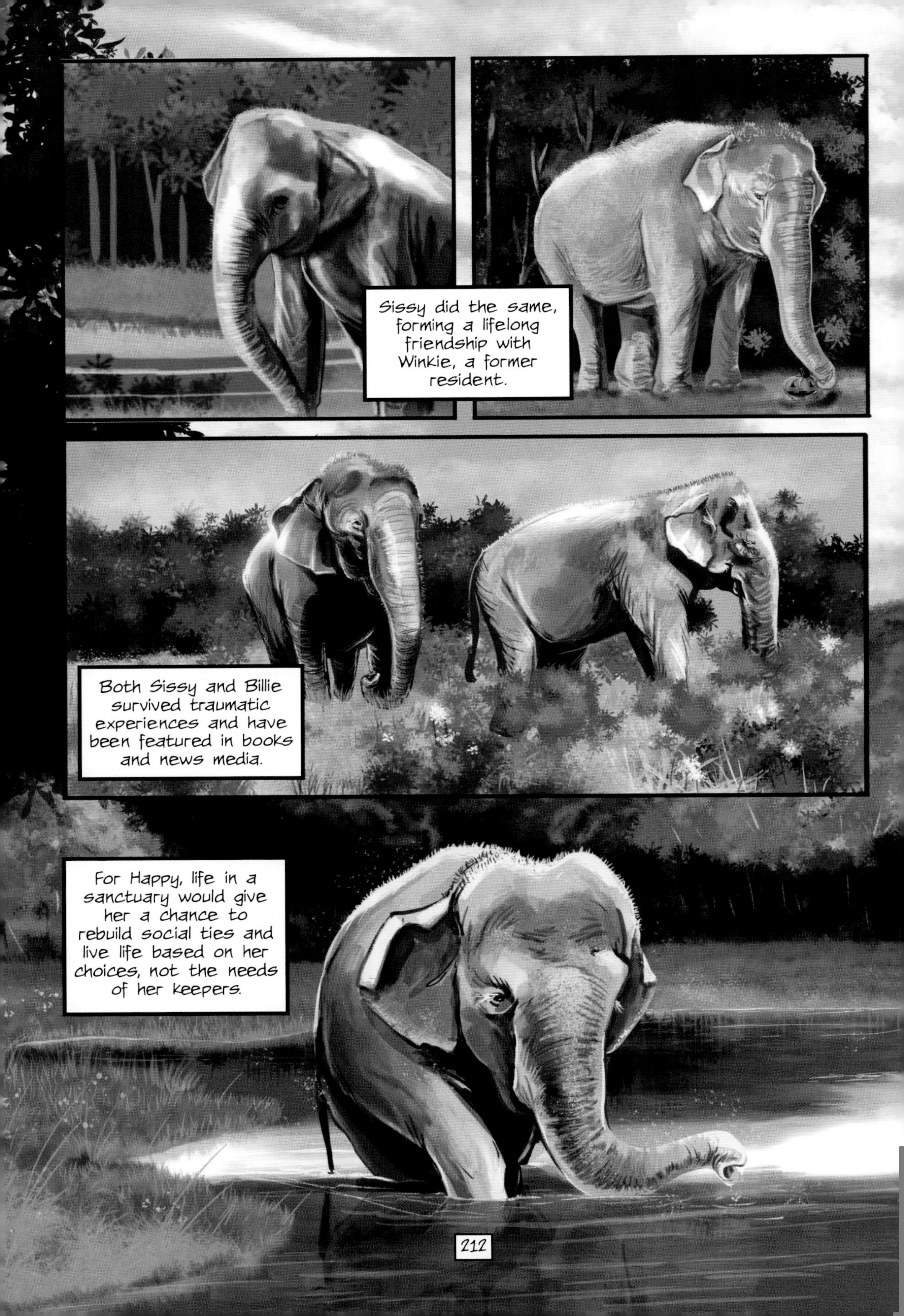
Sissy did the same, forming a lifelong friendship with Winkie, a former resident.
Both Sissy and Billie survived traumatic experiences and have been featured in books and news media.
For Happy, life in a sanctuary would give her a chance to rebuild social ties and live life based on her choices, not the needs of her keepers.

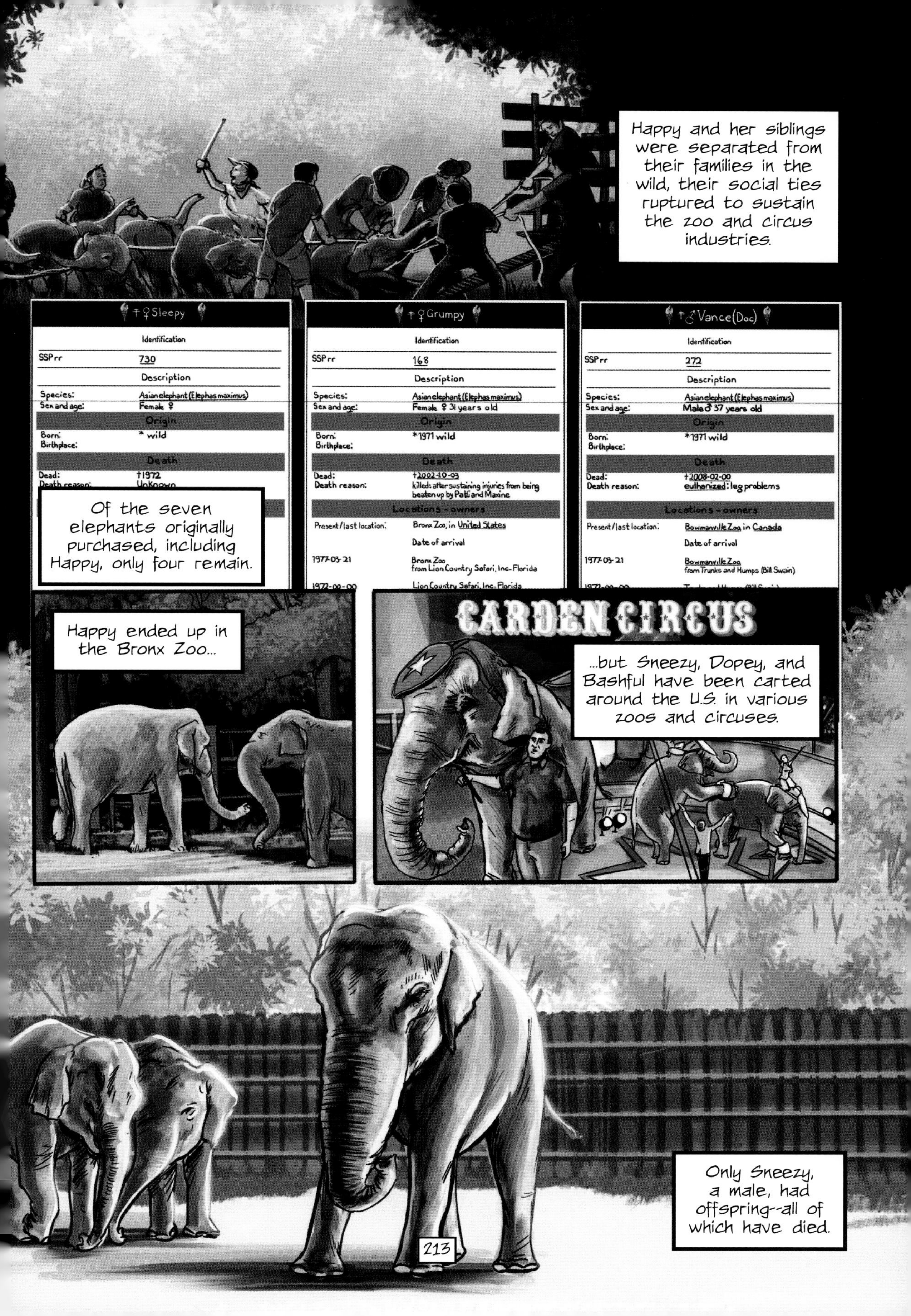

Happy and her siblings were separated from their families in the wild, their social ties ruptured to sustain the zoo and circus industries.

‡ ♀ Sleepy	
Identification	
SSP rr	730
Description	
Species:	Asian elephant (Elephas maximus)
Sex and age:	Female ♀
Origin	
Born:	* wild
Birthplace:	
Death	
Dead:	†1972
Death reason:	Unknown

‡ ♀ Grumpy	
Identification	
SSP rr	168
Description	
Species:	Asian elephant (Elephas maximus)
Sex and age:	Female ♀ 31 years old
Origin	
Born:	*1971 wild
Birthplace:	
Death	
Dead:	†2002-10-03
Death reason:	killed: after sustaining injuries from being beaten up by Patti and Maxine
Locations – owners	
Present/last location:	Bronx Zoo, in United States
	Date of arrival
1977-03-21	Bronx Zoo from Lion Country Safari, Inc-Florida
1972-00-00	Lion Country Safari, Inc-Florida

‡ ♂ Vance (Doc)	
Identification	
SSP rr	272
Description	
Species:	Asian elephant (Elephas maximus)
Sex and age:	Male ♂ 37 years old
Origin	
Born:	*1971 wild
Birthplace:	
Death	
Dead:	†2008-02-00
Death reason:	euthanized: leg problems
Locations – owners	
Present/last location:	Bowmanville Zoo in Canada
	Date of arrival
1977-03-21	Bowmanville Zoo from Trunks and Humps (Bill Swain)
1972-00-00	Trunks and Humps (Bill Swain)

Of the seven elephants originally purchased, including Happy, only four remain.

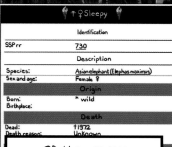

Happy ended up in the Bronx Zoo...

CARDEN CIRCUS

...but Sneezy, Dopey, and Bashful have been carted around the U.S. in various zoos and circuses.

Only Sneezy, a male, had offspring--all of which have died.

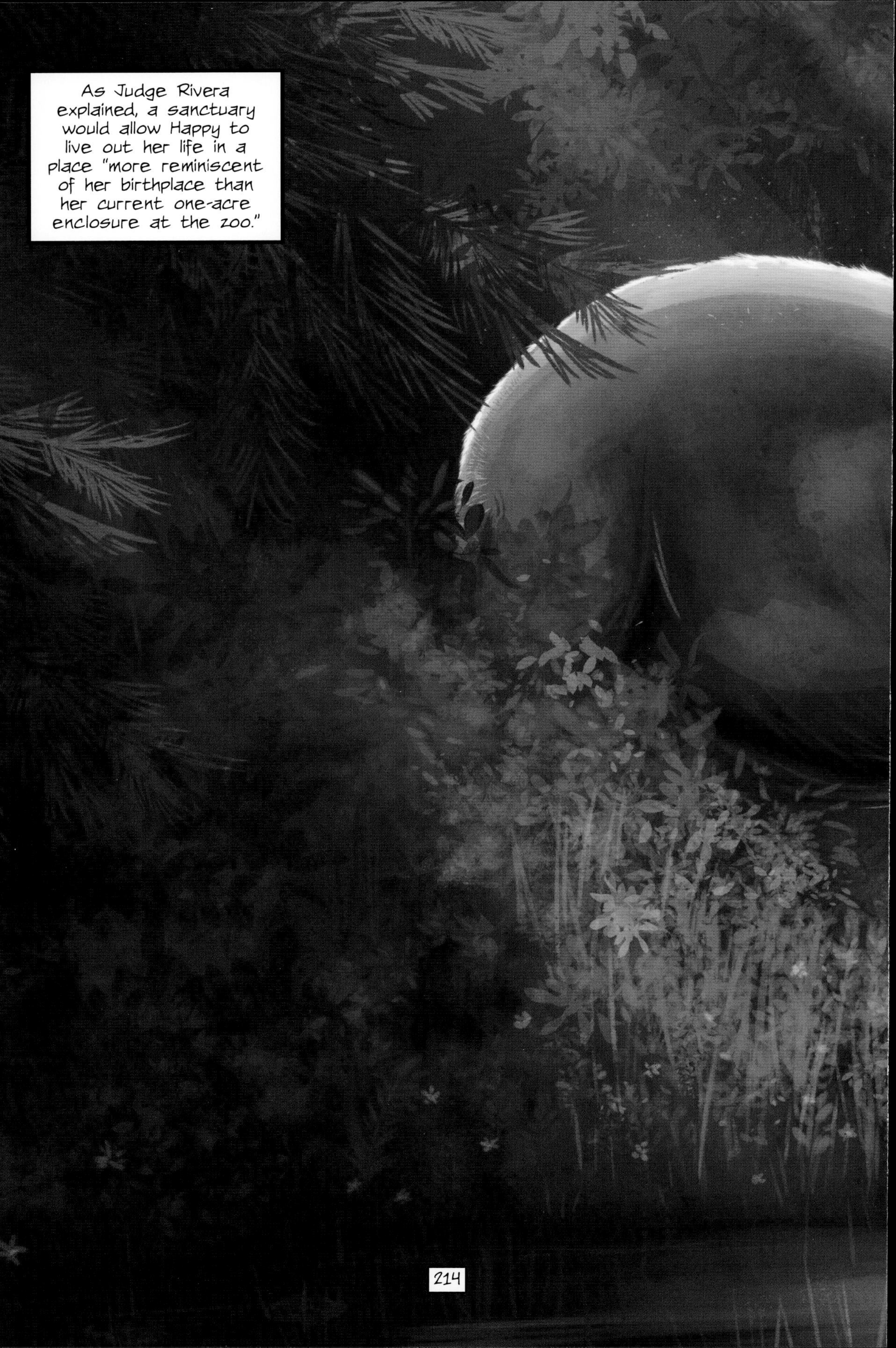
As Judge Rivera explained, a sanctuary would allow Happy to live out her life in a place "more reminiscent of her birthplace than her current one-acre enclosure at the zoo."

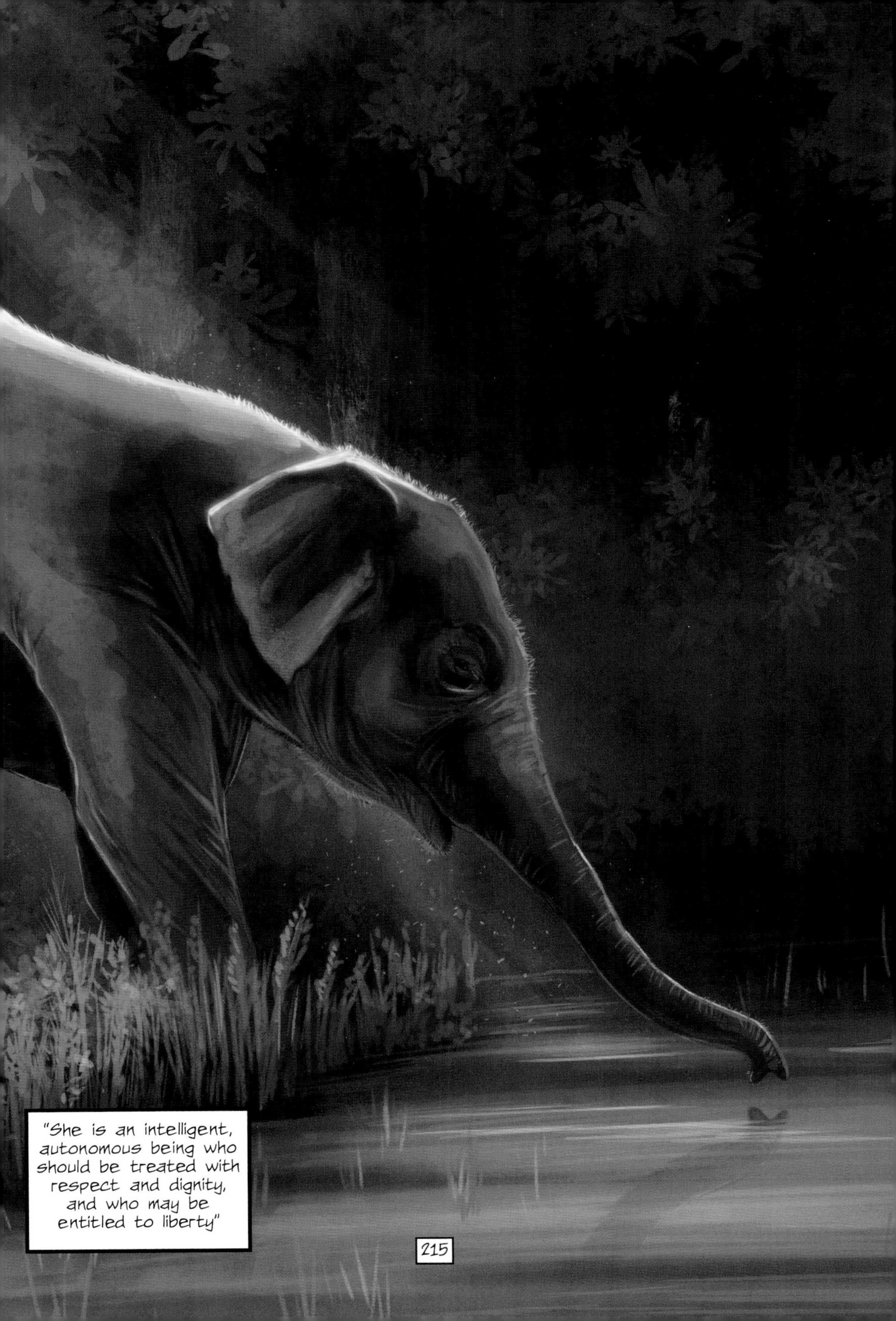
"She is an intelligent, autonomous being who should be treated with respect and dignity, and who may be entitled to liberty"

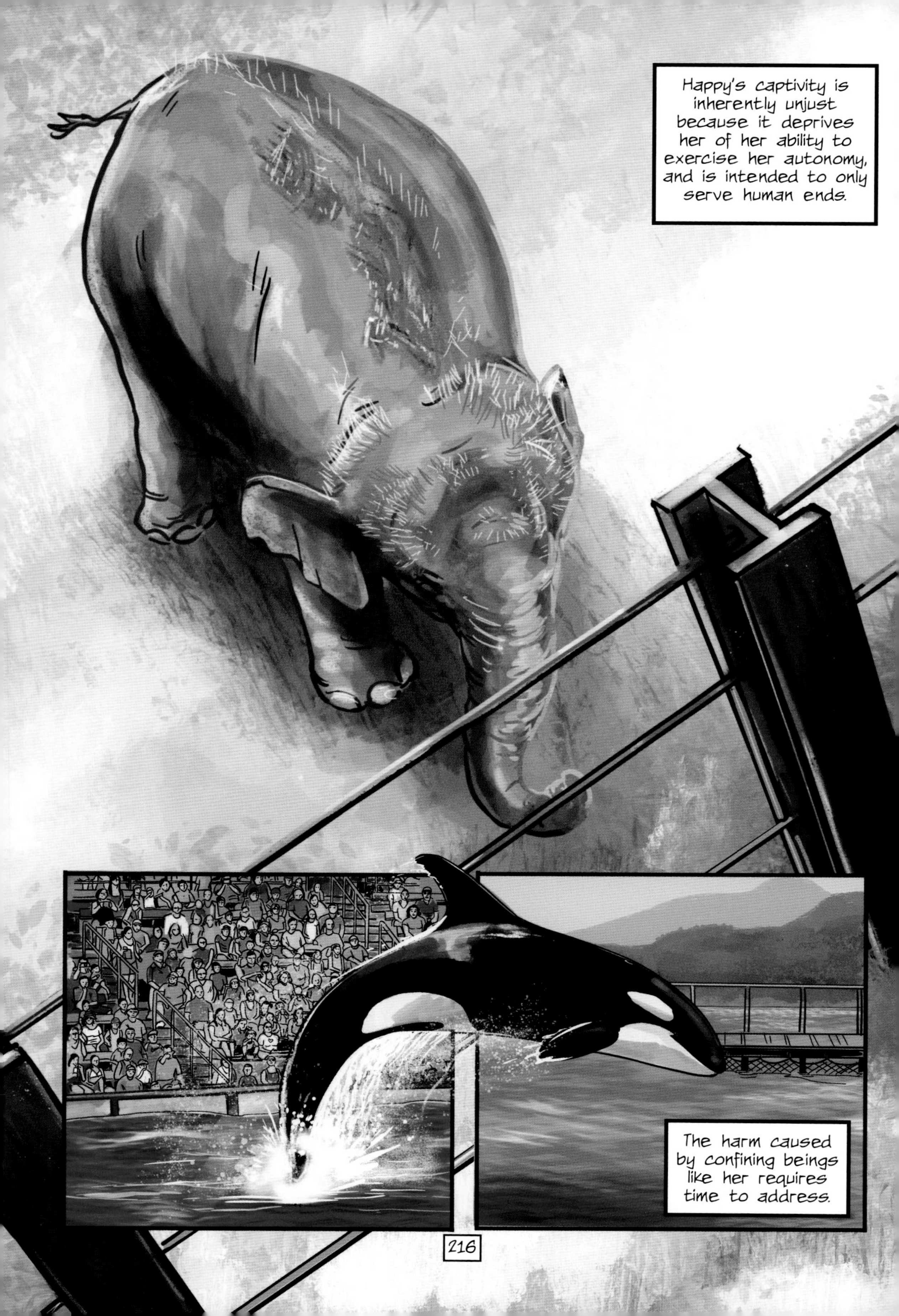
Happy's captivity is inherently unjust because it deprives her of her ability to exercise her autonomy, and is intended to only serve human ends.
The harm caused by confining beings like her requires time to address.

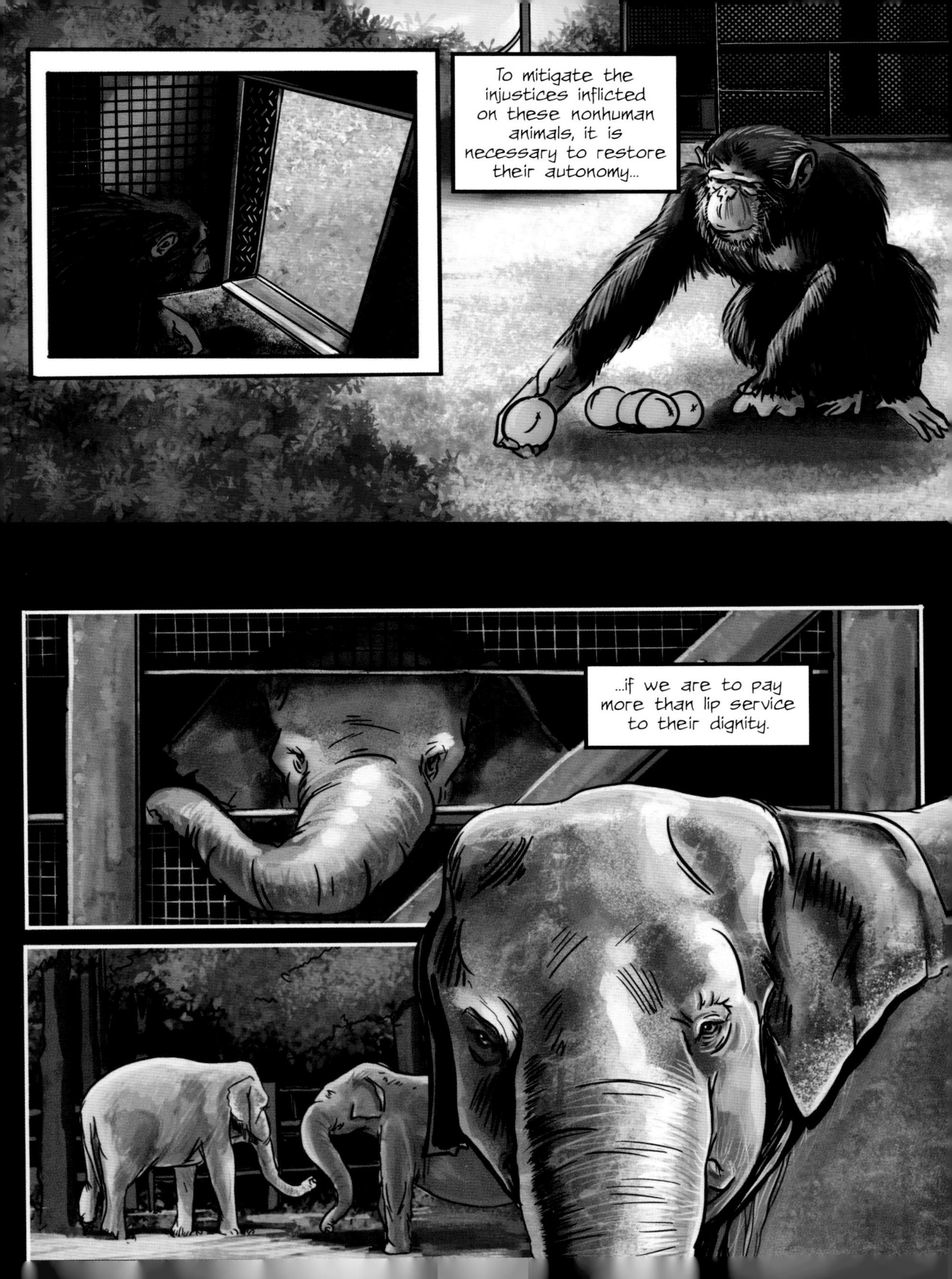
To mitigate the injustices inflicted on these nonhuman animals, it is necessary to restore their autonomy...
...if we are to pay more than lip service to their dignity.

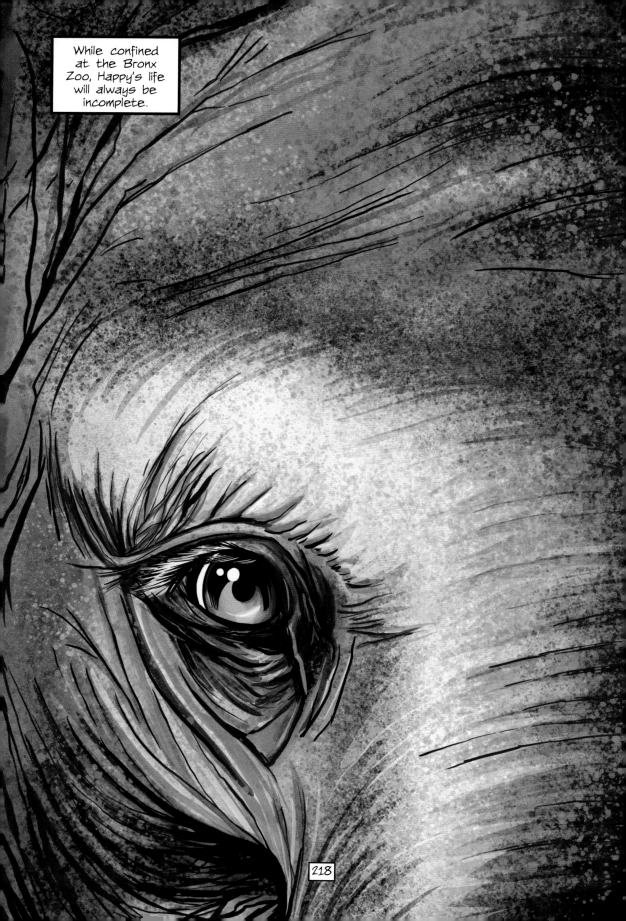

While confined at the Bronx Zoo, Happy's life will always be incomplete.

Throughout history, our society has used the writ of habeas corpus to remedy the harms caused by relegating autonomous beings to property.

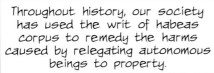

by the founders of this country as the highest safeguard of liberty. People

45 Cal.4th 1063, 1068 (hereafter *Villa*) (internal quotation marks and

tatic, narrow, formalisti

protection of individua

PUS

27   107.   "[H]abeas corpus 'is not now and never has been a static, narrow, formalistic

28 remedy; its scope has grown to achieve its grand purpose—the protection of individuals

1 against erosion of their right to be free from wrongful restraints upon their liberty.'" *Villa*,

2 45 Cal.4th at 1073 (citation omitted). "The very nature of the writ demands that it be

Happy is more than a thing.

WILDLIFE CONSERVATION SOCIETY

BRONX ZOO

In every way that counts, Happy is a person who deserves to be free.

219

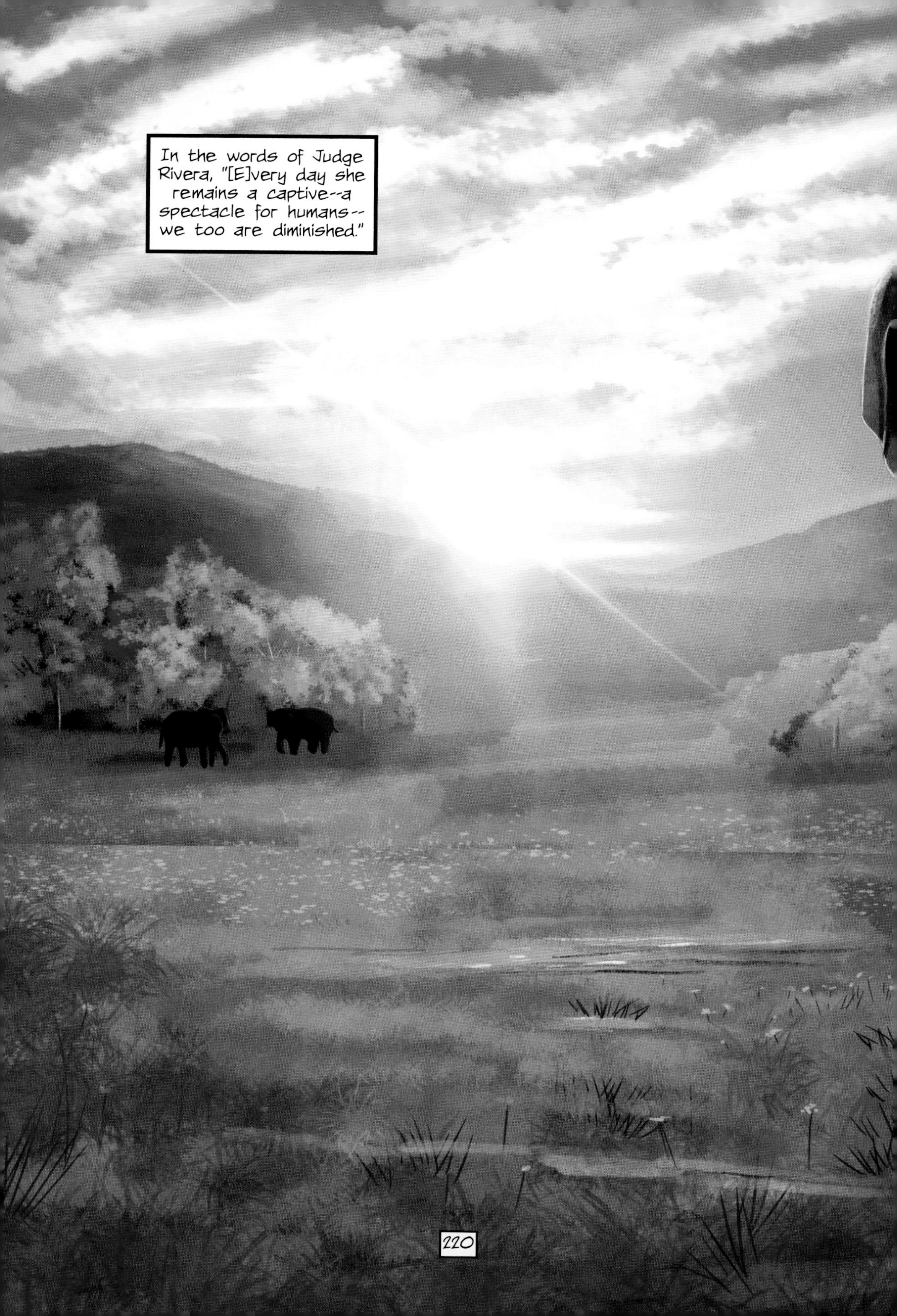
In the words of Judge Rivera, "[E]very day she remains a captive--a spectacle for humans--we too are diminished."

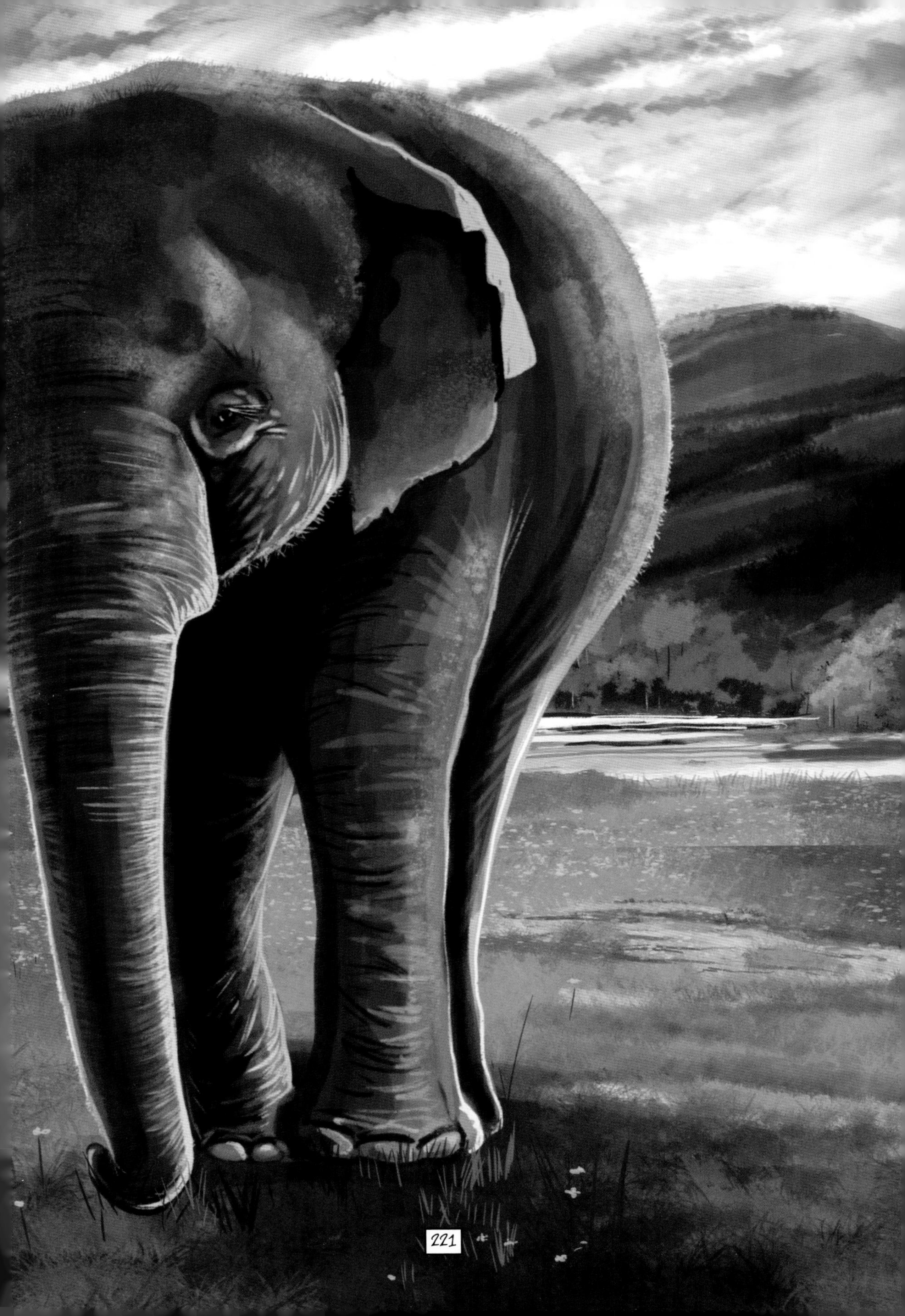

Afterword

by Steven M. Wise

Everything decent, everything righteous, evolves, sometimes quickly, sometimes slowly. For centuries, many humans were considered to be legal things, not legal persons with any rights. It took numerous struggles, for example, to finally decide that all humans were "persons" with rights, irrespective of race or color.

But a million species of nonhuman animals have been considered things, not persons. They were never rationally distinguished nor carefully studied, and generally misunderstood. However, less than two centuries ago, human understanding of the cognition, intelligence, and emotions of a light dusting of species of nonhuman animals, such as elephants and chimpanzees, began to evolve. Sixty years ago, scientists began to deeply study them.

These scientists' findings are giving us an increasingly clear look into the minds of many nonhuman animals. Scientists around the world have obtained increasingly powerful elephant data that focus on their natural and intrinsic nature, and steadily reveal who they actually are. They have come to understand that elephants possess complicated learning abilities that allow them to use a wide variety of gestures and calls, to consciously plan how to live, share their lives, knowledge, information, and thoughts just as humans do, cooperate in innovative problem-solving, remember long-term issues, teach, and build coalitions. Researchers have learned that chimpanzees and elephants are remarkably autonomous, wildly cognitively complex, and that each individual is unique.

Throughout this book, you've followed our litigation on behalf of an elephant named Happy. Zoos claim that granting her the right to bodily liberty will result in an "uprooting of the entire social order," will "upend a state's legal system," and inflict "extensive and far-reaching consequences," specifically "economic consequences." New York's highest court agreed with the Bronx Zoo and went one step further, claiming that nonhuman animals like Happy cannot be legal persons because only humans can have rights.

Yet, this case is not an end to our work on behalf of nonhuman animals, but a beginning. Victory, once so far back as to be unthinkable, is creeping up. In 2018, a thoughtful Court of Appeals judge suggested that one of our chimpanzee clients should have won their fight. As the judge said, "I continue to question whether the Court was right to deny leave in the first instance. The issue whether a nonhuman animal has a fundamental right to liberty protected by the writ of habeas corpus is profound and far-reaching. It speaks to our relationship with all the life around us. Ultimately, we will not be able to ignore it."

And in 2020, when ruling in Happy's trial, Justice Tuitt admitted that she felt "regrettably" bound by previous decisions. The trial court stated that it "recognizes that Happy is an extraordinary animal with complex cognitive abilities, an intelligent being with advanced analytical abilities akin to human beings....Happy is more than just a legal thing. She is an intelligent autonomous being who should be treated with respect and dignity, and who may be entitled to liberty."

Fifty state high courts exist in the United States. Only one has decided whether a nonhuman animal who thinks and plans and appreciates life as human beings do has a right to be protected against enforced detentions. Happy's case is America's first on the foremost nonhuman animal rights issue of our time. There will be many more.

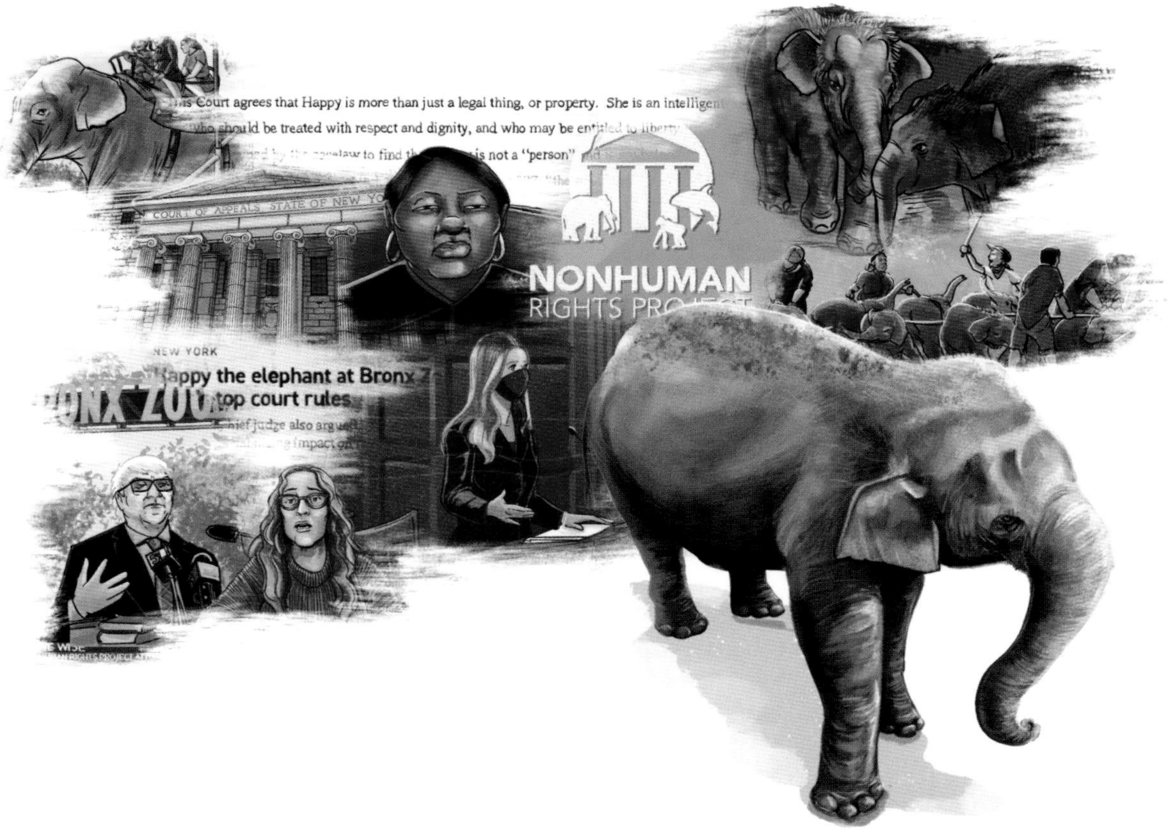

Court agrees that Happy is more than just a legal thing, or property. She is an intelligen
should be treated with respect and dignity, and who may be entitled to liberty
is not a "person"
COURT OF APPEALS STATE OF NEW YO
NONHUMAN
RIGHTS PRO
NEW YORK
appy the elephant at Bronx
top court rules

Notebook

by Samuel Machado

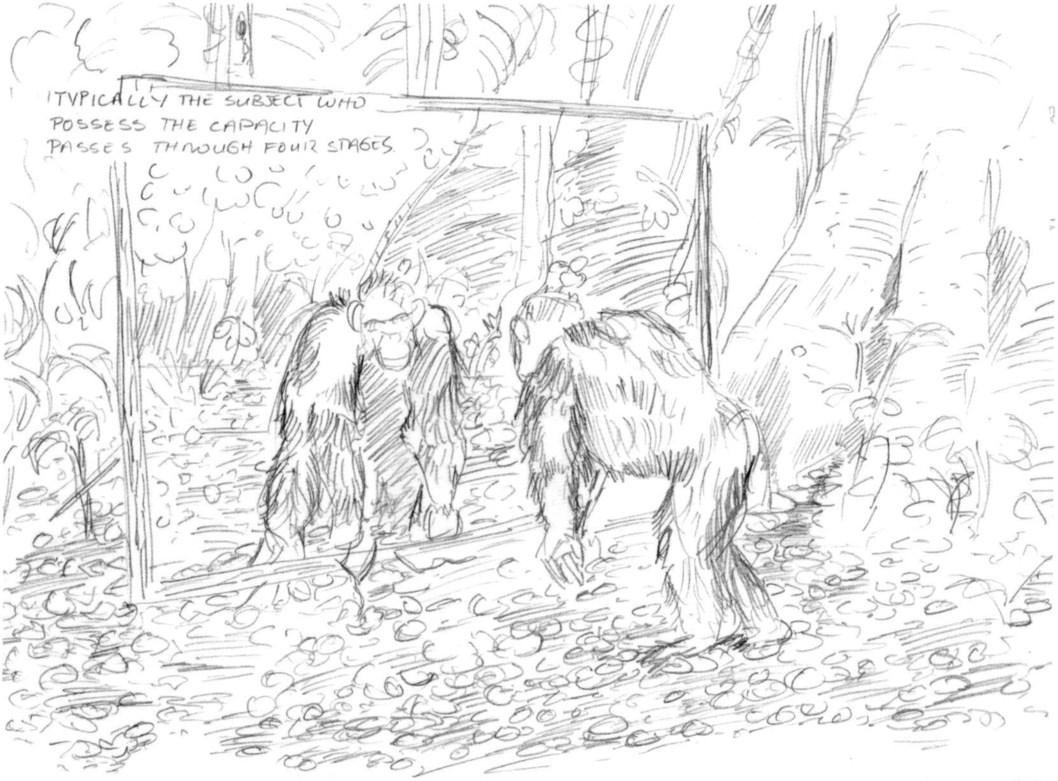
TYPICALLY THE SUBJECT WHO
POSSESS THE CAPACITY
PASSES THROUGH FOUR STAGES.

When Cynthia and I first imagined *Thing*, we were inspired by Scott McCloud's vision of a medium bigger than its historical genre, the superhero serial. We believed in a comic or cartoon's capacity to distill and articulate nuanced meaning through the conventions of sequence, space, and the power of illustration and cartooning. Marshalling those conventions for a nonfiction comic remains a challenge. In this medium, we can easily compare what we create to live-action-documentary films, photographic journalism, and nonfiction books.

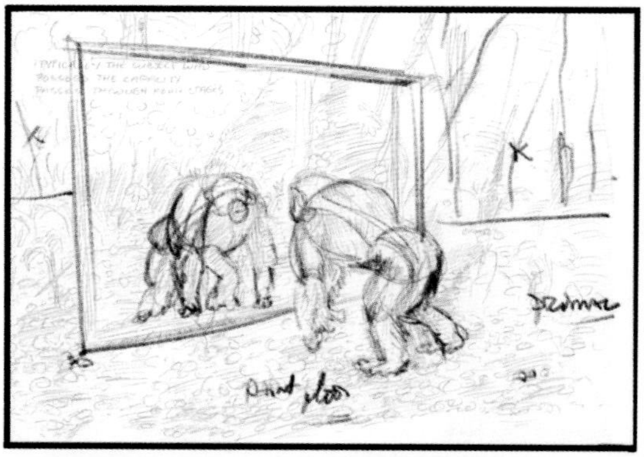

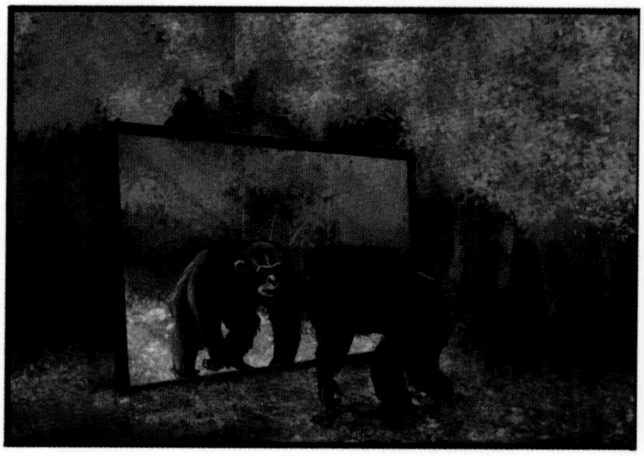

As creators, we make those comparisons, too. When we began our collaboration with Island Press and our editor, Rebecca Bright, we established an idea that this book would have the feel of a documentary with a touch of editorialization. (See our image reference at https://islandpress.org/books/thing for a complete list of credits for images).

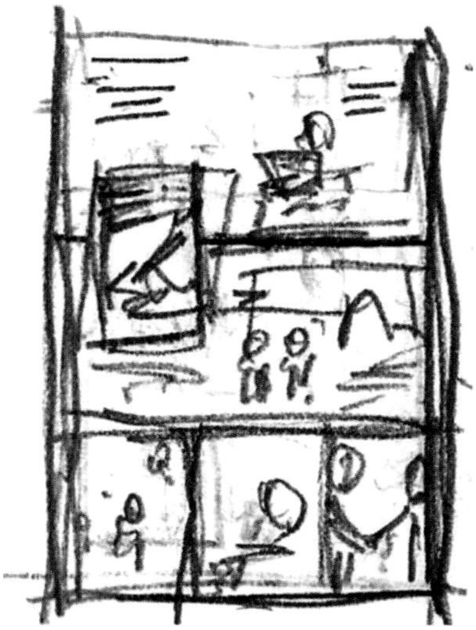

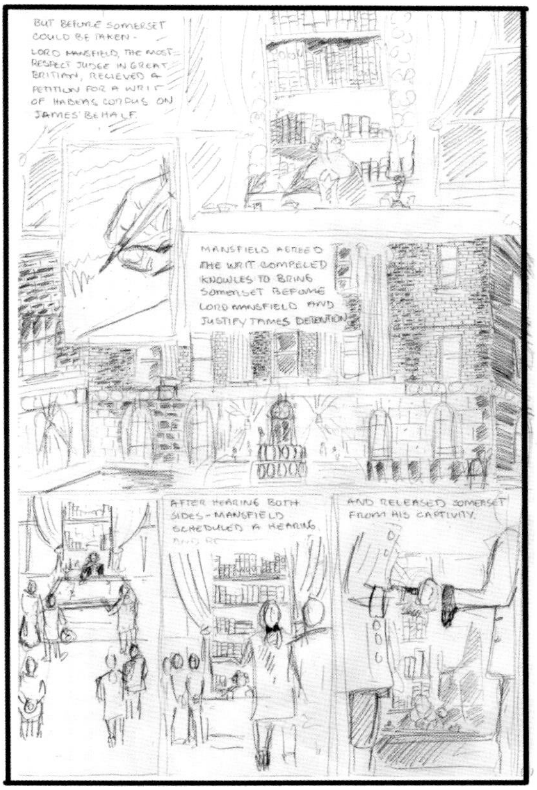
BUT BEFORE SOMERSET COULD BE TAKEN -
LORD MANSFIELD, THE MOST RESPECT JUDGE IN GREAT BRITIAN, RECIEVED A PETITION FOR A WRIT OF HABEAS CORPUS ON JAMES' BEHALF.
MANSFIELD AGREED THE WRIT COMPELED KNOWLES TO BRING SOMERSET BEFORE LORD MANSFIELD AND JUSTIFY JAMES DETENTION
AFTER HEARING BOTH SIDES - MANSFIELD SCHEDULED A HEARING.
AND RELEASED SOMERSET FROM HIS CAPTIVITY.

But, whether we are recreating a photo or building a page, our process has always been the same. I begin with a thumbnail and an idea of what I'd like to say on the page. I usually do a hand-drawn rough with some word placement before sending it to Cynthia. She illustrates and renders the final look and panel design.

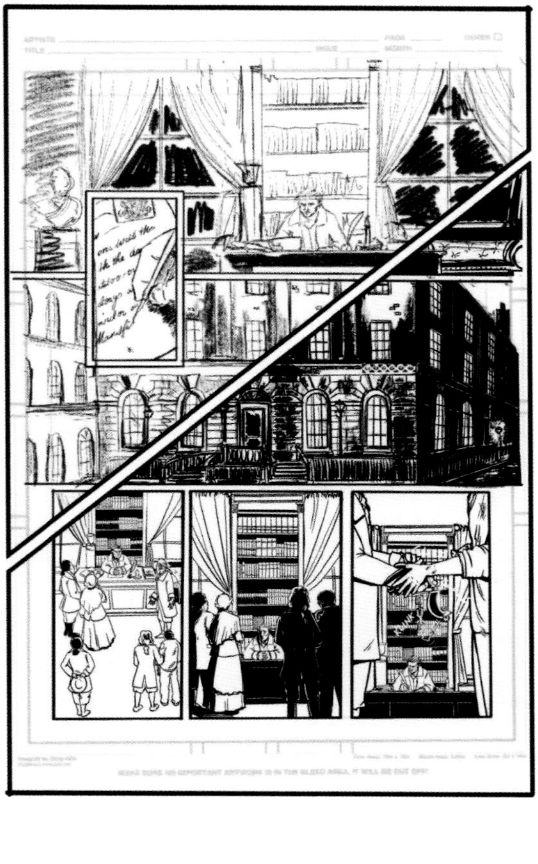

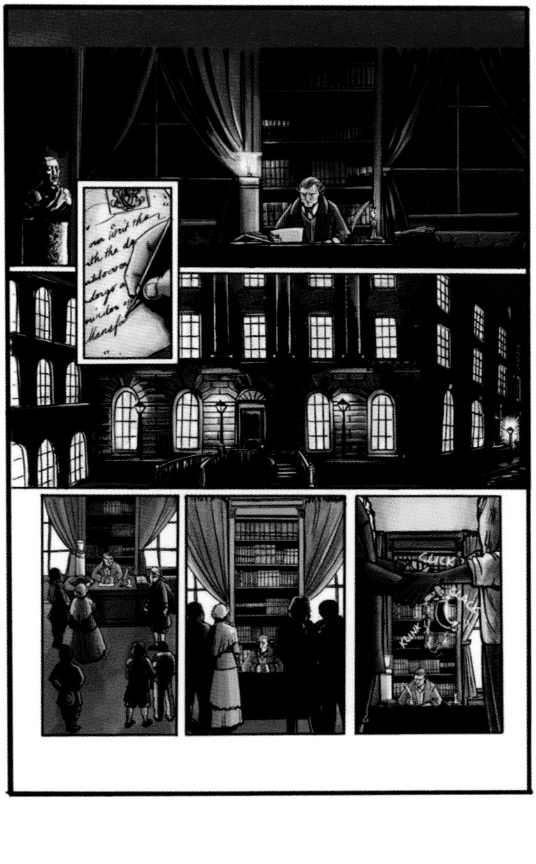

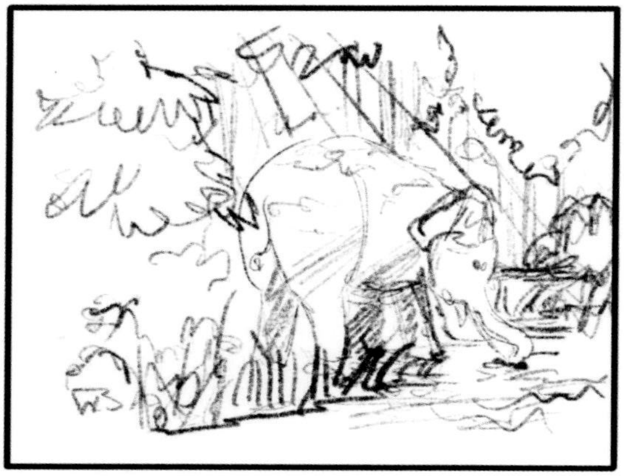
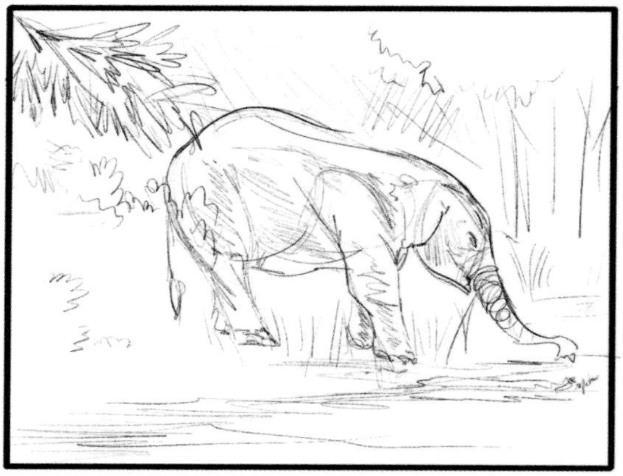
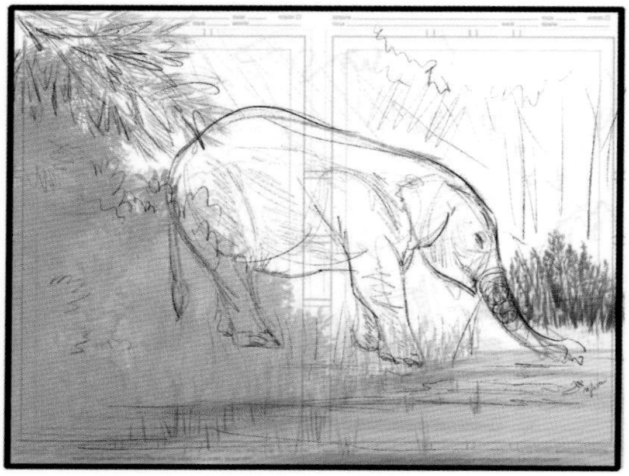
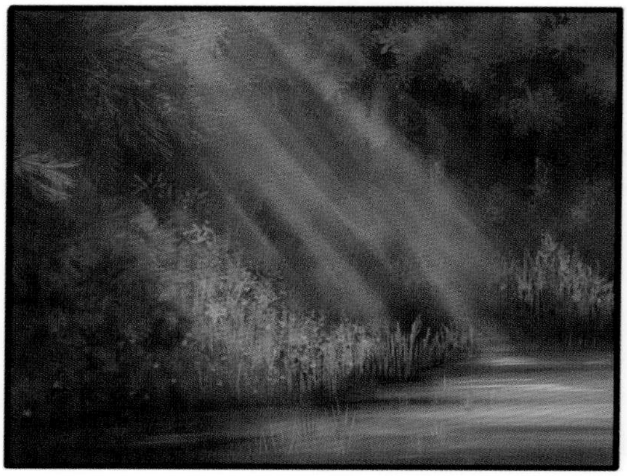

Cynthia works digitally, compositing imagery and choosing perspectives to communicate her intentions for each page or panel. In this case, Cynthia took a painted approach in one of the final double-page splashes of the book. She wanted the reader to imagine an alternate future—where Happy takes her first sip of water from a pond located on the grounds of the Elephant Sanctuary in Tennessee. Cynthia hopes that the reader can place themselves in that moment with Happy, and empathize with the experience of fear and excitement Happy might feel while exerting her autonomy for the first time.

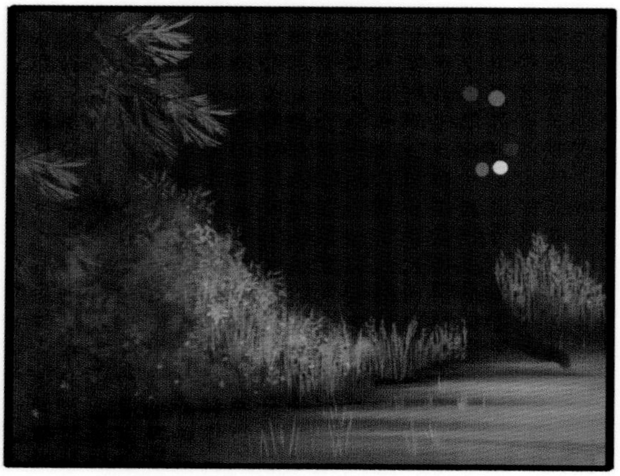
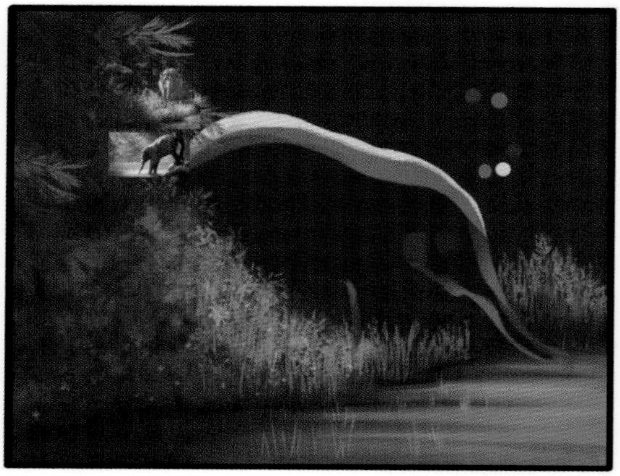

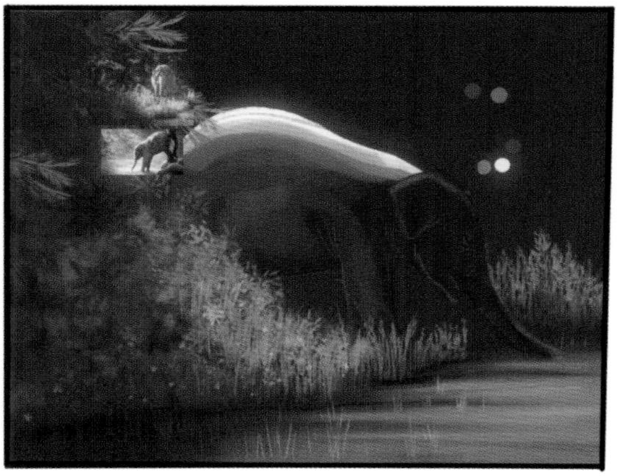

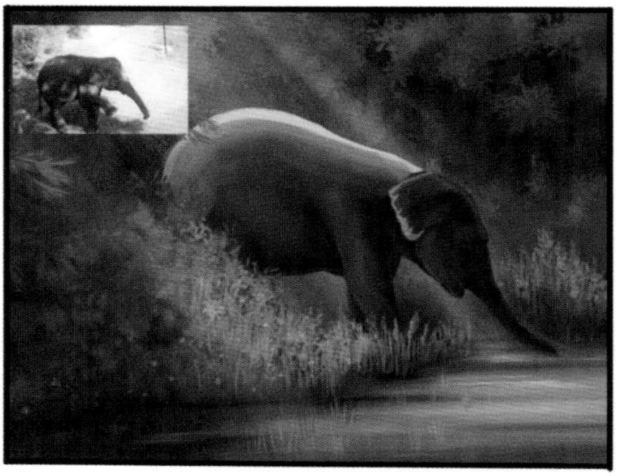

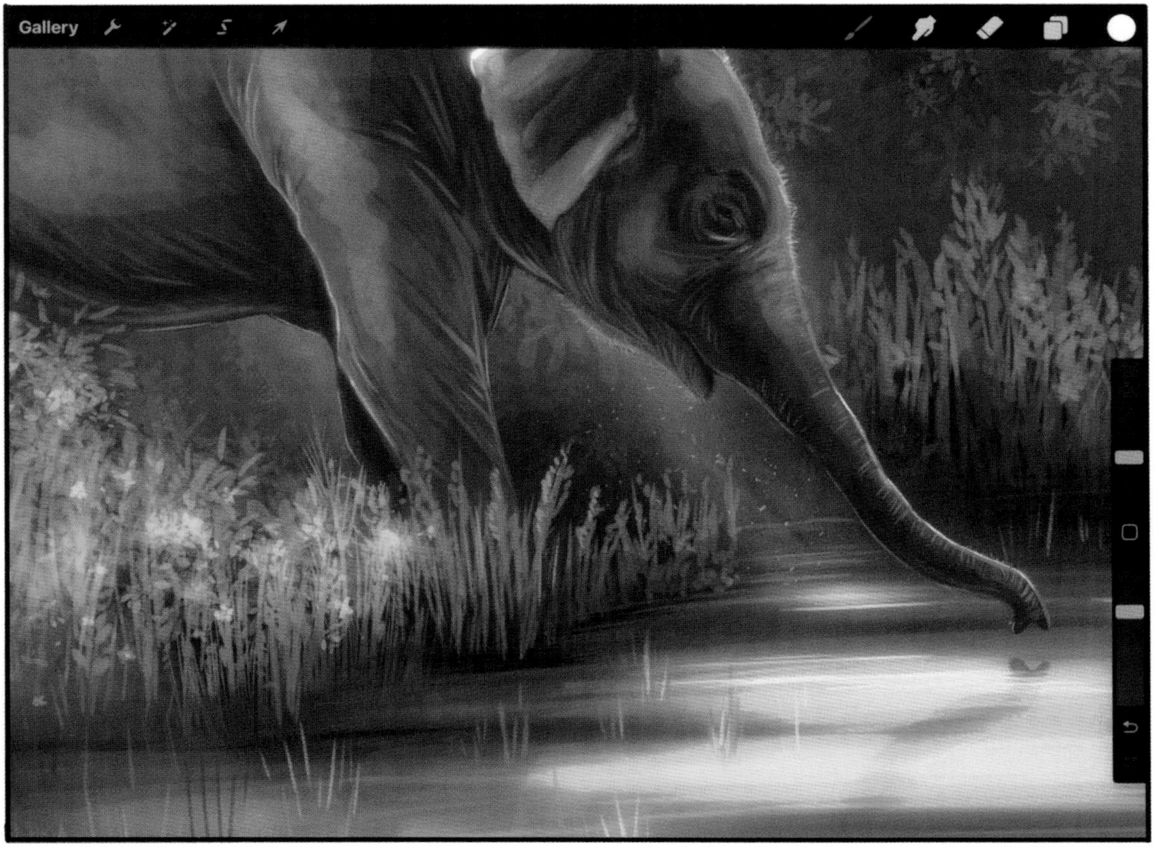
Gallery

Like Happy's first sip, our process produces anxiety and joy. There is no one way to create a comic or cartoon, let alone a single way to approach a genre that is in flux. More often than not, Cynthia's work inspires me to return to the writing and match the art's tone, similar to the method popularized by Stan Lee, Jack Kirby, and Steve Ditko. We take direction from each other to speak in one voice and harmonize our page. Throughout our partnership, the themes of liberty and independence always guide our collaboration.

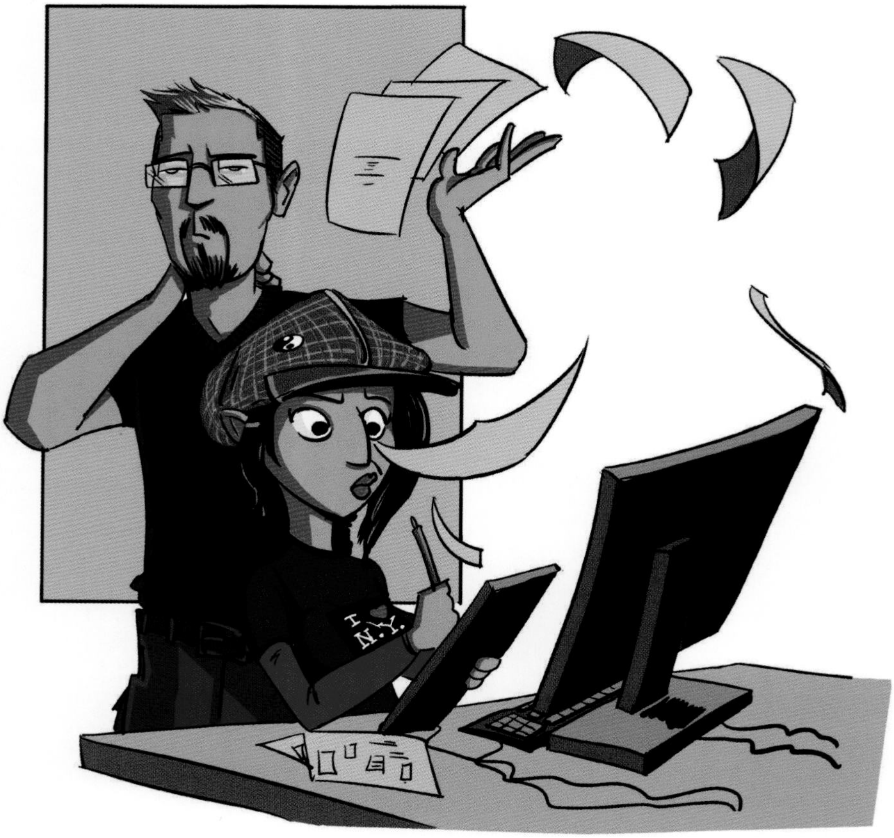
I
N.Y.

About the Authors

Cynthia Sousa Machado and Sam Machado are cartoonists whose work explore politics, law, identity, rights, and social justice issues. They are best known for their editorial cartoons "I Got This" and "If I Don't Get Pants." You can find their webcomic *Cyberbunk* on LINE Webtoon. Cynthia and Sam live and illustrate in Miami, Florida.

Steven M. Wise is Founder and President of the Nonhuman Rights Project. He has practiced animal protection law for 30 years throughout the United States and is the author of four books: *Rattling the Cage: Toward Legal Rights for Animals; Drawing the Line: Science and the Case for Animal Rights; Though the Heavens May Fall: The Landmark Trial That Led to the End of Human Slavery; and An American Trilogy: Death, Slavery, and Dominion Along the Banks of the Cape Fear River*. Wise has taught Animal Rights Law at Harvard, Stanford, and seven other law schools. He holds a J.D. from Boston University Law School and a B.S. in Chemistry from the College of William and Mary.

About Island Press

Since 1984, the nonprofit organization Island Press has been stimulating, shaping and communicating ideas that are essential for solving environmental problems worldwide. With more than 1,000 titles in print and some 30 new releases each year, we are the nation's leading publisher on environmental issues. We identify innovative thinkers and emerging trends in the environmental field. We work with world-renowned experts and authors to develop cross-disciplinary solutions to environmental challenges.

Island Press designs and executes educational campaigns, in conjunction with our authors, to communicate their critical messages in print, in person, and online using the latest technologies, innovative programs, and the media. Our goal is to reach targeted audiences—scientists, policy makers, environmental advocates, urban planners, the media, and concerned citizens—with information that can be used to create the framework for long-term ecological health and human well-being.

Island Press gratefully acknowledges major support from Bobolink Foundation, Caldera Foundation, The Curtis and Edith Munson Foundation, The Forrest C. and Frances H. Lattner Foundation, The JPB Foundation, The Kresge Foundation, the Summit Charitable Foundation, Inc., and many other generous organizations and individuals.

Generous support for this publication was provided by Merloyd Lawrence.

The opinions expressed in this book are those of the author(s) and do not necessarily reflect the views of our supporters.